SOIL FERTILITY DECLINE IN THE TROPICS WITH CASE STUDIES ON PLANTATIONS

Soil Fertility Decline in the Tropics with Case Studies on Plantations

Alfred E. Hartemink
International Soil Reference and Information Centre (ISRIC)
Wageningen
The Netherlands

CABI Publishing

CABI Publishing is a division of CAB International

CABI Publishing
CAB International
Wallingford
Oxon OX10 8DE
UK
Tel: +44 (0)1491 832111
Fax: +44 (0)1491 833508
E-mail: info@cabi.org
Website: www.cabi-publishing.org

CABI Publishing
745 Atlantic Avenue
8th Floor
Boston, MA 02111
USA
Tel: +1 (617)682-9015

E-mail: cabi-nao@cabi.org

A catalogue record for this book is available from the British Library, London, UK.

The Library of Congress has cataloged the hardcover edition as follows:
Hartemink, Alfred E., 1964-
Soil fertility decline in the tropics : with case studies on plantations / Alfred E. Hartemink.
p. cm.
Includes bibliographical references and index.
ISBN 0-85199-670-1 (alk. paper)
1. Soil fertility--Tropics. 2. Soil degradation--Tropics. 3. Agriculture--Tropics--Case studies. I. Title.
S633.5.T76H37 2003
631.4′22′0913--dc21
2003004851

ISBN 978 0 85199 670 7 (HB)

Typeset in 10pt Melior by Columns Design Ltd, Reading.
Printed and bound in the UK by CPI Group (UK) Ltd, Croydon, CR0 4YY.

Contents

Preface

In 1987, I was employed as a pedologist-trainee at the National Soil Service in Tanzania as part of my undergraduate training. Detailed soil surveys and land evaluations were carried out on several sisal plantations and an oil palm plantation. After my BSc graduation, I conducted a soil survey of a tea plantation in Congo (Zaire) and returned to Tanzania for detailed soil surveys of sisal plantations followed by short assignments on tea, cocoa, rubber and oil palm plantations in Indonesia. Some years later I worked in Papua New Guinea, where I fostered my interest in on-plantation research, and a study on soil changes under sugarcane was carried out.

During those years I developed three important passions which helped me in the writing of this book and which also form its cornerstones: soils, the tropics and plantation cropping. I was intrigued by how these were linked and although the conversion of primary forest into agricultural plantations is often regarded as destructive and politically and morally incorrect, the beauty of large-scale plantations, the history, neatness and the efficiency in mass production fascinated me intensely. At the same time, I have often noticed the ease with which natural resources were taken for granted and wondered how soil conditions were affected under permanent cropping. After working in Papua New Guinea, I felt the time had come to contemplate on the years in the field, and that reflection forms the body of this book. Although there are books focusing on plantation agriculture, there is no book solely dedicated to the soil under plantation cropping. This, in combination with the need for hard data on soil fertility decline in the tropics, are the main gaps that this book aims to fill.

The groundwork for this book formed the data and literature brought together in my PhD thesis. I owe a great deal of gratitude to Professor Dennis Greenland and Dr Stephen Nortcliff of the University of Reading for their supervision and guidance during the preparation of my thesis. A special word of thanks goes to my friend and colleague Dr Hans van Baren for his overall assistance and suggestions to improve the text. My director, Dr David Dent, is thanked for critically reading the text and his support. Most of all I would like to thank Ariane who has been with me during the years in which my passions developed. She was part of it and made it all possible. I dedicate this book to her.

Alfred E. Hartemink
Amsterdam
February 2003

Foreword

Soil fertility decline remains one of the most serious problems facing the world. In many developing countries nutrient depletion already threatens food production, so that food shortages in Africa are again a serious problem. Methods for controlling soil degradation and improving productivity are well known, but in many parts of Africa and some countries in Asia the economic and political factors determining the acceptability of improved practices has limited their adoption. Hence this study by Dr Hartemink is important and opportune, particularly because of the emphasis on plantation crops, which have received less attention than their economic importance deserves.

In spite of criticisms, the 'green revolution' methods of greater fertilizer use with responsive crop varieties have mostly ensured that food production has kept pace with the rapid population growth of the past decades. Nevertheless the economic problems determining the costs of importing, manufacturing and distributing fertilizers, key factors in countering soil fertility decline and ensuring yield improvements, have not been solved. This book emphasizes that crop production cannot be sustained unless nutrient removals are balanced by replenishment, and soil erosion is controlled.

In spite of long-term fertilizer experiments with perennials and plantation crops, some of which started more than a century ago, yields have not always been maintained or improved. In this book reports of soil chemical changes under plantation and perennial crops are analysed. In many of these reports the inadequacies of the data limit the value of the results. These inadequacies include failure to provide data on the nutrient status of the soil before cultivation or establishment of the plantation, failure to define the soil type, failure to provide sufficient data regarding the soil management practices used, and failure to give other essential information about crop variety and climate.

On sisal estates in Tanzania serious yield decline occurred between the 1960s and 1980s, and many estates had to be abandoned. Sisal production from Tanzania fell from over 200,000 to 30,000 t/year. The decline was attributed to a fall in the world market price, and the key factor of soil fertility decline was largely ignored. Soil fertility decline has also been an important factor making sugarcane production uneconomic in many developing countries. Yield decline is often due to acidification caused by continued use of ammonium-based fertilizers, without correction by liming.

Dr Hartemink has worked in Kenya, Tanzania, Congo, Indonesia and Papua New Guinea, so that he has first-hand experience of both crop production and plantation developments in the tropics, as well as the relation between soil conditions and productivity.

This book should help to refocus attention on the dangers of continuing soil degradation, and on the need to ensure that positive nutrient balances are maintained for plantation as well as for food crops.

D.J. Greenland
Visiting Professor
Department of Soil Science
University of Reading, UK

Acknowledgements

The author is grateful to the following publishers for the permission to reproduce figures from their publications: CAB International (Fig. 2.5 from: *Soil Use and Management* (1997) 13, 167–177); Cambridge University Press (Fig. 9.4 from: *Experimental Agriculture* (1998) 34, 231–243; Fig. 2.4 from: *Crop Evolution, Adaptation and Yield* by L.T. Evans (1993)); Elsevier Science (Fig. 2.9 from: *Agriculture Ecosystems and the Environment* (1999) 75, 93–100; Fig. 7.3 from: *Forest Ecology and Management* (2001) 140, 227–238; Fig. 8.1 from *Geoderma* (1997) 79, 187–225); IP Publishing (Fig. 10.4 from: *Outlook on Agriculture* (1995) 24, 91–96); John Wiley & Sons (Fig. 2.8 from: *More People, Less Erosion* by M. Tiffen, M. Mortimore, F. Gichuki (1994)); Kluwer Academic Publishers (Fig. 4.5 from: *Plant and Soil* (1999) 209, 137–156).

All photographs are by the author.

1 Introduction

> No scientific investigation is complete until its results can be expressed quantitatively. Only when this is done can the investigators feel reasonably certain that they have gained the right perspective and that they know how nearly their hypotheses approximate to the truth.
>
> Sir E.J. Russell (1926)

General

If all resources are harnessed, and adequate measures taken to minimize soil degradation, sufficient food to feed the population in 2020 can be produced, and probably sufficient for a few billion more. These were the key conclusions from a Discussion Meeting held at the Royal Society in London in December 1996. The meeting entitled 'Land resources: on the edge of the Malthusian precipice?' attempted to make a rigorous scientific assessment of the evidence regarding the availability of land resources to meet future demands, and to determine what further research is needed to strengthen the scientific base on which such assessments depend (Greenland *et al.*, 1997).

At the meeting global data-sets and models were widely used by specialists with expertise in the area of population growth, crop production, climatology, water, soil and land resources, or the environment. Some of the key points to emerge from the meeting were:

- It is expected that most of the extra food will be produced by those countries with a greater extent and better quality of land resources.
- There are many countries where serious food problems will be experienced in the next two decades and that transfers of food from countries with greater resources to food-deficit countries will be necessary.
- Increased understanding of the basic principles of crop production and sustainable land management is needed for a sustainable basis of production increases in resource-poor countries.
- Research efforts of the developed countries must be not only restored but increased, and collaboration with developing countries enhanced, so that the fall in the rate of increase in cereal yields from a high of almost 3% per annum in 1965 to close to 1.3% per annum in 1997 can be reversed.
- In the developing world, research into soil, water and nutrient management must be intensified.
- Long-term trials are essential if the sustainability and environmental acceptability of various management practices are to be properly assessed (Greenland *et al.*, 1997).

Key topics such as human population growth, soil resources and soil degradation, land-use intensification and integrated nutrient management, agricultural production and sustainable land management were widely discussed during the Royal Society meeting. The development of technically feasible and socio-economically acceptable solutions is among the biggest challenges that agricultural sciences are facing. They form the starting point for this book and the general background against which the information is presented.

This Book and its Rationale

Currently, more than 95% of the human population increase is in the tropics, which puts existing agricultural systems under stress. In order to produce sufficient food and to curb land degradation, permanent and productive cropping systems[1] need to be developed. The sustainability of permanent cropping systems is largely effected by the judicious management of the soil chemical fertility. This has been recognized for many decades, but there is a need for hard data on soil

[1] Permanent cropping systems are systems in which the land is continuously cropped year after year with annual crops (e.g. maize, beans, rice) or perennial crops (e.g. cocoa, rubber). There is usually no fallow or resting period in permanent cropping systems.

changes and nutrient management strategies in order to improve our understanding of agricultural systems and to design sustainable cropping systems in the tropics. Although there is also a need for more sustainable land management practices in agriculture in the temperate regions, there are diametrically opposite problems compared with soils under agriculture in the tropics. Many of the soils under agriculture in the temperate regions are very high in plant nutrients owing to excess use of inorganic fertilizers and animal manure, whereas it is often perceived that soils in the tropics suffer from plant nutrient deficiencies (Hartemink, 2002).

So far the discussion on soil productivity decline, nutrient mining and sustainable land management in the tropical regions has focused on low-external-input agriculture by subsistence farmers (Pieri, 1989; Henao and Baanante, 1999; Scoones and Toulmin, 1999; Smaling *et al.*, 1999). This book has its main focus on agricultural plantations, which have been largely neglected in the discussion on soil fertility decline[2] and sustainable land management. Another major difference with the existing studies is that soil fertility decline is assessed using soil chemical data whereas other studies have used nutrient balances as the main tool to evaluate the sustainability of the systems.

The rationale for the focus on plantation agriculture is that it is an important form of land use in the tropics, and in many countries the area under plantation crops has expanded rapidly in the past decades. For example, in Indonesia the area under oil palm expanded from 133,000 ha in 1970 to almost 1.8 million ha by the mid-1990s (Fairhurst, 1996). In Malaysia the extent of oil palm increased from about 150,000 ha in the early 1970s to over 3 million ha at the end of 1998. Plantation agriculture is contributing to the macroeconomies in many tropical countries and provides much employment. Even in middle-income countries such as Malaysia, total export earnings from oil palm plantations are 6% of the Gross National Product (GNP) (Jalani, 1998). In Ivory Coast a group of plantation crops produce 22% of GNP (Tiffen and Mortimore, 1988). As yields are usually higher on plantations than on smallholder farms, they may contribute proportionally more to GNP than the area they occupy, for example in Kenya tea plantations comprise 35% of the area under tea but they produce more than 60% of the total output (Tiffen and Mortimore, 1988).

[2] Soil fertility decline is defined in this book as: the decline in soil chemical fertility, or a decrease in the levels of soil organic C, pH, CEC and plant nutrients. Biological and physical factors are not considered. Soil fertility decline thus includes nutrient depletion (larger removal than addition of nutrients), nutrient mining (large removal of nutrients and no inputs), acidification (decline in pH and/or an increase in exchangeable Al) and the loss of organic matter – see also page 80.

Plantation crops are sometimes referred to as non-CGIAR (Consultative Group for International Agricultural Research) crops (Smith, 2000). Despite the importance of plantation agriculture, long-term effects of plantation cropping on the soil have received little research attention. No systematic effort has been undertaken to prove that plantation agriculture is a more sustainable form of land use than arable cropping. However, it has long been assumed that a perennial plant cover protects the soil better than an annual crop (Jacks and Whyte, 1939), and it has also been stated that land degradation under perennial crops is usually less than in arable farming under similar conditions (Ruthenberg, 1972).

Set-up of the book

The book focuses on soil fertility decline in the tropics. The information is presented in three parts, which have been partly interwoven: (i) a literature review (Chapters 2 to 4); (ii) review of soil changes under annual and perennial crops including two detailed case studies (Chapters 5 to 10); and (iii) an integrative part, in which the literature and case studies are combined (Chapter 11).

Chapter 2 embraces a global literature review on human population growth, soil resources of the tropics, tropical land use and management, soil degradation, and sustainable land management. A general review of the problems in the use and management of soils in the tropics is given with particular emphasis on soil chemical fertility. Chapter 3 reviews historical and productivity aspects of agricultural plantations in the tropics. Chapter 4 covers theoretical considerations of soil fertility decline and includes sections on data requirements, spatial and temporal variation, soil tests and interpretation of soil fertility decline studies. The information forms the theoretical framework for the subsequent chapters, in which evidence for soil fertility decline is presented.

Chapter 5 focuses on soil fertility decline of annual cropping systems and brings together a wide range of data and studies. Soil changes under plantation crops are critically examined in Chapters 6 to 10 using the published literature on perennial crops, sugarcane plantations and forest plantations. Two detailed case studies are included, based on research conducted at plantations in Tanzania and Papua New Guinea. The data of these studies have been re-evaluated and re-interpretated according to schemes developed in Chapter 4. For both case studies a fair amount of data was available and basic statistics were used. In Chapter 11, the information is synthesized. A summary of the soil changes is presented followed by some of the implications for plantation cropping and a set of conclusions is given in the final chapter.

Thus, this book starts with a global view of the pressing matters in the tropics (wide zoom), then focuses on the soil and the decline in soil fertility under plantation cropping (zoom-in phase), and subsequently the data are aggregated again and the wider implications are reviewed (zoom-out again).

Structuring of chapters

This book deals with evidence of soil fertility decline under different land-use systems in the tropics: annual crops (Chapter 5), perennial crop and forest plantations (Chapters 6 and 7), sugarcane (Chapters 8 and 9) and sisal (Chapter 10). Each chapter is structured in a similar way. After a brief history and introduction, soil erosion is discussed followed by a review of changes in soil chemical properties and calculations on the rates of change. Hereafter semiquantitative studies (nutrient balances), soil process oriented and environmental impact studies are reviewed. Each chapter ends with a discussion and conclusions. Although the primary focus is on changes in soil chemical fertility, brief reviews on soil erosion and soil environmental studies are included in order to embed the soil chemical data in a broader setting, and to supplement the soil chemical data with other important soil data.

Literature and own data

This book is based on data from the literature and on my own research in Tanzania and Papua New Guinea and to a lesser extent on my experiences in Congo and Indonesia. Literature data were used to complement and balance detailed case studies. In the literature review sections, a slight preference has been given to some of my own data as far as they were not presented in the case studies. For example, in the discussion on bulk density in relation to soil nutrient depletion, soil physical data from the sugarcane plantation in Papua New Guinea were used to demonstrate changes under permanent cropping. Likewise, I have used my research data on nutrient content of roots to illustrate variation in nutrient uptake over time, and to compare fertilized and unfertilized plants. There are other sections where my own work was used to exemplify relevant soil properties and processes, and although this might appear as a bias, it has been done in combination with other literature.

An analogy

This book has been structured in a particular way and it can best be explained by an analogy. Suppose a patient named X, suffering from an unknown disease, visits a medical doctor named M. The doctor will firstly ask X what his problem is, followed by some general background information on X. Successively, M will measure or observe the problem. In some cases, M may be able to directly diagnose the problem ('You have a broken leg, sir') but in many cases M may request tests to have his diagnosis verified or to obtain some general impression of the medical status of X. If the disease appears complicated, M may need to do some extra reading and study, provided the patient is not transferred to a more specialist colleague. Once the diagnosis has been ascertained, M will prescribe a treatment to cure X, and will speculate about the future health of his patient.[3]

This book is built along these lines of thought, in which the soils of the tropics under permanent cropping are X and this book can be considered some sort of M. The problem and some background information on the patient are discussed in Chapters 2 to 4, whereas measurements and observations are given in Chapter 5 to 10. The final diagnosis is made in Chapter 11 including a discussion on the future health status in the concluding chapter. This book requires the medical specialist to do a considerable amount of reading, but given the size of the patient and the problem, that may be justified.

What's excluded

This book deals with soils and agriculture in the low-altitude regions of the humid and subhumid tropics. The main focus is on basic or standard soil analytical measurements (pH, organic C, total N, available P, CEC and exchangeable Ca, Mg and K). No information is included on micronutrients, or on soil biological or physical properties. Although a lot of ground is covered, virtually no attention is given to livestock, grazing and pastures. In some plantations, livestock and cropping are intensely mixed (i.e. cows and coconuts) but as animals introduce another complicating factor in the relationship between soils and crops, such studies were excluded. It needs to be mentioned that the literature review is largely based on Anglo-Saxon and Dutch literature and virtually no French, Spanish, German or

---

[3] Recently I noticed that this analogy was also used by Doran, J.W. and Parkin, T.B. (1996) in their chapter: 'Quantitative indicators of soil quality: a minimum data-set.' In: Doran, J.W. and Jones, A.J. (eds) *Methods for Assessing Soil Quality.* SSSA, Madison, Wisconsin, pp. 25–37. The analogy was used for the discussion of 'ecosystem health' based on other work in the USA. I found that a bit discomforting, but after careful deliberation the analogy was kept as it had facilitated the way in which the book was structured.

Portuguese literature has been used. None the less, it is assumed that the Anglo-Saxon and Dutch literature covers the main trends and developments in the subjects discussed in this book. It should be mentioned that no attention is given to socio-economic aspects of soil fertility decline. Also, the literature data and the results from the soil investigations in Tanzania and Papua New Guinea were not modelled. Although models can be of great help in the understanding of factors driving certain processes and can also show what data are lacking, a disquieting aspect of computer-based modelling is the gap between the model and the real-world events (Phillip, 1991). In soil science, computer modelling has also partly supplanted laboratory experimentation and field observations (Hartemink *et al.*, 2001) and the focus of the book has been placed on bringing together and critically analysing real-world data. Such data are much needed now our understanding of soil processes and the advances in computer modelling have made great progress. Lastly, no geostatistics were used because the data were too few.

Aims and Approach

The main aims are to quantify soil fertility decline under permanent cropping systems in the tropics and to resolve what strategies are needed to assess the rates of change in soil chemical properties. More specifically the book aims to:

- Review the published literature on human population growth, soil resources and soil degradation, food production, land-use change and sustainable land management in the tropics.
- Review data types and boundary conditions for the assessment and measurements of soil fertility decline.
- Review the literature on the history and importance of agriculture and forest plantations in the tropics.
- Analyse and assess the effects of plantation crops on soil chemical properties.
- Re-evaluate the soil chemical data from sisal and sugarcane plantations in Tanzania and Papua New Guinea, and compare the results with those in the published literature.
- Investigate how soil fertility decline differs between different land-use systems and soils.

Three aspects deserve mentioning up front as they have determined why the information is presented as it is. First, the focus of this book is *change*: changes in human population, changes in crop yield and production, and changes in soil chemical properties. Change is defined here as: *to become different, altered or modified.* Studying changes means a comparison of observations made at different periods or different locations. Historical knowledge is an essential element of learning about change.

Therefore, in the book a fair degree of *historical* information is given, which is the second aspect that has influenced the presentation of the information. Wherever possible and relevant, emphasis is placed on historical developments in biophysical knowledge rather than socio-economic aspects of the subjects treated. In addition to the historical aspects of changes, this book aims to be *quantitative* or in other words, it puts a strong emphasis on reports of measurable properties. Although qualitative studies in soil science have improved our understanding and often encouraged quantitative studies, major advances in soil science have come from quantitative studies. Most studies focusing on the subject of soil fertility decline contain few data. In this book a wide collection of data has been brought together and critically analysed. The aim was for completeness, but only factual and quantitative studies have been included.

In summary: this book investigates changes by critically analysing historical and published information in a quantitative manner supplemented with detailed case studies, so as to improve our understanding of the complex relation between permanent cropping and the chemical fertility of soils in the tropics.

References

Fairhurst, T. (1996) Management of nutrients for efficient use in smallholder oil palm plantations. PhD thesis, Department of Biological Sciences, Wye College, University of London.

Greenland, D.J., Gregory, P.J. and Nye, P.H. (1997) Introduction and conclusions. In: Greenland, D.J., Gregory, P.J. and Nye, P.H. (eds) *Land Resources: On the Edge of the Malthusian Precipice?* The Royal Society, London, pp. 861–867.

Hartemink, A.E. (2002) Soil science in tropical and temperate regions – some differences and similarities. *Advances in Agronomy* 77, 269–292.

Hartemink, A.E., McBratney, A.B. and Cattle, J.A. (2001) Developments and trends in soil science: 100 volumes of Geoderma (1967–2001). *Geoderma* 100, 217–268.

Henao, J. and Baanante, C. (1999) *Estimating Rates of Nutrient Depletion in Soils of Agricultural Lands of Africa*. IFDC, Muscle Shoals, Alabama.

Jacks, G.V. and Whyte, R.O. (1939) *The Rape of the Earth – a World Survey of Soil Erosion*. Faber and Faber, London.

Jalani, B.S. (1998) Oil palm research and development in Malaysia: progress and trends. *The Planter (Kuala Lumpur)* 74, 73–92.

Phillip, J. (1991) Soils, natural science, and models. *Soil Science* 151, 91–98.

Pieri, C. (1989) *Fertilité des Terres de Savanes*. Ministère de la Cooperation et CIRAD-IRAT, Paris.

Russell, E.J. (1926) *Plant Nutrition and Crop Production.* University of California Press, Berkeley.

Ruthenberg, H. (1972) *Farming Systems in the Tropics.* Clarendon Press, Oxford.

Scoones, I. and Toulmin, C. (1999) *Policies for Soil Fertility Management in Africa.* IIED, London.

Smaling, E.M.A., Oenema, O. and Fresco, L.O. (eds) (1999) *Nutrient Disequilibria in Agroecosystems: Concepts and Case Studies.* CAB International, Wallingford.

Smith, R.W. (2000) Tree Crops: Essential Components of Sustainable Tropical Agricultural Systems. *TAA Newsletter* March, 17–19.

Tiffen, M. and Mortimore, M. (1988) *The Role of Plantations in Development: a Literature Review.* Overseas Development Institute, London.

2

Human Population and Soil Degradation

> The peoples of the earth, whether they will it so or not, are bound together today by common interests and needs, the most basic of which are, of course, food supply and other primary living requirements. Those come, all of them, from nature and from nature alone – from the forests, the soils and the waterways.
>
> Fairfield Osborn (1948)

The world's population has doubled since 1960. The developing world accounts for about 95% of the population growth, with Africa as the world's fastest growing region. The growing population has many implications but most of all it requires an increase in agricultural production to meet food demand. Conservation and improvement of the natural resources on which agricultural production depends is essential. Soil degradation, and in particular the decline of soil chemical fertility, has been a concern in relation to food production and the sustainable management of land resources. These subjects form the general background setting of this book and are briefly discussed in this chapter – soil fertility decline in the tropics is reviewed in Chapters 5 to 10.

Malthus and his Followers

In 1998 it was exactly 200 years since Reverend Thomas Malthus (1766–1834) wrote *An essay on the principle of population, as it affects the future improvement of society.* The stimulus for writing this polemic was his concern about the unwarranted euphoria of his colleagues who, in the aftermath of the French revolution, saw mankind progressing ever upwards to a world of universal abundance, peace and prosperity where all would be equal in health, wealth and happiness (Short, 1998). He wished to demystify this utopian fantasy, and used his numeracy to point out a simple truth: population, when unchecked, increases geometrically whereas subsistence increases arithmetically (Malthus, 1826).

Malthus' theory on everlasting food shortages and poverty had three basic assumptions: (i) food was considered to be necessary for the existence of man and the sole limiting factor on human population growth; (ii) human population increases exponentially; and (iii) food production could only be increased linearly. His theory explained the scarcities and misery observed in England and he declared food paucity to be 'checks' to population growth imposed by the prescribed bounds of nature (Seidl and Tisdell, 1999).

Malthus' assumption on exponential population growth was not entirely new and similar views were discussed in demographical research of the 17th and 18th centuries. His essay was, however, written in a brilliant way, which facilitated the widespread acceptance of his theory. Initially he was abused as his essay was held for unholy, atheistic and subversive of social order (Bettany, 1890). A major criticism is that the idea of exponential growth was deduced from growth in North America and it was not observed elsewhere at this time. Population growth in North America was mainly due to immigration, a confounding factor that was initially ignored by Malthus (Seidl and Tisdell, 1999). In later editions of the book he slightly altered that view (Trewavas, 2002).

Malthus has been named founder of the social demographic discipline, but, more importantly, he was one of the first to see the importance of the environmental limiting factors on human material progress. His essay inaugurated a grand tradition of pessimistic environmentalism (Anon., 1997), which probably found its heyday in the 1960s and 1970s. Although many people in the 19th century and at the beginning of the 20th century thought Malthus was right, he was wrong, for he did not foresee the industrial age and the geometric effect of technology upon economic growth (Jensen, 1978).

When C. Darwin read Malthus' essay in 1838 he saw the struggle for existence, which inspired him for the *Origin of Species*, published in 1859. There is thus a substantial influence of Malthus on the most influential biological theory (Bettany, 1890; Seidl and Tisdell, 1999).

The population bomb

After Darwin the most renowned biologist and follower of Malthus is P. Ehrlich, who published in 1968 the book *The Population Bomb*.

The book became an instant bestseller and ran through several editions bearing slightly different names but the same message. Ehrlich's message was similar to Malthus': unchecked population growth will outstrip food production and destroy the earth's environment (Ehrlich, 1968). The book contains a detailed and pessimistic account of what will happen when the population growth continues. Inevitably there will be mass starvation and the 3.5 million who starved to death in 1968 would only be a handful compared with the numbers that will be starving in a decade or so, in addition to the massive environmental degradation. In the foreword, Ehrlich states 'In a book about population there is a temptation to stun the reader with an avalanche of statistics. I'll spare you most, but not all, of that'. The few hard figures and projections given are indeed all wrong. For example, it was projected that there would be over seven billion people in 2000, a figure which was also used by the Club of Rome (there were six billion people in 2000), and the population in Calcutta would have reached 66 million by 2000, whereas the actual population in 2000 was around 11 million.

Ehrlich, being a scientist, included a small section in his book entitled 'What if I'm wrong?' in which he states that the possibility exists that ' ... technology or a miraculous change in human behavior or a totally unanticipated miracle in some other form will save the day' (Ehrlich, 1968). He found that highly unlikely but played it safe: 'If I'm right we will save the world. If I'm wrong, people will be better fed, better housed, and happier, thanks to our efforts.' Not a modest view and impossible to substantiate.

It is likely that Ehrlich's books inspired groups like the Club of Rome, which was formed and headed by the Fiat director A. Peccei. Their study *The Limits to Growth*, which was published in 1972, detailed what would happen if economic growth and population growth continued. It had the following supposition: 'The basic behaviour mode of the world system is exponential growth of population and capital, followed by collapse.' A model was built to investigate five major trends of global concern: accelerating industrialization, rapid population growth, widespread malnutrition, depletion of non-renewable resources and a deteriorating environment. Calculations were made by a team of the Massachusetts Institute of Technology (USA) and the results were shocking as most natural resources would be depleted within 100 years or sooner (Meadows *et al.*, 1972). Moreover the study was pessimistic about the future of the land resources and advocated that the Green Revolution only caused widening inequalities and disruptions of stable societies. It is no exaggeration to note that *The Limits to Growth* study was momentous – partly as it was conducted by computer (so it had to be right) and initiated by an industrialist (a capitalist so not a person from the Green movement). Despite the various predictions made in *The Limits to Growth*, growth continued exponentially and many of the projections proved false – with the comment on the increase of CO_2 in relation to climate change as the most noteworthy exception.

A different sound from Malthus and his more or less faithful followers came from E. Boserup's study entitled *The Conditions of Agricultural Growth* (1965). Contrary to the common reasoning, which is that the supply of food for the human race is inherently inelastic and that this lack of elasticity is the main factor governing the rate of population growth, she argued that population growth is an independent variable, which in its turn is a major factor determining agricultural developments (Boserup, 1965). The growth rate of food production will accelerate when population grows since it forces the population to intensify land use. This holds as long as fallow land is available and the population density threshold value has not been reached. In summary: agricultural developments are caused by population trends rather than the other way around. Her viewpoints were largely ignored by both Ehrlich and the Club of Rome, whose studies fell in much more fertile public grounds. The reasons hereto may be related to the fact that there was ample food in Europe and the USA and that the post-war baby-boom generation felt a need to change the world.

Population Growth

Estimates of world population are published at the website of the US Government on population census (www.census.gov/ipc/www/worldhis.html). The estimates are based on various sources and the mean of the upper and lower boundary for the period 10,000 BC to AD 2000 and the period 0 to AD 2000 is shown in Fig. 2.1. Global population hardly changed up to 1000 BC and slightly decreased in medieval times. The real increase started from 1650 onwards, when global population passed through the 'J-bend' of the exponential growth curve. Population growth remained below 0.5% up to 1800 and was about 0.6% in the 19th century. In the first half of the 20th century growth was 1%, but the largest rate occurred in the second half of the 20th century, when the world population grew over 2% in some years (Lutz and Qiang, 2002).

What has caused the exponential increase in human population since the 1600s? The main reason is science and technology – in particular medical, industrial and agricultural sciences. The conquest of infectious diseases in infancy and childhood and other medical inventions are the main contributors to the exponential growth of the human population. Another factor is the decline in traditional breastfeeding practices by urbanization and by the premature introduction of animal

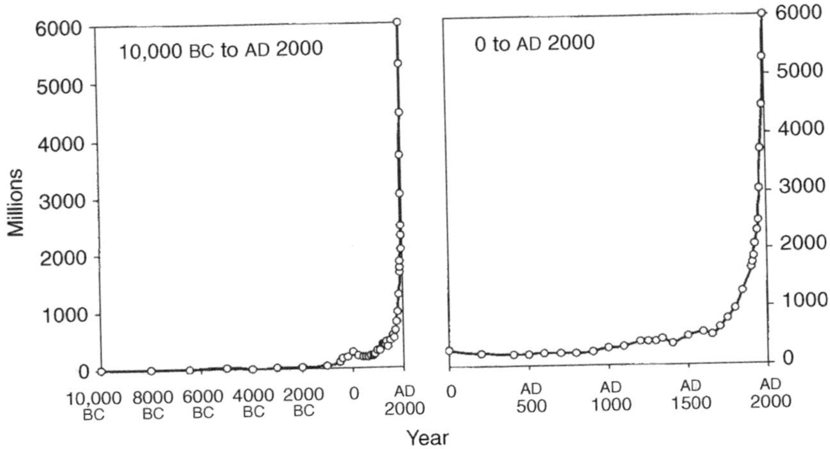

Fig. 2.1. World population estimates for the years 10,000 BC to AD 2000 and the years 0 to AD 2000. Based on reconciliation of published data.

milk or infant milk (Short, 1998). Also the increase in food production in Europe in the 17th and 18th centuries due to advanced cultural techniques (ploughing, liming) and more stable societies resulted in an increase in human population. Important inventions like the acidulation of bones and the invention of superphosphate by J.B. Lawes and technological marvels like the Haber–Bosch process, which allowed the industrial production of urea, indirectly caused a large increase in the European population. These were highly important and the invention and extensive use of superphosphate probably resulted in the economic domination of the world by Western Europe, according to Greenland (1994). Factors explaining the strong population growth in tropical regions were the abolishing of the slave trade, the suppression of tribal wars by European colonists, improvements in the health systems and the relief of famine (Nye and Greenland, 1960). Most of these factors only became significant in the 20th century.

Recent estimates

Since 1950, the world population has grown almost linearly (Fig. 2.2). A linear regression through the 1950 to 2000 data showed that the average increase was 73 million people per year. The regression was highly significant and a high correlation was obtained (r^2 = 0.994). Official statistics have shown that the annual increase in human population was 85 million in the late 1980s and had decreased to 80 million per year in 1995 (Smil, 1999). Currently, the world population is growing by 1.3% per year, which is significantly less than the 2.0%

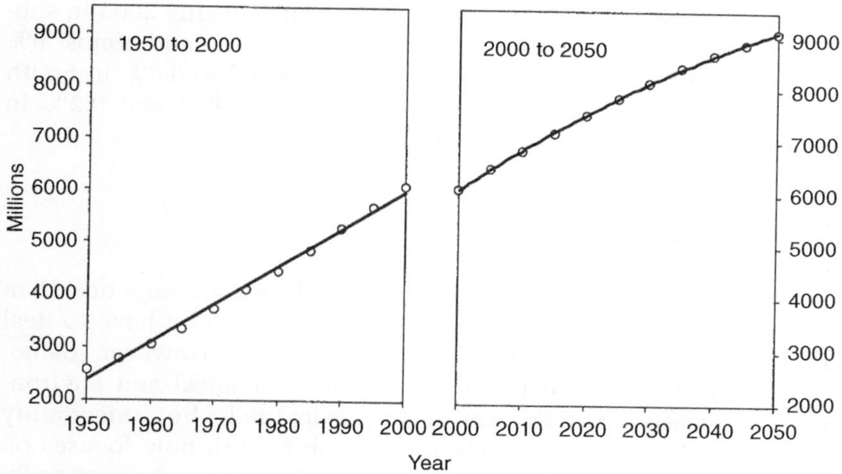

Fig. 2.2. World population estimates for the years 1950 to 2000 and projections for the years 2000 to 2050. Based on reconciliation of published data.

growth rate of the late 1960s. Population growth has been different in different regions. More than 80% of the population lives in developing regions, and Asia accounts for 61% of the world total. Two out of five people in the world live in either China or India. According to the population division of the United Nations, Africa's population is now larger than that of Europe but in 1960 Africa had less than half Europe's population.

It has been estimated that the world population will be 9.4 billion by 2050. Fischer and Heilig (1997) estimated that the average population increase between now and 2015 will be 80 million per year, which will decrease to around 50 million per year in 2050. Doubling of the human population by 2050 is therefore unlikely and the UN Department of Social and Economic Affairs has also lowered its forecast to 8.9 billion in 2050 (Lutz *et al.*, 1997; Smil, 1999) and more recently to 8.8 billion (Lutz and Qiang, 2002). About one-third of this drop is due to the unexpectedly dire ravages of AIDS in sub-Saharan Africa and parts of the Indian subcontinent. Despite this decrease in the rate of population growth, an increase in agricultural production remains essential.

Population growth is also slowing down due to a change of attitude in the developing world. In 1969, people in the developing world had an average of six children compared with three today. The population keeps on growing, however, because more babies survive and old people live longer and in Africa each woman has on average five children. By 2050, there will be three times as many people in Africa as in Europe – if AIDS allows it. Of the 34 most AIDS-affected countries, 29 are in Africa and life expectancy has been reduced on average by seven years (Anon., 1999). Deaths due to AIDS during 2000 in sub-Saharan Africa was estimated to be about 2.4 million and almost 9% of adults were infected with HIV/AIDS, as opposed to 0.6% in South and South-East Asia (470,000 million deaths in 2000) and 0.2% in Western Europe (7000 deaths in 2000).

Uncertainty in projections

Population forecasts have a mixed record and they are no worse than forecasts by economists, meteorologists and others who have to deal with complex and partially understood systems. However, demographic projections could point the way for ecological and environmental forecasts, which are also bedevilled by uncertainty (Tuljapurkar, 1997). Much of the demographic work now focuses on the forecasting of uncertainty, *per se*. A recent report uses a new probabilistic approach that makes use of expert opinion on trends in fertility, mortality and migration and on the 90% uncertainty range of those trends in different parts of the world. It was concluded that the chances for a doubling of the world population by 2050 are less than a third (Lutz *et al.*, 1997). This is based on the expert opinion that human fertility will continue to fall everywhere, trailing the decline of mortality by about a half-century (Tuljapurkar, 1997).

A new focus of attention is developing in demographic studies, and in Western Europe and the USA the focus of the public, political and scientific concern is shifting from global population growth to population ageing (Lutz *et al.*, 1997), or as Tuljapurkar (1997) puts it 'for individuals, families and countries everywhere, the largest question of the next few decades will almost surely be, how to age gracefully'. Two hundred years after Malthus' essay that is quite a shift of focus – at least for those parts of the world where food is ample. The fear exists that the issue of ageing will detract the much needed attention from those areas in the world where population keeps on increasing, hunger is widespread and a higher food production is needed to nourish current and future generations. That combination is mostly found in developing countries in tropical regions.

Food Production and Soil Science

Whether agricultural development is governed by population growth (Boserup, 1965) or vice versa like Malthus and his followers proclaimed, the fact remains that more food needs to be produced when the population grows and if starvation is to be avoided (Tilman *et al.*, 2002). In the absence of massive food relocation, the extra food needs to come either from the available land through intensification, better crop husbandry practices and new high-yielding varieties (yield increases) or through taking more land into production (area increases). Both production increase and area increase depend on a thorough knowledge of the soil and technological applications of this knowledge. Soil science, being essentially an interdisciplinary and applied science, has a long tradition of considering increased food production for the growth of the human population. This emerged in the 1920s (e.g. Penck, 1928) and continues to date (e.g. Greenland *et al.*, 1997; Bouma *et al.*, 1998) and in this section some early and recent studies in which soil science, population growth and food production are linked are reviewed.

At the first Congress of the International Society of Soil Science (ISSS) in 1927, it was suggested that the world could feed at a maximum 15.9 billion people although at that time 7.7 billion was considered a more likely figure (Penck, 1928). The estimate was largely based on the climatic maps of the world by Köppen as soil maps of the world were not available. The human population in 1927 was 1.8 billion, of which 72% lived in the temperate zone. Penck (1928) correctly foresaw a dramatic increase in the human population and that most of the increase would occur in the tropics, which was first expressed in the early 1900s.

Soil science and food production became closely linked in the 1960s and the motto for the Seventh International Congress of Soil Science in 1960 in Madison was 'Alleviate Hunger, Promote Peace through Soil Science.' In his presidential address R. Bradfield mentioned that he could think of no single group of scientists who have more to contribute to feed the world than this group (Bradfield, 1960). He also mentioned that agriculturists including soil scientists have had more experience and in general more success in increasing food production than population experts have had in population control. The growing human population and the adequacy of food production were a point of concern in both the British Empire and the Dutch East Indies, although commercial developments usually had higher priorities than smallholder agriculture in the tropical colonies.

British studies

Before the Second World War, British administrators felt responsible for the feeding of the increasing population that had followed the cessation of war and raiding in many of their colonies and territories. Sir A.D. Hall, the first director at Rothamsted Experimental Station after J.B. Lawes, summarized the situation in the mid 1930s as follows '...native agriculture especially in those vast regions of Africa for which we are responsible, is inadequate to provide for the growing population, that is leading to land hunger and political unrest, that is wasting and will eventually destroy even the present limited production from the land'. Hall stated that the increase of population in Africa had become very marked since the advent of European government, and in many tribes land hunger had developed already to an alarming degree. 'Unless remedial measures are taken, a state of general congestion is threatened within 30 years and famine is never far away' (Hall, 1936). He strongly believed that an increase in the amount of available food and the raising of living standards would be accompanied by an automatic reduction in the rate of increase of the population. As we know now, that point has not been reached yet in Africa, and in effect quite the opposite occurred: a rapidly increasing population because of better living standards.

Hall's successor Sir E.J. Russell, who was Rothamsted's director for 31 years, showed great interest in the relation between human population growth and soil science and published a thorough book on the subject (Russell, 1954). In the book's Preface he mentioned that there have always been great inequalities in the food supplies of different countries. Before the Second World War such inequalities were accepted as part of the natural order of things, with which it was not for us (i.e. northwest Europeans) to interfere. He added that in the 1950s and 1960s many people in Europe and their descendants overseas had a growing feeling that they must do something to mitigate the hunger that oppresses so many in the undeveloped countries. That argument, which has deeply rooted, is still with us today, particular in soil science.

Dutch East Indies

The link between population growth and soil science was recognized in the Dutch East Indies (Indonesia) and in particular on densely populated Java. Based on earlier work, E.C.J. Mohr showed that Indonesia had a mean population density of 32 people per km^2 in 1930 but with large regional differences: Java carried 316 people per km^2 whereas population density in some parts of Sumatra and Borneo (Kalimantan) was only 11 people per km^2. The soil made all the difference: Java is largely volcanic and most fertile soils are derived from volcanic ejecta, which also affects the quality of the irrigation water, which is highly important in the rice-based farming systems of Java (van Baren, 1960). Mohr (1947) compared population densities for different districts near the active Merapi Volcano in Central Java (Indonesia) on volcanic soils and non-volcanic soils derived from limestone (Fig. 2.3). Much higher population densities were found in the area where regular ash deposits are made by the volcano.

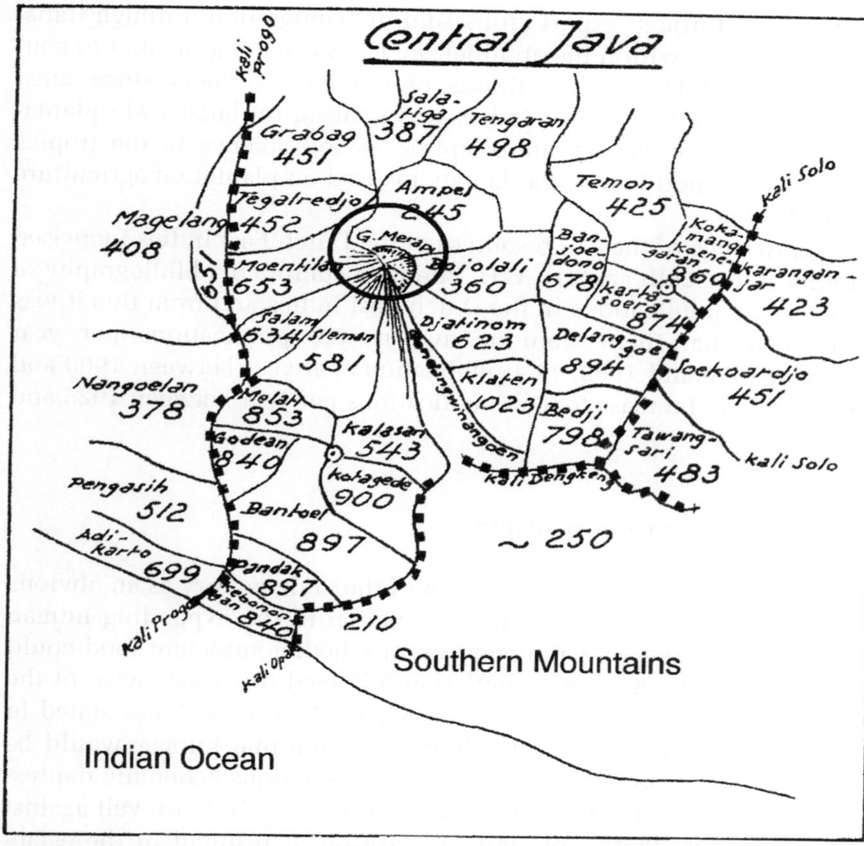

Fig. 2.3. Population density (persons/km^2) around the Merapi Volcano (Central Java) in the 1930s. The Merapi Volcano (encircled) is an active volcano that regularly ejects lava and ashes that are deposited between the Pogo River in the East and the Opa and Solo in West (dotted lines). Soils in the Southern Mountains are generally poor. Modified after Mohr (1947).

Further studies relating soil properties to population density showed that they were closely linked and the problem of the pressure on the land from an ever increasing growth of the population should take this into account (van Baren, 1960). Transmigration of people from Java to sparsely populated regions of Sumatra, Papua or Kalimantan was a way in which both the Dutch colonizers and subsequently the Indonesian government dealt with the growing population on Java's fertile soils.

White (1941) considered soil fertility as a matter of national concern in relation to the rapidly growing population on Java. It is interesting that he considered two types of soil fertility loss: 'emigration' of soil fertility through export crops, and 'transmigration' through transport of produce within the archipelago. He was among the first to consider the soil fertility conditions of the local farmers since most research in the Dutch East Indies was focused on large-scale plantations. The situation was no exception as soil science in the tropics before the Second World War largely focused on plantation agriculture (Hall, 1936).

Knowledge of the soil resources of the Dutch East Indies increased rapidly in the 20th century. C.H. Edelman compiled a bibliography of soil science publications in the Dutch East Indies and from this it was calculated that there were on average 5.0 publications per year between 1850 and 1900, 44.4 publications per year between 1900 and 1925, with an increase to 63.3 publications per year between 1925 and 1940.

Soil erosion and environmentalism

Soil erosion emerged in the first half of the 20th century as an obvious factor affecting food production in relation to the expanding human population. In the USA the question of whether sufficient food could produced for the growing population followed the 'dustbowls' of the 1930s caused by severe wind erosion. The *New York Times* stated in 1937 that if soil erosion was allowed to continue hunger would be common in the USA. This coincided with a serious economic depression in the USA, and an active fight of President F.D. Roosevelt against both economic depression and soil erosion. It resulted in the establishment of the Soil Conservation Service and the belief that productive soils can be maintained through centuries of farming if serious erosion is prevented (Bennett, 1939).

A global overview of soil erosion was prepared by Jacks and Whyte (1939) titled *The Rape of the Earth – a World Survey of Soil Erosion* (the American expurgated title was *Vanishing Lands*). They arrived at similar conclusions: world food production would be seriously affected if erosion remained unchecked. Depressing views on the future of the earth were also expressed in *An Agricultural Testament* (Howard, 1940), *Our Developing World* (Stamp, 1960) and *Our Plundered Planet* (Osborn, 1948), containing quotes like '...It is easy to understand in present times, with the world so crowded and in need of food, how any overpopulated country might deplete its land in a desperate effort to feed its crowding millions.' The cover of *Our Plundered Planet* from 1948 stated '...we are more likely to destroy ourselves in our persistent and world-wide conflict with nature than in any war of weapons yet devised'.

It is certainly not the case that the soil scientific community embraced the conclusions of all these books and the well-known Dutch soil scientist C.H. Edelman rightly called them ‘scare books’ (Edelman, 1951). He was also convinced that man is inventive so that a much larger human population would be possible provided modern agricultural techniques were available in developing countries.

Although environmentalism is generally associated with the 1960s and 1970s, these ‘scare books’ make clear that the seeds for a pessimistic environmental outlook were sown in the first half of the 20th century (e.g. Hall, 1936; Jacks and Whyte, 1939; Howard, 1940; Osborn, 1948). As explained earlier, it was not until the 1960s that these views gained widespread attention and acceptance.

After the Second World War, when international organizations such as the United Nations Food and Agriculture Organization (FAO) were established and many countries were aiming at independence, the feeding of the growing population became an important area of research. Increasing food production was a concern in Western Europe because of the devastation after the war and the baby boom. Fortunately, science came out of the war with high status and was overall respected (Tinker, 1985). There was great optimism and positivism in the 1950s and agricultural research rapidly expanded. Most, if not all, agricultural research was directed towards agricultural production, which increased dramatically thanks to technological developments and major investments in agricultural infrastructure. Even though the term Green Revolution – a term coined in 1968 for the agricultural changes that began to spread through developing countries in the mid-1960s – is mostly being reserved for agricultural production in developing countries, it could apply as well to post-war agriculture in Western Europe (Bouma and Hartemink, 2003). There is no doubt that soil science played an important role in the increase of agricultural productivity, and Malthus would have been correct in predicting that population growth would outstrip food supplies, but for the discoveries of soil scientists (Greenland, 1991).

Some recent studies

Nearly all studies from the 1930s to the mid-1970s focused on soil physical and chemical properties limiting agricultural production (e.g. Mohr, 1947; Pendleton, 1954; Bradfield, 1960). These studies were largely qualitative. The first quantitative study estimating the world food production was conducted in the mid-1970s. Following the publication of the report of the Club of Rome (Meadows *et al.*, 1972), the Dutch Nobel laureate J. Tinbergen requested a group of Wageningen researchers headed by the soil scientist P. Buringh to estimate the maximum food production of the world if all suitable agricultural land were cropped (Buringh *et al.*, 1975). As a result, an assessment was made of land resources and productivity of more than 200 regions of the world, introducing the regional aspects of food production and productivity. The absolute maximum production was expressed in grain equivalents of a standard cereal crop and estimated to be 40 times the production in 1975, when there were 4.1 billion people. It was assumed that less than half of the potential agricultural land of the world was cultivated (Buringh *et al.*, 1975). Although the authors admitted that the results have only a theoretical and scientific value, the study showed locations where most productive land is available and that the highest possible land productivity is in the tropics, where double or triple cropping can be practised. Characteristics of these early quantitative studies on global food production are the lack of scenarios and uncertainty in the calculations. Analogous to demographic studies which now largely focus on uncertainty *per se* (Lutz *et al.*, 1997), such studies started to appear in the 1990s (Gallopin and Raskin, 1998).

A study by Kendall and Pimentel (1994) included three scenarios: continuation of present trends whereby population is to reach 10 billion in 2050, a pessimistic scenario with 13 billion by 2050 and climatic changes, and an optimistic scenario assuming a population of 7.8 billion by 2050 and improved agricultural practices. In the first scenario food production cannot keep pace with population growth. The pessimistic scenario assumed considerable climatic change causing 15% yield loss and gave little hope of providing adequate food for the majority of humanity by 2050. Only the optimistic scenario considered that grain production might be adequate for the growing population by 2050 but it would require a doubling of the production and the implementation of soil and water conservation programmes (Kendall and Pimentel, 1994). Similar views were expressed by the International Food Policy Research Institute (IFPRI) (Pinstrup-Andersen, 1998).

Another estimate showed that a two- to fourfold increase in food production could be easily achieved to satisfy the growing population (Penning de Vries *et al.*, 1997). The estimate was based on the potential production limited by radiation and temperature, by moisture in rain-fed areas, and it was further assumed that all surface water was available for irrigation. The authors conclude that the actual (or attainable) level of agricultural production will be much lower because land is limited, water use is inefficient and there is loss of productivity because of soil degradation.

Soil degradation was taken into account in a study by Bouma *et al.* (1998). They explored the effects of different types of soil degradation (compaction, erosion, acidification) on agricultural production. Although the calculated values should not be considered as absolute, the study showed that the effects of different forms of degradation cannot simply be extrapolated. So far no study has been conducted which quantifies the effects of land degradation on global agricultural production. This is because no accurate data exist on the extent and types of land degradation (see 'Soil Degradation' on page 43).

Conclusions

The human population grew very fast in the past century and much of this growth occurred in tropical regions. There is no doubt that the concentration of people has had environmental implications and in many cases it is likely that the environment has degraded. There are also cases where the environment is improved. Various studies in the past predicted gloom: more people, less to eat, scarcity, starvation, misery, war, environmental devastation etc. Obviously, these studies need to be examined against the available information in their times, but it could be argued that the political and emotional content often exceeds the scientific content including the uncertainties in the predictions. But did they help? It is impossible to appraise whether Ehrlich's efforts in predicting gloom have been helping Earth, but we know that most of his predictions from the 1960s were wrong.

At the extreme, there are two groups in the world, which either believe that food production and yield increases have reached a plateau (the pessimists) or argue that sufficient food can be produced for many billions to come (the optimists). The pessimists, and followers of Malthus, believe that the world is approaching its carrying capacity, that no more cultivable land is available and land degradation is widespread, and that production cannot be sufficiently increased to decrease the 800 million or so who are chronically malnourished (Pinstrup-Andersen, 1998). They also believe that socio-economic constraints limit the adoption and spreading of improved cultural practices. The optimists believe that there is room to grow more food by taking new land under cultivation, that the green revolution has not run out of steam and that biotechnology has great potential to feed the growing population. The optimists believe that future generations will produce enough geniuses to solve the problems that more people would cause. Both pessimists and optimists have non-scientific motives in their baggage, and apart from political or emotional motives, they largely base their conclusions on the projections of agricultural production.

It can be argued that the preaching of gloom is fruitless unless it is underpinned with science, and is harmful as it encourages fatalism instead of much-needed determinism. Given the many unknowns it is fortunate that the discussion on the carrying capacity of the world continues. Like any important subject, the discussion should be based on the collection and careful interpretation of facts in which research plays a major role. Science can provide much-needed answers and guide the future focus of the political and research agenda (Greenland *et al.*, 1997). Since agricultural production is largely dependent on the soil's productive capacity, soil science should be upfront in providing the much-needed data on soil resources and scenario studies of how soil and land-use changes affect food production. However, it seems that the interest of the developed world is moving towards human ageing and obesity instead of population growth and hunger (Hartemink, 2002). This, in combination with reduced funding for soil research (Mermut and Eswaran, 1997) and the inability of the soil science community to clearly demonstrate the importance of soil science (Greenland, 1991), leaves little room for optimism despite some favourable projections for food production and hunger alleviation (FAO, 2000).

Trends in Crop Yields

Feeding the growing population is largely dependent on crop production. Yields of most crops have increased in the past century (Fig. 2.4). The increase has been very rapid since the 1960s, and ever since that time the rise in yields has taken over from extension of the cultivated area as the major source of greater food production (Evans, 1993;

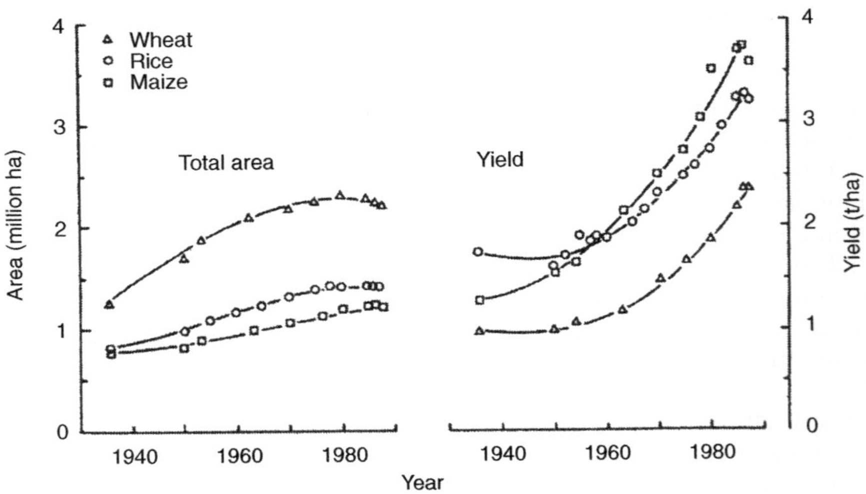

Fig. 2.4. Trends in total area and average world yield of the three major cereals. From Evans (1993) based on FAO production yearbooks and earlier publications.

Greenland, 1997; Tilman *et al.*, 2002). The increase in crop yield has mainly arisen from the fruits of the Green Revolution (Huang *et al.*, 2002). Apart from a doubling or tripling of crop yields, these changes also resulted in less yield variability. Coefficients of variation for world production of wheat, rice and maize have fallen from 6–8% in the 1950s to 3–4% in the 1990s, indicating that the greater reliance on yield enhancement has not increased the instability of food supplies at the global level (Evans, 1993).

Yield increases in the developing regions have remained below the world average. Figure 2.5 shows that worldwide between 1950 and 1995 yields have increased from 1.1 to 2.8 Mg/ha but yields were lower in the developing regions of the world. Evans (1993) had shown that there had been little increase in the area under cultivation. Hence Greenland (1997) concluded that food production kept pace with the huge growth in the population due to the improvement of yields. These yield increases were made possible by improved soil conditions and the availability of crop varieties able to respond to the better conditions. The improvements were the use of fertilizers and lime, irrigation and drainage, and the effects of these amendments outweighed soil physical and chemical degradation (Greenland, 1997).

There is a decreasing trend in the growth rates of cereal production. Staff from IFPRI calculated the growth rates of wheat, rice and maize for different regions in the 1970s, 1980s and 1990s (Table 2.1). The question arises whether crops have reached their genetic plateau

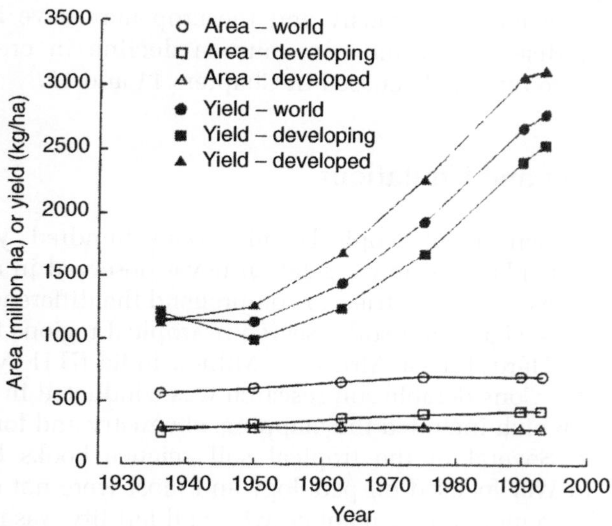

Fig. 2.5. Average area and cereal yield in the developed and developing world. From Greenland (1997) based on FAO production yearbooks.

Table 2.1. Annual growth rates (%) of wheat, rice and maize yields in the 1970s, 1980s and 1990s. Modified from Gruhn *et al.* (2000) based on FAO databases.

Crop	Region	1970s	1980s	1990s
Wheat	Asia	4.3	3.7	0.7
	Latin America	0.6	3.4	2.4
	Sub-Saharan Africa	3.5	0.9	−0.8
	World	2.1	2.8	0.4
Rice	Asia	1.6	2.4	1.6
	Latin America	0.7	3.0	3.7
	Sub-Saharan Africa	<0.1	2.5	−0.6
	World	1.5	2.4	1.5
Maize	Asia	3.4	2.8	1.6
	Latin America	1.5	0.6	3.8
	Sub-Saharan Africa	2.3	1.7	2.1
	World	3.2	0.6	1.8

so that further improvements in soil conditions have no more effect, or whether the stagnation and decline in growth rates is related to soil degradation (Gruhn *et al.*, 2000).

Crop yield decline (or a negative growth rate) may be caused by a range of factors, including weed infestation, the invasion of pests and diseases, genetic decline of the crop, or deterioration of soil chemical, biological or physical properties. The factors may be interrelated. If crops grow poorly because of a declining level in available P, weeds or striga may grow more vigorously and the crop may have less resistance against diseases, which can cause a decline in crop yields. These factors are further discussed in Chapters 11 and 12.

Soil Resources and Limitations

Little was known about tropical soils some hundred years ago. Travellers saw landscapes and vegetation never observed in any of the temperate regions and many tried to comprehend the differences. In the 1920s and 1930s, significant soil research in tropical regions took place in Trinidad (F. Hardy), East Africa (G. Milne), India (H.H. Mann) and Congo (INEAC). Considerable soil research was conducted in Indonesia (E.C.J. Mohr), which included the mapping, chemistry and formation of tropical soils. Several of the tropical soil science books before the Second World War focused on pedology and most were not concerned with the soil as a medium for plant growth. Soil fertility was mainly the research terrain of the agronomist and books focusing on soils as medium for plant growth started to appear in the 1960s (Nye and

Greenland, 1960). Systematic soil inventories started mostly after the Second World War following rapid developments in soil surveying and soil chemistry, and an overall increased interest in the natural resources of the tropics, which was, amongst other things, driven by the fact that many in northwest Europe felt they must do something to mitigate the hunger in developing countries (Russell, 1954). Owing to tremendous research efforts the knowledge base about tropical soils has increased dramatically in the past decades and excellent but slightly dated reviews are given by Sanchez (1976) and van Wambeke (1992).

Soil distribution

Soil inventories were conducted in a number of countries before the Second World War and a brief overview of the advent of soil surveys is given by Simonson (1989) and by Dalal-Clayton and Dent (2001). Soil surveys started particularly in sparsely populated areas where there was ample land for farm extension and a clear need for soil inventories (Kellogg, 1974). This was, for example, the case in the Russian Empire and the USA. In densely populated Europe, where land was relatively scarce, research efforts were much more devoted to maintaining and improving the soil conditions. Farmers in Western Europe had also learned much about their soils by trial and error (Kellogg, 1974). In most European countries soil survey organizations were only established after the Second World War.

In the 1930s and 1940s soil surveys in British tropical territories were much neglected and many of the so-called surveys produced between the two world wars were general accounts, mainly chemical, of the soils occurring within areas covered by the formations portrayed on geological maps. Very little formal soil mapping was done which could be termed soil surveying (Charter, 1957). In the 1950s, soil surveying became popular and was recognized as essential for the planned development of the tropics (Charter, 1957). Each country followed more or less its own system of mapping and classifying soils. As result there was a need for standardized methods of soil observations to allow cross-border correlations of soil information.

At the Sixth International Society of Soil Science (ISSS) Congress in Paris in 1956, a committee was formed with the aim of developing the classification and correlation of the soils of great regions of the world (Dudal and Batisse, 1978). In 1961, the Food and Agriculture Organization (FAO) and United Nations Education Scientific and Cultural Organization (UNESCO) agreed jointly to prepare a soil map of the world at a scale of 1:5 million in association with the ISSS. This international effort encouraged soil scientists from all over the world to cooperate in order to develop a common map legend. In many countries the map compilers could use the existing small-scale maps but there were also considerable gaps, for which extrapolation was required. The complete set of the Soil Map of the World was presented at the Tenth ISSS Congress in 1974 (Dudal and Batisse, 1978) and marked one of the most successful efforts by the international soil scientific community (van Baren *et al.*, 2000).

The FAO–UNESCO soil map has been translated into Soil Taxonomy and the comparative distribution of the soil orders in the world and tropics is shown in Table 2.2. Predominant soils of the tropics are Oxisols, Aridisols, Alfisols, Ultisols and Entisols but the relative distribution of these soils varies between regions (Lal, 1990). The data show that most of the Oxisols and Ultisols occur in tropical regions whereas most of the Mollisols and Spodosols are found in other areas of the world.

The FAO–UNESCO soil map of the world has, since its completion, found wide applications, such as for the assessment of desertification, delineation of major agroecological zones, evaluation of global land degradation, calculation of population-supporting capacity, and the creation of a digital global soils and terrain database (SOTER).

Limitations

With increasing knowledge about the extent and distribution of soils in tropical regions, limitations for agricultural production and other

Table 2.2. Land area of different soil orders in millions of hectares. Figures from Buringh (1979) based on data from 1:5M FAO–UNESCO soil maps and Sanchez (1976).

	World land area		Land area in the tropics	
Soil order	Million ha	%	Million ha	%
Alfisols	1,730	13.1	800	16.2
Aridisols	2,480	18.8	900	18.4
Entisols	1,090	8.2	400	8.2
Histosols	120	0.9	–	–
Inceptisols	1,170	8.9	400	8.3
Mollisols	1,130	8.6	50	1.0
Oxisols	1,120	8.5	1,100	22.5
Spodosols	560	4.3	–	–
Ultisols	730	5.6	550	11.2
Vertisols	230	1.8	100	2.0
Mountains	2,810	21.3	600	12.2
Total	13,170		4,900	

purposes could be quantified. This was important because it enabled many global studies including estimates on the food production capacity of the world. It was found that tropical soils are as diverse and varied as those of temperate regions (Moormann, 1972), which makes it difficult to generalize about their distribution and limitations for agriculture and other uses (Lal, 1990). Table 2.3 presents some major limitations for agriculture in the arable soils of different continents. Mineral stress, defined as nutritional deficiencies or toxicities related to chemical composition or mode of origin (Larson, 1986), affects about one-fifth of the soils in Africa, 59% of the soils in South-East Asia, and almost half of the soils in South America. Soils in Africa and South Asia are prone to drought, whereas 38% of the soils in North and Central Asia have limited soil depth.

Sanchez *et al.* (1982) estimated that two-thirds of the soils of the humid tropics have low nutrient reserves, nearly 60% suffer from Al toxicity and 38% of the land has soils with high P fixation. Estimates were based on the Fertility Capability Soil Classification System. The system was developed to bridge the gap between the subdisciplines of soil classification and soil fertility (Sanchez *et al.*, 1982). At the continental level, Eswaran *et al.* (1997) calculated, based on the FAO–UNESCO maps, that one-quarter of the soils in Africa have medium to low potential and that more than 50% of the land is fragile and not productive.

The problem with such small-scale studies and their interpretation is that soil limitations are crop specific. For example, tea tolerates or even requires soils high in exchangeable aluminium, whereas other crops are often affected at low aluminium concentrations. In other words, what is favourable for one crop might be detrimental for another. Therefore in many crop simulation models a 'generic crop' with average and non-excessive soil requirements is used, which is usually a cereal (e.g. Timlin *et al.*, 1996; Eijsackers, 1998). A general

Table 2.3. Soil resources and their major limitations for agriculture as a percentage of the total land area. Adapted from Larson (1986).

	Drought	Mineral stress	Shallow depth	Water excess	Permafrost	No serious limitation
Africa	44	18	13	9	–	16
South Asia	43	5	23	11	–	18
North and Central Asia	17	9	38	13	13	10
South-East Asia	2	59	6	19	–	14
Central America	32	16	17	10	–	25
South America	17	47	11	10	–	15
World	28	23	22	10	6	11

overview of soil limitations provides, among other things, guidelines for the direction and focus of soil fertility research, even though it is largely based on soil analytical data from the 1950s to 1970s. It also allows studies focusing on land productivity and human population growth. The small-scale studies may also be used in larger-scale studies which take into consideration crop requirements. An example of such a model is QUEFTS (Quantitative Evaluation of the Fertility of Tropical Soils) developed by Janssen *et al.* (1990). It has been used quite extensively in Kenya (Nandwa and Bekunda, 1998) and has recently been adopted for rice (Witt *et al.*, 1999). Overall, model studies for the quantitative evaluation of tropical soil fertility have not been used extensively and this is probably because of lack of sufficient data and the high spatial and temporal variation in soil fertility parameters.

Conclusions

Although soils in tropical regions are highly diverse, with some soils having a high production potential, there are many areas where the soil resources suffer from serious limitations hindering agricultural production and development. Some soils have a very low chemical fertility, are extremely acid or contain toxic substances, but the exact extent of low-fertility soils and the extent of spare land is, despite advanced mapping techniques, still open for debate (Young, 1999). It is therefore unfortunate that surveying and mapping was curtailed or even stopped in many countries in the 1980s (Bouma, 1989), especially because the need for reliable spatial soil information is increasing with the sophistication of soil and crop models. Many of the current global modelling efforts are largely based on soil information from the 1950s to 1970s. The evaluation of soil fertility in tropical regions is also largely based on old data and several generic crop models have been developed. There is a great need for updated soil survey and soil fertility information to monitor the effects of current and past land management on soil properties.

Soil Fertility

The chemical fertility of tropical soils has been the cause for much debate and uncertainty in the past century. This is largely because of the lack of adequate research and because of persistent ideas, which were hard to eliminate. As this book mainly focuses on the chemical soil fertility, a summary is given here of how the theory on the fertility of tropical soils has changed through the years.

In the late 1800s and early 1900s, it was assumed that soil fertility in the humid tropics must be very high because it supports such abundant vegetation as rainforest. The point of view was fairly popular amongst tropical agriculturists, and was prominently mentioned in the book of J.C. Willis (Willis, 1909), which ran through several editions during the first two decades of the 1900s. The American soil scientist E.W. Hilgard thought that soils of the humid tropics must be rich in humus because of the abundant vegetation supplying plant material (Hilgard, 1906). Continuous and rapid rock and soil decomposition was thought to be very high under the prevailing climatic conditions, hence providing a constant supply of minerals for plant growth (Hilgard, 1906). Also Shantz and Marbut (1923) stated that the soil under tropical rainforest is relatively fertile.

The high fertility theory of tropical soils was dispelled when the forest was cut, crops planted and it was discovered that yield levels were disappointingly low or rapidly declining. The idea emerged that soil fertility in the tropics was uniformly low and easily lost by cultivation (Jacks and Whyte, 1939). This was substantiated by the fact that travellers in the tropics noted that soils were lighter in colour and hence assumed that such soils had lower organic matter contents and chemical fertility. It was also assumed that many tropical soils would inversibly change to laterite (ironstone) upon cultivation.

After the Second World War, research emphasis was put on the improvement of soil fertility by the judicious application of inorganic fertilizers. This followed the introduction and widespread use of inorganic fertilizers in the temperate regions (see Table 2.4). A large number of inorganic fertilizer experiments was conducted from the 1950s onwards and the experiments focused on the search for balanced nutrition, the economics of fertilizers, credit, subsidies and marketing of fertilizers, and fertilizer training programmes and extension.

The increased use of inorganic fertilizers was deemed necessary: (i) to increase production per unit of land in the face of a growing

Table 2.4. Fertilizer use in some selected European countries and the USA in different periods. Values in kg nutrients (N, P_2O_5, K_2O)/hectare/year. Modified after Knibbe (2000).

	1913	1936	1986
Germany	47	64	427
The Netherlands	146	320	784
United Kingdom	26	44	356
USA	6	8	94

shortage of arable land in many developing countries; (ii) to increase marketed food supplies or exports; and (iii) to raise incomes and return to labour (FAO, 1987). Furthermore inorganic fertilizers were needed to make full use of the new high-yielding varieties (Greenland, 1975). The combined package of new crop varieties, pest and disease control and the use of inorganic fertilizers caused a dramatic increase in crop yields in many parts of the tropics. This was particularly the case in Asian countries (Table 2.5).

Following the food production decline in the 1960s, the FAO launched in 1961 the Freedom From Hunger Campaign (FFHC), which was partly financed by the world fertilizer industry. The FFHC's main target was to encourage the use of fertilizers by small-scale farmers through education, effective means of distribution and credit. The overall idea was that agricultural production cannot be significantly increased in the developing countries of the world without improving the nutrient status of most soils (Olson, 1970).

Fertilizer use in African countries remained low and was below that necessary to meet future production (Donovan and Casey, 1998). The annual growth rate in fertilizer use in East Africa was about 10% in the 1960s but had decreased to 0.7% in the 1990s. In southern Africa the annual growth rate in fertilizer use decreased from 10% in the 1970s to −3.4% in the 1990s. Overall growth rates in sub-Saharan Africa remained behind the developing world from the 1960s although the gap widened in the 1980s and 1990s (Gruhn *et al.*, 2000). In sub-Saharan Africa, average fertilizer use in 1990 was 8 kg nutrients per hectare arable land and land under permanent crops, as compared to the world average of 93 kg/ha and 81 kg/ha for developing countries (Mwangi, 1997). In addition, most of the inorganic fertilizer in sub-Saharan Africa was used on plantation crops and the majority of the small-scale farmers did not use fertilizers (Donovan and Casey, 1998).

Table 2.5. Fertilizer use in some selected Asian countries in different periods. Values in kg nutrients (N, P_2O_5, K_2O)/hectare/year. Modified from Hossain and Singh (2000) based on FAO databases.

	1968–1970	1983–1985	1993–1995
India	16	61	105
Indonesia	16	111	135
Bangladesh	12	49	93
Thailand	7	20	70
Vietnam	36	62	170
Pakistan	19	79	124

With a few exceptions, large-scale and widespread inorganic fertilizer trials are no longer conducted. Instead of advocating the use of inorganic fertilizers, studies in the late 1980s and early 1990s focused on new arguments to justify the use of inorganic fertilizers. These arguments were developed when nutrient balances were re-introduced as a research tool and widespread soil fertility decline and nutrient mining were being reported for sub-Saharan Africa (Stoorvogel and Smaling, 1990). Not only inorganic fertilizers are being advocated but also integrated nutrient management is being promoted to improve the overall negative nutrient balance and the efficiency of nutrient use (Sanchez, 1994; Vanlauwe *et al.*, 2002).

Decreasing inorganic fertilizer use

Locally it was noted that inorganic fertilizers had little or no effect owing to crop husbandry practices (poor seedbed preparation, improper seeding, delay in sowing, etc.) or because of wrong fertilizer placement, unbalanced nutrient application, incorrect identification of nutrient limitations, or weed and insect problems. These factors were mostly eliminated when fertilizer trials were conducted on a research station but came to the surface when fertilizers were used by smallholders. As an overall result inorganic fertilizers gave a poor profitability, which affected their widespread use in smallholder farming systems in many tropical regions.

Some of the fertilizers being used in tropical regions were given as aid by European countries. This was meant to stimulate the use of fertilizers in tropical regions and increase crop production but also meant that aid funds were retained in Europe. Moreover, European countries could maintain their fertilizer industry, which suffered from the declining rate of fertilizer use by European farmers. The decline in inorganic fertilizer use started in the 1980s when environmental concerns about inorganic fertilizers were rising, for example the eutrophication of surface water and the nitrate content of drinking water, which is said to create health hazards for humans under specific conditions (Addiscott *et al.*, 1991). Inorganic fertilizers have also been associated with the destruction of the ozone layer, as nitrous oxides resulting from denitrification can give rise to products which catalyse ozone destruction (Bouwman, 1998). In other words, fertilizers were regarded as environmentally damaging and as such their use was reduced.

The negative image of inorganic fertilizers in the temperate regions had probably some indirect effects on the use of fertilizers in the tropical regions – although the environmental consequences of continued low use of fertilizers are more devastating than those anticipated from increased fertilizer use in the tropics (Dudal and Byrnes,

1993). The FFHC, which was replaced in the late 1970s by FAO's Fertilizer Programme, gradually ceased in the 1990s and currently the FAO has no such programme.

Summary

Views and opinions about soil fertility in tropical regions have gone through various stages. In the late 1800s and early 1900s it was assumed that tropical soils were uniformly rich. This was followed by a period in which it was believed that tropical soils were of inherent low fertility that was quickly lost by cultivation. From the 1950s onwards research efforts focused on the use of inorganic fertilizers to overcome low soil fertility and a large number of field trials were conducted. In the subsequent period it was found that inorganic fertilizers were not widely used and might be damaging the environment. As a result of the low use of inorganic fertilizers, soil fertility is being mined, leading to a decline in agricultural production. Since the early 1990s a combination of organic and inorganic nutrient inputs has been advocated to maintain and improve the soil fertility under smallholder agriculture in tropical regions.

Tropical Land Use and Management

Land in tropical regions is being used for the same reasons as in temperate regions, i.e. to grow trees, crops and animals for food, as building sites for houses and roads, or for recreational purposes. Sanchez (1976) distinguished the following land management systems in the tropics: nomadic herding, livestock ranching, shifting cultivation, subsistence tillage, and plantation agriculture. In this section, the focus is on arable farming by smallholders and plantation agriculture. Most land in the tropics is being used by smallholders who farm for subsistence but also may grow some cash crops. When in the 16th century Europeans started to sail the world oceans and settled in new continents, a new type of tropical agriculture developed: plantation agriculture including perennial tree crops and annual crops. Plantation agriculture is discussed in Chapter 3.

Types of land use

Smallholder rain-fed agriculture is differently practised in different parts of the world but in essence it has the following characteristics: small scale, subsistence or semi-subsistence with cash crops, little or no external inputs, low level of mechanization, and relatively low yields. Farm sizes largely depend on the intensity of the farming system, which is determined by both population pressure and agroecological conditions. Farming systems in the tropics are as diverse as tropical soils, which makes it difficult to generalize.

Not so long ago smallholder agriculture in the tropics was regarded as backward and wasteful or destructive of natural resources (Hall, 1936; FAO-Staff, 1957). More recent investigations showed that most smallholders are dynamic and innovative and that the farming systems are often well adapted to the local environment (Reijntjes *et al.*, 1992). Some smallholder farming systems are to be regarded as a sustainable form of land use (Torquebiau, 1992). Part of the misconception about smallholder farming systems was the lack of research. Up to the Second World War much of the agronomic research in tropical regions was devoted to plantation crops (Webster and Wilson, 1980). Agricultural development in the smallholder sector was deemed necessary to produce more food for domestic consumption but also to provide exports to earn foreign exchange for the purchase of resources which must be imported for industrial development (Webster and Wilson, 1980). Research was often station-based but from the 1950s onwards considerable attention was given to shifting cultivation (swidden cultivation, slash-and-burn systems), which was the dominant form of land use in many parts of the tropics. As explained earlier this shift of attention was due to changes in political thinking, mainly in Western Europe, on poverty alleviation in the developing countries (Russell, 1954).

Up to the mid-1800s, shifting cultivation had no serious effect on the farmland since the soil and vegetation were given adequate time to regenerate after a period of cropping (Nye and Greenland, 1960). Shifting cultivation systems broke down because of an increase in population, which was caused by abolishing the slave trade, the suppression of tribal wars, medical health improvement, and the relief of famine (Nye and Greenland, 1960). This resulted in a larger demand for crop land, which was aggravated in some countries by European settlers using subsistence farming land to establish agricultural plantations. The pattern is illustrated in the evolution of farming systems in the Kikuyu District of Kenya, in which originally shifting cultivation prevailed but increasing land pressure led to semi-permanent and permanent farming (Table 2.6). Coffee and other cash crops were introduced following the Swynnerton plan, and in the mid-1960s, tea production, maize, potatoes and the application of inorganic fertilizers were spreading in the Kikuyu District (Ruthenberg, 1972). The driving force for changes in the Kikuyu farming systems was population pressure, which is the main force for land-use changes in most of the tropics (Nye and Greenland, 1960; Fischer and Heilig, 1997). Such

Table 2.6. Evolution of the farming system in the Kikuyu Districts of Kenya. Modified after Ruthenberg (1972).

Year (approximate)	Type of farming	Cropping pattern
1860	Shifting cultivation	Maize, beans, mixed cropping
1920	Semi-permanent cultivation	Maize, beans, sweet potatoes, mixed cropping
1950	Permanent cultivation	Maize, beans, sweet potatoes, banana, wattle
1960	Permanent crops + permanent cultivation + some ley farming	Coffee, maize, beans, sweet potatoes, banana, potato, vegetable, pineapple, leys

land-use changes are the results of *in situ* changes in the population. Exceptions occur in the Amazon, where city dwellers become farmers, and in the transmigration areas of Indonesia, where people from densely populated islands like Java and Bali are moved to sparsely populated islands like Sumatra, Kalimantan or Papua.

Changes in land use

Changes in land use and land cover are central to the study of global environmental change including soil degradation, and reflect the rapid population growth in tropical regions. Changes in land use are relatively easy to observe and quantify and several recent studies have focused on the prediction of land-use changes (Veldkamp and Lambin, 2001; Jansen and Di Gregorio, 2002). Simple techniques include a comparison of on-the-ground photographs of the same site at different periods and aerial photographs combined with satellite imagery. Table 2.7 summarizes some of the main data types in land-use and land-cover changes studies, which is followed by a brief review of some major studies on changes in land use in tropical regions.

Table 2.7. Data types in land-use change studies.

Data type	Description	Typical example
Expert knowledge	Oral history of land-use patterns	(Jones, 1999)
Photographs (on the ground)	Pictures from landscapes at different times	(Shantz and Turner, 1958)
Photographs (aerial)	Aerial pictures from the same site and scale taken at different times	(Abbot and Homewood, 1999)
Satellite imagery	Images from satellites, sometimes in combination with a GIS	(McAlpine *et al.*, 2001)

Expert knowledge

Endfield and O'Hara (1999) used historical archival and political data to observe environmental change in west central Mexico. A similar approach was used by Conte (1999) studying forest use and environmental change in West Usambara mountains in Tanzania, whereas oral history was used to understand land degradation in the Uluguru mountains in Tanzania (Jones, 1999). Although these studies yield useful information on the patterns of change, they contain little or no hard data.

On-the-ground photographs – Africa

One of the first studies on soils and vegetation in Africa was conducted by the American botanist H.L. Shantz and soil scientist C.F. Marbut in the early 1920s (Shantz and Marbut, 1923). As part of the fieldwork they toured from Cairo to Capetown in 1919/20 and 1924, during which time they took about 5000 4×5 photographs, mostly in South and East Africa. In 1956 and 1957, a large number of sites were revisited and pictures were taken at approximately the same time of the year. The time interval of 33 and 37 years enabled a broad picture of vegetational and land-use changes (Shantz and Turner, 1958). Almost 80% of the original sites could not be revisited because of urban development or agricultural extensions over the original sites, which reflects some of the major land-use changes that occurred in Africa in the first half of the 20th century. The major conclusions of the report were that in the desert grasslands of South Africa and Kenya, succulent bushes and thorny shrubs have increased at the expense of grasses, and that in the dry forest, clearing and burning transformed woodland into savannah, whereas rainforest was transformed into high grass and low tree savannah. The increase of population has placed a great strain on the more productive lands and in many instances submarginal areas have been taken into cultivation (Shantz and Turner, 1958).

An example of some of the changes in the forest zone in Congo is depicted in Fig. 2.6 showing clearing of forest for cultivation. In Kenya (Fig. 2.7) it was found that the area under tree cover actually had increased, probably as a result of decreased occupancy and cultivation (Shantz and Turner, 1958). At the same time new tree plantings (*Eucalyptus*) have been made. It would be of interest if the sites of Shantz and Turner (1958) were revisited now – more than three-quarters of a century later.

A similar approach was taken by Tiffen *et al.* (1994), who explored the relationship between increasing population density, productivity and environmental degradation, through a case study of

Fig. 2.6. Tree vegetation on hill slope in Kuala Nord (Congo) in 1920 (left picture). Note dense cut-over tree (*Brachystegia* spp., *Isoberlinia* sp., etc.) and liana growth. The same site 36 years later showed that the hill slope on the right has been cleared, probably twice, as indicated by the scrubby vegetation shown in the centre of the picture. Modified after Shantz and Turner (1958).

Fig. 2.7. Changes in vegetation and land use at Lumbwa railway station between Kisumu and Nairobi (Kenya). The left picture was taken in 1920 – note the very open savannah-type vegetation on the hill in the background. The site was photographed again in 1957 – note increase of *Acacia* tree cover on hill in background. Larger trees behind buildings are plantings of *Eucalyptus*. Modified after Shantz and Turner (1958).

Machakos District in Kenya, over the period 1930 to 1990 (Fig. 2.8). They argued that despite a fivefold increase in population the environment in 1990 was in a much better condition than in the 1930s (Tiffen *et al.*, 1994). This was due to a number of reasons. Machakos is

Fig. 2.8. Kiima Kimwe Hill in Machakos District in Kenya in 1937 and 1991. In 1937, the slope (>55%) was almost treeless but in 1991 woodlots and bananas and other fruits were planted on the contour. Modified after Tiffen *et al.* (1994).

close to Nairobi, which meant both a market for agricultural produce and the possibility of earning cash income through off-farm labour. Investments were made in roads and other infrastructure necessary for ready access to urban and overseas markets and to local processing facilities. The farming systems changed with increased soil conservation practices and the growing of cash crops. As a result of off-farm income farmers in Machakos could improve the land (Laegreid *et al.*, 1999). Similar conclusions were derived from a study on increasing population density and the availability of land resources in Sukumuland near Lake Victoria in Tanzania (Meertens *et al.*, 1996).

The study of Tiffen *et al.* provided evidence for Boserup's model on agricultural intensification (Boserup, 1965). The findings were at odds with the common assumptions, generally made on the basis of a shorter period, which assume that there has been little increase in agricultural productivity in Africa, that increased commercial production harms food supplies, that out-migration is all negative, that development depends overwhelmingly on government initiatives and aid support, and that population growth means fewer trees and harms the environment (Tiffen *et al.*, 1994). Various methods and data-sets have been used to arrive at these conclusions. An important method extensively used by Tiffen *et al.* has been landscape pictures from different periods, although the more extensive study of Shantz and Turner (1958), which arrived at different conclusions, had escaped their attention.

Aerial photographs and satellite images – Africa

Several studies on land use in the tropics have used aerial photographs and satellite images from different periods and Geographic Information Systems (GIS) to observe and quantify changes. Holmgren *et al.* (1994) presented surveys of woody biomass on farmland in Kenya using aerial photographs and field measurements in a subsample of the photos. A rapid increase in planted woody biomass was observed between 1986 and 1992, and the annual increase was estimated to be 4.7%. Population density was positively correlated with the volume of planted woody biomass instead of increasing fuel-wood deficit and land degradation following rapid population growth. The results imply that some pessimistic opinions on land-use development in Kenya are false (Holmgren *et al.*, 1994) and confirm some of the observations of Tiffen *et al.* (1994).

A study in Tanzania using normalized difference vegetation index (NDVI) imagery from NASA showed that the overall greenness increased between 1982 and 1994 (Pelkey *et al.*, 2000). Woodland and forest pixels increased in greenness but swamp pixels showed a marked decline in vegetative cover. A detailed study in the Usambara Mountains in Tanga Region, Tanzania, showed a drastic reduction in forest cover from 53,000 ha in 1965 to 30,000 ha in 1991 (Kaoneka and Solberg, 1994). About one-third of the natural forests was used for the conversion to forest plantations and the main cause of the deforestation was the expansion of farmlands and settlements as the population increased (Kaoneka and Solberg, 1994).

Detailed studies from several parts of Kenya arrived at similar conclusions. In the Embu region in Kenya, Imbernon (1999b) studied change in land use in a semiarid and humid area of Mount Kenya. The tree cover decreased from 26% in 1956 to 24% in 1995. The extent of perennial crops (tea, coffee) increased from 1% to 33% over the same period. Bushland, which covered about one-quarter of the area in 1958, was no longer in existence by 1995. Land use in the humid area changed significantly before 1985 and has since stabilized. In the lower arid zone, drastic changes occurred after 1985 and these changes were related to those in the humid region (Imbernon, 1999b). In the highlands of Murang'a District, north of Nairobi, Ovuka (2000) observed substantial land-use changes between 1960 and 1996. In 1960 there was 15% fallow land in one area but this had decreased to 6% in 1996. Woodlots had increased from 1% to 3% and coffee gardens from 0.2% to 12% over the same period. Areas without soil and water conservation practices increased from about 25% in 1960 to 70% in 1996. Most farmers depended on income from the land and thought that livelihood was better in 1996 than in 1960 (Ovuka, 2000).

A study in a protected area in Lake Malawi National Park, Malawi, using aerial photographs showed measurable conversion of closed *miombo* to sparse woodlands (Abbot and Homewood, 1999). Between 1982 and 1990 closed canopy woodland decreased by 7% whereas sparse woodland increased by 342%. The human population more than doubled between 1977 and 1992 and the authors consider increased fuel-wood harvesting the main cause for the decline in *miombo* woodland (Abbot and Homewood, 1999). In southwestern

Burkina Faso, Gray (1999) showed that between 1981 and 1993 the area under cultivation roughly doubled at the expense of scrub savannah. The changes in land use mirrored changes in population, which doubled between 1971 and 1985. In another area, few changes were found as there were nearly no new areas to be cultivated (Gray, 1999).

Politics and national land-use policies are major factors affecting land use. Omiti *et al.* (1999) studied changes in Ethiopia, where since 1975 agriculture land tenure arrangements have changed from a feudal system to a socialist model with semi-collective villages, and from 1991 onwards to a smallholder system based on private ownership. It seems that there was excessive fragmentation of the land in the new system but that more than half of the farmers claimed to have started soil conservation efforts since the collapse of the socialist agrarian structure. More than three-quarters of the farmers are now planting trees and expressed the wish to plant more trees than during the socialist rule. About 20% of the farmers now use fertilizers and at higher rates than before, when only 10% of the farmers used inorganic fertilizers. The authors concluded that there are some grounds for optimism as the freeing-up of the market will bring more sustainable land-use practices (Omiti *et al.*, 1999).

A different conclusion was reached by Tekle and Hedlund (2000), who compared aerial photographs from 1958 and 1986 in the highlands of Kalu District, Ethiopia. Two maps were made using GIS, and a decrease in coverage by shrublands, riverine vegetation and forests was observed. Areas under cultivation remained more or less unchanged so it was concluded that land-cover changes were the result of clearing of vegetation for fuel wood and grazing (Tekle and Hedlund, 2000). The difference from the study of Omiti *et al.* (1999) could possibly be explained by the period of observation, i.e. Tekle and Hedlund (2000) observed changes during the feudal and socialist period, whereas the current free-market system may have yielded a different picture of land cover in the area studied.

Aerial photographs and satellite images – Asia and Central America

Lumbanraja *et al.* (1998) and Syam *et al.* (1997) described changes in land use in West Lampung (south Sumatra), where low fertility soils (Dystropepts and Kanhapludults) are dominant. Between 1970 and 1990, the area under primary forest decreased from 57% to 13%. In 1970, 9% of the area was under slash-and-burn agriculture but in 1990 there was no land left under shifting cultivation. Monocropping plantations, mainly lowland coffee, were absent in 1970 but occupied 40% of West Lampung in 1990. In North Lampung, where Oxisols are predominant, Imbernon (1999a) described land-use changes between

1930 and 1996. Dense forest covered about 80% of the area in 1930, but no more forest was left in 1996. Most changes occurred between 1969 and 1985 following the transmigration programme and the development of agricultural plantations.

In densely populated Java, which had a long history of population pressure causing agricultural land use to expand and intensify, much of the prime agricultural land is converted into residential and industrial areas (Verburg *et al.*, 1999). This is a major factor affecting food production in rapidly growing areas, which will influence the production capacity of existing agricultural areas and occurs in many parts of the world.

In Papua New Guinea, which has an average population of less than 10 persons/km^2, McAlpine *et al.* (2001) compared aerial photographs from the early 1970s with LandSat^{TM} imagery from 1996 and found that there was an increase of approximately 10% in the area of land used for food production. Rural population increased, however, by more than 40%, which indicates that land use intensified significantly. This took place in areas that were already classified in 1975 as having significant intensity of use. Considerable intensification rather than expansion of the area of land appears to have occurred in response to population growth and movement (McAlpine *et al.*, 2001).

In the hillside region of central Honduras it was found that forest cover was reduced from 56% in 1955 to 36% in 1995 (Fig. 2.9). This was largely due to increasing population pressure and agricultural activities, and the largest reduction occurred in land which had slopes of less than 30% (Kammerbauer and Ardon, 1999).

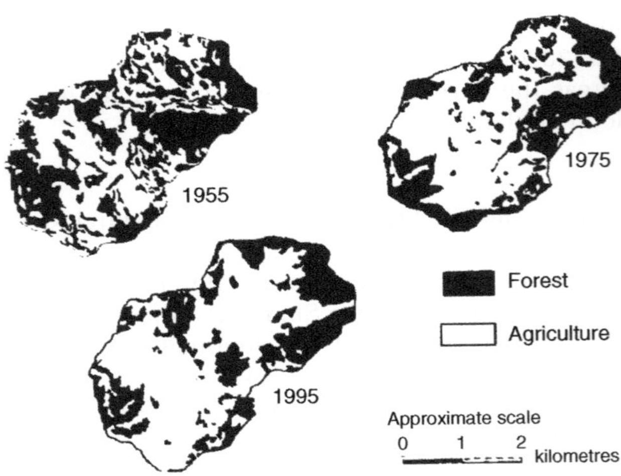

Fig. 2.9. Forest cover maps of the La Lima watershed in the central region of Honduras at different times. From Kammerbauer and Ardon (1999).

Conclusions – land-use change

The remote-sensing studies all showed major changes in land use which could be quantified since the 1930s (North Lampung), 1950s (Honduras, Kenya) or 1970s (Burkina Faso, south Sumatra, Tanzania, Papua New Guinea). Changes in land use reflect the rapid population growth, which occurred almost universally in the areas under study, although the relation between agricultural expansion and population growth is unlikely to be linear (Meertens *et al.*, 1996). No universal pattern in land-use change following population growth could be traced and some interesting trends were observed. The Papua New Guinea study showed that population increase was much higher than the increase of the area under cultivation and little changes were observed in the area under rainforest despite a doubling of the population (McAlpine *et al.*, 2001). In most countries the area under natural forest declined when the population increased. A national survey in Kenya revealed, however, an increase in planted tree biomass in many smallholder agricultural systems (Holmgren *et al.*, 1994) and this was confirmed in large-scale surveys (Tiffen *et al.*, 1994; Imbernon, 1999b). On the other hand, in Malawi and in parts of Tanzania no increase in planted tree biomass was observed (Kaoneka and Solberg, 1997; Abbot and Homewood, 1999). Part of the biomass increase in the Embu Region (Kenya) was caused by the increase in perennial tree crops in smallholder farming systems, which was also observed in Sumatra, where plantation agriculture increased at the expense of rainforest.

All studies showed that land-use changes occur and the quantification of agricultural land has been greatly facilitated by aerial photographs and satellite images (Seto *et al.*, 2000; Jansen and Di Gregorio, 2002). Studying changes in land use is not a goal in itself and such studies could be used for land-use planning or to analyse spatial patterns of soil fertility changes, or regional nutrient budgets. However, the scale at which the observations are made may not be directly applicable to the scale at which the factors that drive land-use change can be predicted or influenced. From the studies reviewed here, there is no overwhelming evidence that land-use changes lead to massive environmental degradation although it can be anticipated that land-use intensification without external inputs or the encroachment of marginal areas may lead to soil degradation.

Soil Degradation

Soil degradation is not a new problem and many of the ancient cultures broke down and disintegrated because of soil degradation problems such as erosion and salinization (Hillel, 1991). Soil degradation caused by agricultural intensification started to receive research interest from the 1940s onwards (e.g. Jacks and Whyte, 1939; Howard, 1940; Osborn, 1948) but it took until the 1970s before it received serious international attention. In 1971, FAO prepared a report entitled *Land degradation* intended to be reviewed at the UN Conference on the Human Environment at Stockholm in 1972. The report contained guidelines for the appraisal and evaluation of land degradation. Information was included on the excess use of inorganic fertilizers but soil fertility decline was not mentioned (FAO, 1971). In the late 1970s, FAO and UNEP prepared a joint report on the methodology for assessing soil degradation (FAO–UNEP, 1978). Chemical degradation was included in the methodology, and pH and base saturation were chosen as attributes to assess soil degradation. Classes were established to evaluate and appraise changes in these attributes in the top 0.30 m of the soil. Although the proposed methodology was fairly detailed, it has never been applied in subsequent studies on soil degradation. In Europe in the 1980s, it was realized that soil degradation was a problem in much of the land under agriculture, although the focus was chemical overloading, groundwater pollution, and soil physical degradation (Boels *et al.*, 1981). None of the studies in the 1970s and 1980s included spatial distribution of soil degradation.

Many soils have been changed in their chemical, physical or biological properties through the agricultural activities of man, including cultivation, tillage, weeding, terracing, subsoiling, deep ploughing, manure and fertilizer addition, liming, draining, irrigation and impoldering (Bridges and de Bakker, 1997). Up to the 1970s, results of human activities on soils were not given much prominence by soil surveyors. Jenny (1941) discussed changes in the nutritional status of the soil resulting from cropping, but deliberately left out the effects of modern mankind in his later book 'in order to establish benchmark for our activities' (Jenny, 1980). Yaalon and Yaron (1966) introduced a framework for man-made soil changes. *Soil Taxonomy* included an anthropic and plaggen epipedon at the Seventh Approximation in 1960 but does not recognize human activities at the highest system level (soil order) in the second edition of *Soil Taxonomy* (Soil Survey Staff, 1999). Anthrosols and fimic horizons were only introduced in the 1988 revised legend of the FAO–UNESCO soil classification (1988). Man as a soil-forming factor has been a difficult issue in pedology, whereas many soils in the world have been drastically altered or degraded as a result of human interference. Soil degradation also occurs in natural environments, but in this book only soil degradation resulting from agricultural activities is considered.

Definition

Although many agricultural activities have been beneficial for the soil, some have caused degradation, i.e. pollution, accelerating erosion, organic matter reduction and nutrient mining, salinization, soil physical degradation and compaction (Bridges and de Bakker, 1997). According to Lal (1997) soil degradation happens when soil cannot perform one or several of the following principal functions:

1. Sustain biomass production and biodiversity including preservation and enhancement of gene pool.
2. Regulate water and air quality by filtering, buffering, detoxification, and regulating geochemical cycles.
3. Preserve archaeological, geological and astronomical records.
4. Support socio-economic structure, cultural and aesthetic values and provide engineering foundation.

Soil degradation is the loss of actual or potential productivity and utility, and soil degradation implies a decline in the soil's inherent capacity to produce economic goods and perform environmental regulatory functions (Lal, 1997). It is difficult to sustain all functions at the same time because some are mutually exclusive. Soil degradation is not the same as land degradation, which embraces the degradation of the overall capacity of the land to produce economic goods and to perform environmental regulating functions. Although there are many definitions of land degradation, they all emphasize the degradation of the soil as a key aspect, and there is probably no form of land degradation that does not include degradation of soil resources.

Soil erosion is obviously the most visible and sometimes most destructive form of soil degradation and it has received considerable research attention. Although salinization, acidification, pollution or nutrient depletion are also important forms of soil degradation, they are far less visible, particularly to the layman and have not received the same amount of research attention as soil erosion, which can be difficult to reverse. Soil erosion can be relatively easily assessed whereas the assessment of nutrient depletion may require long-term monitoring, complicated and expensive laboratory analyses, and is prone to considerable difficulties for its assessment (see Chapter 4).

Assessment

Assessing soil degradation is not an easy task. Soil degradation happens when soil cannot perform one or more of the principal functions discussed above (Lal, 1997). Recently formed gullies in the land or a newly formed salt crust on the topsoil makes it easier to confirm that soil degradation takes place. What if such clear signals are absent and crop yields decrease each season and soil degradation is suspected? The question arises of what are appropriate indicators to assess soil degradation, how should they be measured and analysed, and how is their temporal and spatial variability related to the trends in these indicators? For example, the capacity of the soil to regulate geochemical cycles as given in Lal's definition (1997) is not only difficult to grasp and thus subject to debate, but for proper assessment it requires a thorough understanding of the processes involved and measurements of many properties, each with its unique characteristics and fluctuations. Moreover, many processes interact and threshold values of the soil to sustain a certain function may vary over time. In addition, some soil properties may show resilience whereas others are irreversibly changed within relatively short periods. The great many factors and processes involved make any such assessment a scientifically challenging undertaking. Given the importance of soil degradation, problems with its assessment should not discourage such studies.

There are some guidelines that can be of help in assessing soil degradation:

- Clear signs of soil degradation that can be observed in the field. These could be erosion features (Fig. 2.10), sealing or slaking of the soil surface, salt accumulation at the surface, or compacted and dense soil layers. These features may be accompanied by poor crop growth but that may also be caused by other, less visible, symptoms, such as drought or the outbreak of pests and diseases.

Fig. 2.10. Clear signs of soil degradation in Eastern Congo following forest clearance and cattle ranching.

- Trends in soil properties, which include, for example, a declining pH. The assessment and interpretation of soil chemical properties is discussed in Chapter 4.
- Trends in crop yield. This is probably the best indicator of soil degradation although a number of confounding factors exist, such as a build-up of pests and diseases over time, increased weed infestation, or weather fluctuations.

Soil degradation can be: (i) observed, but that might subjective and susceptible to personal biases; or it can be (ii) measured and monitored, which is subject to a range of methodological problems. Provided the personal biases and the methodological problems in the measurements are understood and quantified, soil degradation can be properly assessed.

Global Assessment of Soil Degradation (GLASOD)

The first approximation to assess and map soil degradation at a global scale was made by Oldeman *et al.* (1991). In this study, soil degradation was defined as a process that lowers the current and/or future capacity of the soils to produce goods or services. Two categories of soil degradation are distinguished: the first group relates to displacement of soil material (soil erosion by water and wind), and the second group deals with soil deterioration *in situ*, such as chemical and physical degradation. More than 200 soil and environmental scientists worldwide were asked to give their expert opinion on the types, degrees, areal coverage and human-induced causes of soil degradation in their regions. The GLASOD map recognizes five different types of human intervention that have caused soil to degrade to its present status: deforestation, overgrazing, agricultural practices, overexploitation of the vegetative cover, and bioindustrial and industrial activities.

GLASOD indicated that almost 40% of the agricultural land in the world has been affected by human-induced soil degradation, and that more than 6% is degraded to such a degree that rehabilitation is only possible through large capital investments. In Oceania about 16% of the agricultural land, 19% of the permanent pasture and 8% of the forest and woodland is degraded. In total 104 million ha of the total land area of 644 million ha is affected by human-induced soil degradation. The GLASOD study has shown that soil chemical degradation (loss of nutrients and/or organic matter, salinization, acidification, pollution) is believed to be important in many parts of the tropics. The loss of nutrients (i.e. soil fertility decline) was severe in Africa and South America, but less of a problem in the upland soils of Asia (Oldeman *et al.*, 1991). Although the GLASOD data are more than 10 years old now and the approach is questionable from a scientific point of view, there is no recent and accurate overview of soil degradation at the global scale. As with accurate information on the distribution of soil types and the soil chemical fertility, there is a need to update and further quantify the information.

Sustainable Land Management

After the Bruntland report of 1987 on 'Our Common Future' and the 'Earth Summit' in Rio de Janeiro in 1992, there has been a wide debate on the issue of sustainability,[1] particularly in relation to soils, land and agriculture. When in the future the history of soil science of the 20th century is written, the 1980s and 1990s will be remembered as eras in which the term 'sustainability' was overwhelmingly present in the soil science literature. What does it mean?

Definition

Overall, sustainability is an inherently vague concept whose scientific definition and measurement still lack wide acceptance (Phillis and Andriantiatsaholiniaina, 2001). In the past decade a plethora of definitions of sustainable land management have been produced (Greenland, 1994). Some have been very lengthy, but in essence, it refers to the combination of production and conservation of the natural resources on which the production depends (Young, 1997). Furthermore, sustainable agricultural production should not release any products into the environment that make the environment less desirable for human occupation and that cannot readily be removed (Greenland, 1994).

The soil is the most important component in sustainable land management, which has been indicated by pedologists (Bouma, 1994), soil fertility experts (Scholes *et al.*, 1994) as well as soil biologists (Swift, 1994). Sustainable land management is deemed necessary for both the developed world, where high-external-input agriculture dominates, and the less developed countries, where the agricultural production is locally dominated by low yields, little or no nutrient inputs, and inadequate soil conservation practices. Most studies dealing with sustainable land management have focused on subsistence

[1] 'Sustainablity' is widely used in economy and development studies, but in this book 'sustainability' is only used in a biophysical and ecological context, with an emphasis on the sustainability of the soil resource for agricultural production.

agriculture. Less attention has been given to high-external-input or plantation agriculture, although it has been indicated that soil scientists have an important role to play in the sustained productivity of plantation forests by developing soil-based sustainability indicators (Nambiar, 1996).

Assessment

Assessing sustainable land management is as difficult as defining it (Fig. 2.11). Key problems are the spatial and temporal borders that need to be chosen for its assessment (Fresco and Kroonenberg, 1992; Bockstaller *et al.*, 1997; Heilig, 1997) and the selection of indicators to evaluate sustainability in a given locality (Pieri *et al.*, 1995; Smyth and Dumanski, 1995). Long-term data are imperative to evaluate the sustainability of land management practices but they are scarce for the tropics (Greenland, 1994). In essence, the problems in the assessment of sustainable land management are similar to those for the assessment of soil degradation. Comparable to developments in soil science, fuzzy logic and fuzzy set theory are increasingly used in studies on sustainability (Cornelissen *et al.*, 2001; Phillis and Andriantiatsaholiniaina, 2001). These models translate linguistic propositions into numerical data but they are largely theoretical exercises, and thus far have not been successfully used in soil fertility studies.

Fig. 2.11. Unshaded tea fields at a plantation in the East Usambara Mountains of Tanzania. The tea was planted in the 1920s and still produces: Is tea cultivation at this plantation sustainable?

Sustainability, although a dynamic concept, implies some sort of equilibrium or steady state (O'Callaghan and Wyseure, 1994). Indicators, defined as attributes that measure or reflect conditions of sustainability (Smyth and Dumanski, 1995), should therefore not show a significant declining trend (Larson and Pierce, 1994). The indicators could be physical, chemical or biological. The identification of suitable indicators is complicated by the multiplicity of physical, chemical and biological factors that control biogeochemical processes and their variation in intensity over time and space (Doran and Parkin, 1996).

A set of basic indicators of soil fertility decline has not been defined, largely due to the range across which soil chemical properties vary in magnitude and importance. Pedotransfer functions can be used to estimate soil attributes that are lacking, for example the CEC can be estimated from the clay content and type and the organic C content. In forestry, there is a wide discussion on the selection of indicators for sustainable land use (de Montgolfier, 1999; Lawes *et al.*, 1999; Szaro *et al.*, 2000). The discussion is more advanced than in agriculture – possibly as forestry land-use systems are less diverse and complex.

FESLM

Various approaches have been advocated to assess sustainable land management and the Framework for the Evaluation of Sustainable Land Management (FESLM) is best known (Smyth and Dumanski, 1995). The FESLM is an extension of the framework for land evaluation (FAO), except that evaluations are based on indicators of performance over time, rather than suitability (Smyth and Dumanski, 1995). Sustainable land management in the context of FESLM consists of five pillars:

1. Maintain or enhance productivity.
2. Reduce the level of production risk (security).
3. Protect the quality of natural resources and prevent soil and water degradation (protection).
4. Be economically viable (viability).
5. Be socially acceptable (acceptability).

All five pillars or criteria must be satisfied simultaneously for a land use to be considered sustainable and all pillars are rated as being equally important in the achievement of sustainability. Although the approach assists in distinguishing factors that drive land management, the framework requires many socio-economic and biophysical data. No case study exists in which the FESLM has been successfully used in the tropics. Except for the high data requirement, there are conceptual problems with the FESLM. For example, consider an agricultural system in which output and input of nutrients is zero. From a soil nutrient point of view such a system may be termed sustainable but if it implies very little crop production and hungry farmers, it is socially unsustainable (Janssen, 1999). In other words, it is almost impossible simultaneously to satisfy all criteria. A partial approach may be more workable.

The partial approach

Agricultural decision-making does not need to await the rigorous development and endorsement of a full framework that incorporates all of the determinants of sustainability, such as indicators, criteria, standards and threshold values. Analyses often reveal that some, but not all, sustainability requirements are met and hence a system can be evaluated as 'partially' or 'conditionally' sustainable. Being able to track the general process towards the goal may be more useful than setting specific, rigid targets to be achieved (El-Swaify, 2000). This seems a useful alternative to the FESLM approach. It is preferable to evaluate the sustainability of a land-use system solely from a soil perspective, or from a socio-economic point of view. The partial approach to evaluate the sustainability of land-use systems is further discussed in Chapter 11.

Discussion and Conclusions

The human population grew very fast in the past century and much of this growth occurred in tropical regions. The concentration of people had environmental implications and in many cases it is likely that the environment, including the soil, has degraded. There are also cases where the environment has improved. Various studies in the past predicted gloom. At the extreme, there are two groups in the world – those who believe that food production and yield increases have reached a plateau and those who argue that sufficient food can be produced for many billions to come. The pessimists believe that the world is approaching its carrying capacity, that no more cultivable land is available and land degradation is widespread, and that production cannot be sufficiently increased to decrease the 800 million or so who are chronically malnourished (Ehrlich, 1968; Pinstrup-Andersen, 1998). The optimists believe that future generations would produce enough geniuses to solve the problems that more people would cause (Huber, 1999; Lomborg, 2001).

Soil science has played, and continues to play, an important role in global studies on food production in relation to the earth's natural resources. This was done through the provision of information on global soil resources, in particular by the FAO–UNESCO Soil Map of the World. The data formed the baseline for many model studies on food production and more recently in studies on climate change in relation to the soil resources. Soil degradation has also received considerable attention, and global and smaller-scale maps are now available indicating types, extent and severity of human-induced soil degradation. Soil fertility decline and nutrient mining are perceived to be an important form of soil degradation in the agricultural land-use systems of the tropics. These studies are mostly qualitative or semi-quantitative, meaning that there are few hard data. They have largely investigated low-external-input agriculture of subsistence farmers, for that is the focus of attention in many developing country projects. It is also in these areas where the largest expansion of the human population takes place and the risks of environmental degradation are large. High-external-input agriculture (i.e. plantation cropping) has been largely ignored in these studies.

With the increased perception and quantification of soil degradation in the tropics, came the call for the design and study of sustainable land management systems. Systems and models were designed for the evaluation of the sustainability of farmers' practices and these models were data driven and either too large and complex, so that they never could be fed with sufficient data, or too small to be meaningful. Partial assessment of sustainable land management seems a more workable method.

References

Abbot, J.I.O. and Homewood, K. (1999) A history of change: causes of miombo woodland decline in a protected area in Malawi. *Journal of Applied Ecology* 36, 422–433.

Addiscott, T.M., Whitmore, A.P. and Powlson, D.S. (1991) *Farming, Fertilizers and the Nitrate Problem*. CAB International, Wallingford, UK.

Anon. (1997) Environmental scare – plenty of gloom. *The Economist* 20 December, 19–21.

Anon. (1999) Unshapely world, too old or too young. *The Economist* 25 September, 60.

Bennett, H.H. (1939) *Soil Conservation*. McGraw-Hill, New York.

Bettany, G.T. (1890) Introduction. In: Malthus, T.R. *An Essay on the Principle of Population or a View of its Past and Present Effects on Human Happiness with an Inquiry into our Prospects Respecting the Future Removal of Mitigation of the Evils which it Occasions*, 6th and last edn. Ward, Lock and Co., London, New York, and Melbourne.

Bockstaller, C., Girardin, P. and Vanderwerf, H.M.G. (1997) Use of agro-ecological indicators for the evaluation of farming systems. *European Journal of Agronomy* 7, 261–270.

Boels, D., Davies, D.B. and Johnston, A.E. (eds) (1981) *Soil Degradation.* A.A. Balkema, Rotterdam.

Boserup, E. (1965) *The Conditions of Agricultural Growth. The economics of agrarian change under population pressure.* Aldine Publishing, New York.

Bouma, J. (1989) Using soil survey data for quantitative land evaluation. *Advances in Soil Science* 9, 177–213.

Bouma, J. (1994) Sustainable land use as a future focus for pedology. (Guest editorial). *Soil Science Society of America Journal* 58, 645–646.

Bouma, J. and Hartemink, A.E. (2003) Soil science and society. *NJAS, Wageningen Journal of Life Sciences* 50, 133–140.

Bouma, J., Batjes, N.H. and Groot, J.J.R. (1998) Exploring land quality effects on world food supply. *Geoderma* 86, 43–59.

Bouwman, A.F. (1998) Nitrogen oxides and tropical agriculture. *Nature* 392, 866–867.

Bradfield, R. (1960) Opportunities for soil scientists in freeing the world from hunger. In: *7th International Congress of Soil Science.* ISSS, Madison, Wisconsin, pp. 1–10.

Bridges, E.M. and de Bakker, H. (1997) Soils as an artefact: human impacts on the soil resource. *The Land* 1, 197–215.

Buringh, P. (1979) *Introduction to the Study of Soils in Tropical and Subtropical Regions.* Pudoc, Wageningen.

Buringh, P., van Heemst, H.D.J. and Staring, G.J. (1975) *Computation of the Absolute Maximum Food Production of the World.* Department of Tropical Soil Science, Agricultural University, Wageningen.

Charter, C.F. (1957) The aims and objects of tropical soil surveys. *Soils and Fertilizers* XX, 127–158.

Conte, C.A. (1999) The forest becomes desert: forest use and environmental change in Tanzania's West Usambara mountains. *Land Degradation and Development* 10, 291–309.

Cornelissen, A.M.G., van den Berg, J., Koops, W.J., Grossman, M. and Udo, H.M.J. (2001) Assessment of the contribution of sustainability indicators to sustainable development: a novel approach using fuzzy set theory. *Agriculture Ecosystems and Environment* 86, 173–185.

Dalal-Clayton, B. and Dent, D. (2001) *Knowledge of the Land: Land Resources Information and its Use in Rural Development.* Oxford University Press, Oxford.

de Montgolfier, J. (1999) Indicators for sustainable development in forestry. *International Journal of Environment and Pollution* 12, 451–458.

Donovan, G. and Casey, F. (1998) *Soil Fertility Management in Sub-Saharan Africa.* The World Bank, Washington, DC.

Doran, J.W. and Parkin, T.B. (1996) Quantitative indicators of soil quality: a minimum data set. In: Doran J.W. and Jones, A.J. (eds) *Methods for Assessing Soil Quality.* SSSA, Madison, Wisconsin, pp. 25–37.

Dudal, R. and Batisse, M. (1978) The soil map of the world. *Nature and Resources* 14, 2–6.

Dudal, R. and Byrnes, B.H. (1993) The effects of fertilizer use on the environment. In: van Reuler, H. and Prins, W.H. (eds) *The Role of Plant Nutrients for Sustainable Food Crop Production in Sub-Saharan Africa.* Dutch Association of Fertilizer Producers, Leidschendam, pp. 141–162.

Edelman, C.H. (1951) De onuitputtelijke natuur. *Economisch-Statistische Berichten* 12 (Sept.), 1–4.

Ehrlich, P. (1968) *The Population Bomb.* Ballantine, New York.

Eijsackers, H. (1998) Soil quality assessment in an international perspective – generic and land-use based quality standards. *Ambio* 27, 70–77.

El-Swaify, S.A. (2000) Is sustainable agriculture a fallacy? *Journal of Soil and Water Conservation* 55, 2–3.

Endfield, G.H. and O'Hara, S.L. (1999) Perception or deception? Land degradation in post-conquest Michoacan, west central Mexico. *Land Degradation and Development* 10, 383–398.

Eswaran, H., Almaraz, R., Vandenberg, E. and Reich, P. (1997) An assessment of the soil resources of Africa in relation to productivity. *Geoderma* 77, 1–18.

Evans, L.T. (1993) *Crop Evolution, Adaptation and Yield.* Cambridge University Press, Cambridge.

FAO (1971) *Land Degradation.* FAO, Rome.

FAO (1987) *Fertilizer Strategies.* FAO, Rome.

FAO (2000) *Agriculture: Towards 2015/30.* FAO, Rome.

FAO-Staff. (1957) Shifting cultivation. *Tropical Agriculture* 34, 159–164.

FAO–UNEP (1978) *Methodology for Assessing Soil Degradation.* FAO, Rome.

FAO–UNESCO (1988) *Soil Map of the World, Revised Legend.* ISRIC, Wageningen.

Fischer, G. and Heilig, G.K. (1997) Population momentum and the demand on land and water resources. In: Greenland, D.J., Gregory, P.J. and Nye, P.H. (eds) *Land Resources: on the Edge of the Malthusian Precipice?* The Royal Society, London, pp. 869–889.

Fresco, L.O. and Kroonenberg, S.B. (1992) Time and spatial scales in ecological sustainability. *Land Use Policy* 9, 155–168.

Gallopin, G.C. and Raskin, P. (1998) Windows on the future – global scenarios and sustainability. *Environment* 40, 7–11.

Gray, L.C. (1999) Is land being degraded? A multi-scale investigation of landscape change in southwestern Burkina Faso. *Land Degradation and Development* 10, 329–343.

Greenland, D.J. (1975) Bringing the green revolution to the shifting cultivator. *Science* 190, 841–844.

Greenland, D.J. (1991) The contributions of soil science to society – past, present, and future. *Soil Science* 151, 19–23.

Greenland, D.J. (1994) Soil science and sustainable land management. In: Syers, J.K. and Rimmer, D.L. *Soil Science and Sustainable Land Management in the Tropics.* CAB International, Wallingford, pp. 1–15.

Greenland, D.J. (1997) Inaugural Russell-memorial-lecture – soil conditions and plant growth. *Soil Use and Management* 13, 169–177.

Greenland, D.J., Gregory, P.J. and Nye, P.H. (1997) Introduction and conclusions. In: Greenland, D.J., Gregory, P.J. and Nye, P.H. (eds) *Land Resources: on the Edge of the Malthusian Precipice?* The Royal Society, London, pp. 861–867.

Gruhn, P., Goletti, F. and Yudelman, M. (2000) *Integrated Nutrient Management, Soil Fertility, and Sustainable Agriculture: Current Issues and Future Challenges.* IFPRI, Washington, DC.

Hall, A.D. (1936) *The improvement of native agriculture in relation to population and public health.* Oxford University Press, London.

Hartemink, A.E. (2002) Soil science in tropical and temperate regions – some differences and similarities. *Advances in Agronomy* 77, 269–292.

Heilig, G.K. (1997) Sustainable development – ten arguments against a biologistic 'slow-down' philosophy of social and economic development. *The International Journal of Sustainable Development and World Ecology* 4, 1–16.

Hilgard, E.H. (1906) *Soils, their Formation, Properties, Composition, and Relation to Climate, and Plant Growth.* Macmillan, New York.

Hillel, D. (1991) *Out of the Earth. Civilization and the Life of the Soil.* University of California Press, Berkeley.

Holmgren, P., Musakha, E.J. and Sjöholm, H. (1994) Not all African land is being degraded: a recent survey of trees on farms in Kenya reveals rapidly increasing forest resources. *Ambio* 23, 390–395.

Hossain, M. and Singh, V.P. (2000) Fertilizer use in Asian agriculture: implications for sustaining food security and the environment. *Nutrient Cycling in Agroecosystems* 57, 155–169.

Howard, A. (1940) *An Agricultural Testament.* Oxford University Press, New York.

Huang, J.K., Pray, C. and Rozelle, S. (2002) Enhancing the crops to feed the poor. *Nature* 418, 678–684.

Huber, P. (1999) *Hard Green. Saving the Environment from the Environmentalist. A Conservative Manifesto.* Basic Books, New York.

Imbernon, J. (1999a) Changes in agricultural practice and landscape over a 60-year period in North Lampung, Sumatra. *Agriculture Ecosystems and Environment* 76, 61–66.

Imbernon, J. (1999b) Pattern and development of land-use changes in the Kenyan highlands since the 1950s. *Agriculture Ecosystems and Environment* 76, 67–73.

Jacks, G.V. and Whyte, R.O. (1939) *The Rape of the Earth – a World Survey of Soil Erosion.* Faber and Faber, London.

Jansen, L.J.M. and Di Gregorio, A. (2002) Parametric land cover and land-use classification as tools for environmental change detection. *Agriculture Ecosystems and Environment* 91, 89–100.

Janssen, B.H. (1999) Basics of budgets, buffers and balances of nutrients in relation to sustainability of agroecosystems. In: Smaling, E.M.A., Oenema, O. and Fresco, L.O. (eds) *Nutrient Disequilibria in Agroecosystems: Concepts and Case Studies.* CAB International, Wallingford, pp. 27–56.

Janssen, B.H., Guiking, F.C.T., van der Eijk, D., Smaling, E.M.A., Wolf, J. and van Reuler, H. (1990) A system for quantitative evaluation of the fertility of tropical soils (QUEFTS). *Geoderma* 46, 299–318.

Jenny, H. (1941) *Factors of Soil Formation. A System of Quantitative Pedology.* McGraw-Hill, New York.

Jenny, H. (1980) *The Soil Resource. Origin and Behavior*. Springer-Verlag, New York.

Jensen, N.F. (1978) Limits to growth in world food production. *Science* 201, 317–320.

Jones, S. (1999) From meta-narratives to flexible frameworks: an actor analysis of land degradation in Highland Tanzania. *Global Environmental Change* 9, 211–219.

Kammerbauer, J. and Ardon, C. (1999) Land use dynamics and landscape change pattern in a typical watershed in the hillside region of central Honduras. *Agriculture Ecosystems and Environment* 75, 93–100.

Kaoneka, A.R.S. and Solberg, B. (1994) Forestry related land use in the West Usambara mountains, Tanzania. *Agriculture Ecosystems and Environment* 49, 207–215.

Kaoneka, A.R.S. and Solberg, B. (1997) Analysis of deforestation and economically sustainable farming systems under pressure of population growth and income constraints at the village level in Tanzania. *Agriculture Ecosystems and Environment* 62, 59–70.

Kellogg, C.E. (1974) Soil genesis, classification, and cartography: 1924–1974. *Geoderma* 12, 347–362.

Kendall, H.W. and Pimentel, D. (1994) Constraints on the expansion of the global food supply. *Ambio* 23, 198–205.

Knibbe, M.T. (2000) Feed, fertilizer, and agricultural productivity in the Netherlands, 1880–1930. *Agricultural History* 74, 39–57.

Laegreid, M., Bockman, O.C. and Kaarstad, O. (1999) *Agriculture Fertilizers and the Environment*. CAB International, Wallingford, in association with Norsk Hydro ASA, Oslo.

Lal, R. (1990) Tropical soils: distribution, properties and management. *Resource Management and Optimization* 7, 39–52.

Lal, R. (1997) Soil quality and sustainability. In: Lal, R., Blum, W.H., Valentin, C. and Stewart, B.A. (eds) *Methods for Assessment of Soil Degradation*. CRC Press, Boca Raton, pp. 17–30.

Larson, W.E. (1986) The adequacy of soil resources. *Agronomy Journal* 78, 221–225.

Larson, W.E. and Pierce, F.J. (1994) The dynamics of soil quality as a measure of sustainable management. In: Doran, J.W., Coleman, D.C., Bezdicek, D.F. and Stewart, B.A. (eds) *Defining Soil Quality for a Sustainable Management*. SSSA–ASA, Madison, Wisconsin, pp. 37–51.

Lawes, M.J., Everard, D. and Eeley, H.A.C. (1999) Developing environmental criteria and indicators for sustainable plantation management: the South African perspective. *South African Journal of Science* 95, 462–469.

Lomborg, B. (2001) *The Skeptical Environmentalist: Measuring the Real State of the World*. Cambridge University Press, Cambridge.

Lumbanraja, J., Syam, T., Nishide, H., Mahi, A.K., Utomo, M., Sarno and Kimura, M. (1998) Deterioration of soil fertility by land use changes in South Sumatra, Indonesia – from 1970 to 1990. *Hydrological Processes* 12, 2003–2013.

Lutz, W. and Qiang, R. (2002) Determinants of human population growth. *Philosophical Transactions of the Royal Society of London B* 357, 1197–1210.

Lutz, W., Sanderson, W. and Scherbov, S. (1997) Doubling of world population unlikely. *Nature* 387, 803–805.
Malthus, T.R. (1826) *An Essay on the Principle of Population or a View of its Past and Present Effects on Human Happiness with an Inquiry into our Prospects Respecting the Future Removal of Mitigation of the Evils which it Occasions.* Ward, Lock and Co., London.
McAlpine, J., Freyne, D.F. and Keig, G. (2001) Land use and population change in PNG, 1975–1996. In: Bourke, R.M., Allen, M. and Salisbury, J. (eds) *Food Security in Papua New Guinea.* ACIAR, Canberra, pp. 277–283.
Meadows, D.H., Meadows, D.L., Randers, J. and Behrens, W.W. (1972) *Limits to Growth. A Report for the Club of Rome's Project on the Predicament of Mankind.* Universe Books, New York.
Meertens, H.C.C., Fresco, L.O. and Stoop, W.A. (1996) Farming systems dynamics: impact of increasing population density and the availability of land resources on changes in agricultural systems. The case of Sukumaland, Tanzania. *Agriculture, Ecosystems and Environment* 56, 203–215.
Mermut, A.R. and Eswaran, H. (1997) Opportunities for soil science in a milieu of reduced funds. *Canadian Journal of Soil Science* 77, 1–7.
Mohr, E.C.J. (1947) De wereldvoedselproductie en de bodem. *Landbouwkundig Tijdschrift* 59, 1–8.
Moormann, F.R. (1972) Soil microvariability. In: *Soils of the Humid Tropics.* NAS, Washington DC, pp. 45–49.
Mwangi, W.M. (1997) Low use of fertilizers and low productivity in Sub-Saharan Africa. *Nutrient Cycling in Agroecosystems* 47, 135–147.
Nambiar, E.K.S. (1996) Sustained productivity of forests is a continuing challenge to soil science. *Soil Science Society of America Journal* 60, 1629–1642.
Nandwa, S.M. and Bekunda, M.A. (1998) Research on nutrient flows and balances in East and Southern Africa: state-of-the-art. *Agriculture Ecosystems and Environment* 71, 5–18.
Nye, P.H. and Greenland, D.J. (1960) *The Soil under Shifting Cultivation.* Commonwealth Bureau of Soils, Harpenden.
O'Callaghan, J.R. and Wyseure, G.C.L. (1994) Proposals for quantitative criteria in the management of sustainable agricultural systems. In: Syers, J.K. and Rimmer, D.L. (eds) *Soil Science and Sustainable Land Management in the Tropics.* CAB International, Wallingford, pp. 27–39.
Oldeman, L.R., Hakkeling, R.T.A. and Sombroek, W.G. (1991) *World Map of the Status of Human-induced Soil Degradation: an explanatory note,* 2nd revised edn. ISRIC, Wageningen.
Olson, R.A. (1970) FFHC. In: Bunting, A.H. (ed.) *Change in Agriculture.* Gerald Duckworth, London, pp. 599–604.
Omiti, J.M., Parton, K.A., Sinden, J.A. and Ehui, S.K. (1999) Monitoring changes in land-use practices following agrarian de-collectivisation in Ethiopia. *Agriculture Ecosystems and Environment* 72, 111–118.
Osborn, F. (1948) *Our Plundered Planet.* Little, Brown and Company, Boston.
Ovuka, M. (2000) More people, more erosion? Land use, soil erosion and soil productivity in Murang'a District, Kenya. *Land Degradation and Development* 11, 111–124.

Pelkey, N., Stoner, C. and Caro, T. (2000) Vegetation in Tanzania: assessing long term trends and effects of protection using satellite imagery. *Biological Conservation* 94, 297–309.

Penck, A. (1928) Das Hauptproblem der physischen Anthropogeographie. In: *First International Congress of Soil Science.* ISSS, Washington DC, pp. 98–116.

Pendleton, R.L. (1954) The place of tropical soils in feeding the world. *Ceiba* 15, 201–222.

Penning de Vries, F.W.T., Rabbinge, R. and Groot, J.J.R. (1997) Potential and attainable food production and food security in different regions. In: Greenland, D.J., Gregory, P.J. and Nye, P.H. (eds) *Land Resources: on the Edge of the Malthusian Precipice?* The Royal Society, London, pp. 917–928.

Phillis, Y.A. and Andriantiatsaholiniaina, L.A. (2001) Sustainability: an ill-defined concept and its assessment using fuzzy logic. *Ecological Economics* 37, 435–456.

Pieri, C., Dumanski, J., Hamblin, A. and Young, A. (1995) Land quality indicators. *World Bank Discussion Paper* 315, World Bank, Washington, DC.

Pinstrup-Andersen, P. (1998) Food security and sustainable use of natural resources. *Ecological Economics* 26, 1–10.

Reijntjes, C., Haverkort, B. and Water-Bayer, A. (1992) *Farming for the Future. An Introduction to Low-external Input and Sustainable Agriculture.* Macmillan Press, London.

Russell, E.J. (1954) *World Population and World Food Supplies.* George Allen & Unwin, London.

Ruthenberg, H. (1972) *Farming Systems in the Tropics.* Clarendon Press, Oxford.

Sanchez, P.A. (1976) *Properties and Management of Soils in the Tropics.* John Wiley & Sons, New York.

Sanchez, P.A. (1994) Tropical soil fertility research: towards the second paradigm. In: *Transactions 15th World Congress of Soil Science.* ISSS, Acapulco, pp. 65–88.

Sanchez, P.A., Cuoto, W. and Buol, S.W. (1982) The fertility capability soil classification system: interpretation, applicability and modification. *Geoderma* 27, 283–309.

Scholes, M.C., Swift, M.J., Heal, O.W., Sanchez, P.A., Ingram, J.S.I. and Dalal, R. (1994) Soil fertility research in response to the demand for sustainability. In: Woomer, P.L. and Swift, M.J. (eds) *The Biological Management of Tropical Soil Fertility.* John Wiley & Sons, Chichester, pp. 1–14.

Seidl, I. and Tisdell, C.A. (1999) Carrying capacity reconsidered: from Malthus' population theory to cultural carrying capacity. *Ecological Economics* 31, 395–408.

Seto, K.C., Kaufmann, R.K. and Woodcock, C.E. (2000) Landsat reveals China's farmland reserves, but they're vanishing fast. *Nature* 406, 121.

Shantz, H.L. and Marbut, C.F. (1923) *The Vegetation and Soils of Africa.* National Research Council and the American Geographical Society, New York.

Shantz, H.L. and Turner, B.L. (1958) *Photographic Documentation of Vegetational Changes in Africa over a Third of a Century.* University of Arizona, Arizona.

Short, R. (1998) An essay on the principle of population, by T.R. Malthus. *Nature* 395, 456.

Simonson, R.W. (1989) *Historical Highlights of Soil Survey and Soil Classification with Emphasis on the United States, 1899–1979.* ISRIC, Wageningen.

Smil, V. (1999) How many billions to go? *Nature* 401, 429.

Smyth, A.J. and Dumanski, J. (1995) A framework for evaluating sustainable land management. *Canadian Journal of Soil Science* 75, 401–406.

Soil Survey Staff (1999) *Soil Taxonomy*, 2nd edn. USDA National Resources Conservation Services, Washington, DC.

Stamp, L.D. (1960) *Our Developing World.* Faber and Faber, London.

Stoorvogel, J.J. and Smaling, E.M.A. (1990) *Assessment of soil nutrient decline in Sub-Saharan Africa, 1983–2000.* Winand Staring Centre-DLO, Wageningen.

Swift, M.J. (1994) Maintaining the biological status of soil: a key to sustainable land management. In: Greenland, D.J. and Szabolcs, I. (eds) *Soil Resilience and Sustainable Land Use.* CAB International, Wallingford, pp. 235–248.

Syam, T., Nishide, H., Salam, A.K., Utomo, M., Mahi, A.K., Lumbanraja, J., Nugroho, S.G. and Kimura, M. (1997) Land use and cover changes in a hilly area of South Sumatra, Indonesia (from 1970 to 1990) *Soil Science and Plant Nutrition* 43, 587–599.

Szaro, R., Langor, D. and Yapi, A. (2000) Sustainable forest management in the developing world: science challenges and contributions. *Landscape and Urban Planning* 47, 135–142.

Tekle, K. and Hedlund, L. (2000) Land cover changes between 1958 and 1986 in Kalu District, Southern Wello, Ethiopia. *Mountain Research and Development* 20, 42–51.

Tiffen, M., Mortimore, M. and Gichuki, F. (1994) *More People, Less Erosion. Environmental Recovery in Kenya.* John Wiley & Sons, Chichester.

Tilman, D., Cassman, K.G., Matson, P.A., Naylor, R. and Polasky, S. (2002) Agricultural sustainability and intensive production practices. *Nature* 418, 671–677.

Timlin, D., Pachepsky, Y.A. and Acock, B. (1996) A design for a modular, generic soil simulator to interface with plant models. *Agronomy Journal* 88, 162–169.

Tinker, P.B. (1985) Soil science in a changing world. *Journal of Soil Science* 36, 1–8.

Torquebiau, E. (1992) Are tropical agroforestry home gardens sustainable? *Agriculture, Ecosystems and Environment* 41, 189–207.

Trewavas, A. (2002) Malthus foiled again and again. *Nature* 418, 668–670.

Tuljapurkar, S. (1997) Demography – taking the measure of uncertainty. *Nature* 387, 760–761.

van Baren, F.A. (1960) Soils in relation to population in tropical regions. *Tijdschrift voor Economische en Sociale Geografie* 51, 230–234.

van Baren, J.H.V., Hartemink, A.E. and Tinker, P.B. (2000) 75 Years The International Society of Soil Science. *Geoderma* 96, 1–18.

van Wambeke, A. (1992) *Soils of the Tropics – Properties and Appraisal.* McGraw Hill, New York.

Vanlauwe, B., Diels, J., Sanginga, N. and Merckx, R. (2002) (eds) *Integrated Plant Nutrient Management in Sub-Saharan Africa: From Concept to Practice*. CAB International, Wallingford, UK.

Veldkamp, A. and Lambin, E.F. (2001) Predicting land-use change. *Agriculture Ecosystems and Environment* 85, 1–6.

Verburg, P.H., Veldkamp, A. and Bouma, J. (1999) Land use change under conditions of high population pressure: the case of Java. *Global Environmental Change* 9, 303–312.

Webster, C.C. and Wilson, P.N. (1980) *Agriculture in the Tropics*, 2nd edn. Longman, Harlow.

White, J.T. (1941) *Bodemvruchtbaarheid en Landsbelang*. Archipel Drukkerij, Buitenzorg.

Willis, J.C. (1909) *Agriculture in the Tropics. An elementary treatise*. Cambridge University Press, Cambridge.

Witt, C., Dobermann, A., Abdulrachman, S., Gines, H.C., Wang, G.H., Nagarajan, R., Satawatananont, S., Son, T.T., Tan, P.S., Tiem, L.V., Simbahan, G.C. and Olk, D.C. (1999) Internal nutrient efficiencies of irrigated lowland rice in tropical and subtropical Asia. *Field Crops Research* 63, 113–138.

Yaalon, D.H. and Yaron, B. (1966) Framework for man-made soil changes – an outline of metapedogenesis. *Soil Science* 102, 272–277.

Young, A. (1997) *Agroforestry for Soil Management*, 2nd edn. CAB International, Wallingford.

Young, A. (1999) Is there really spare land? A critique of estimates of available cultivable land in developing countries. *Environment, Development and Sustainability* 1, 3–18.

3 Plantation Agriculture

No general judgement is possible on the environmental impact of plantation management *per se*; the literature is insufficient.

M. Tiffen and M. Mortimore (1988)

Plantation agriculture is more than 400 years old and contributes to the regional and national economies in many tropical countries. Throughout its history, plantation agriculture has been a cause for debate because it is associated with the exploitation of land and labour. A British plantation owner remarked in the 1980s: 'A sound respectable case in favour of plantation agriculture has yet to be made, whereas the case against plantation agriculture has been based on real injustices and has a strong emotional appeal' – a clear quest for scientific investigations. However, an assessment of the biophysical and socio-economic impact of plantation agriculture in the tropics would be an arduous task. This chapter focuses on the history and importance of plantation agriculture with an emphasis on soil management aspects. In subsequent chapters, soil changes under plantation cropping are reviewed including detailed case studies on sisal and sugarcane plantations.

Introduction

There are a number of definitions of plantation agriculture but it is often referred to as a large scale, mostly foreign owned and specialized high-input/high-output farming system that is export oriented (Courtenay, 1980; Goldthorpe, 1987, 1994). Tiffen and Mortimore (1988) defined plantation agriculture as a farm over 100 ha in size, with a specialized management team in charge of a labour force with specialized production techniques. A more recent definition is given by Stephens *et al.* (1998): plantations are defined as areas that are typically monocropped with perennials, producing tropical or subtropical products that commonly require prompt initial processing and for which there is an export market. Plantation production systems generally require large amounts of fixed capital investment on planting material, processing/packaging equipment and infrastructure such as roads and housing. They normally generate some activity during most of the year and therefore tend to have a large permanent labour force. These characteristics mean that there is limited opportunity for rapid changes in either product or process, and plantation organizations are therefore vulnerable to fluctuations in the cost of inputs and the price of the commodity (Stephens *et al.*, 1998).

An important motive underlying large-scale production is to obtain a regular supply of goods of constant quality, whether for delivery to a factory (tea, oil palm) or timely delivery to an established market such as export bananas (Ruthenberg, 1972). The term 'plantation' is also used in forestry: timber plantations, plantation forests or forest plantations (Evans, 1986; Parrotta, 1992; Sedjo and Botkin, 1997). Forest plantations are man-made forests and cultivation is generally less intensive and no prompt processing of the produce is required. Forest plantations are defined as: a forest crop or stand raised artificially, by either sowing or planting (Evans, 1992).

History

In order to understand some of the theoretical and political controversies that surround plantation agriculture or the difficulties in defining either the plantation or plantation crops, a brief survey of the historical background is needed (Tiffen and Mortimore, 1988). Plantation agriculture started in the 16th century when the Portuguese settled in coastal parts of Brazil (Courtenay, 1980). The area lacked any known mineral wealth and settlers were involved in the cultivation of sugarcane, which would find a ready market in Portugal. The Portuguese became familiar with the cultivation of sugarcane in India. Slaves were brought in from Portugal's Atlantic island colonies, and during the 16th century Brazil was the world's major supplier of sugar (Courtenay, 1980). It showed that large-scale agriculture in the tropics could be a successful business.

Two types of plantation establishment developed in the 17th century. In densely populated areas like Java the initial focus was on trading settlements leading to intensive exploitation. In the 1800s, a forced cultivation scheme was introduced whereby Javanese farmers were obliged to hand over for export a fixed quantity of agricultural produce. About one-fifth of the land of each village was used for this compulsory cultivation, and in most villages coffee, sugar and indigo and to a lesser extent tobacco, tea and spices were planted. This type of plantation agriculture is usually referred to as *hinterland of conquests* and was practised in the Portuguese and Dutch colonies in South and South-East Asia (Courtenay, 1980). The colonial power was mainly interested in the organization of the people for the production of wealth, but was not interested in the land.

Many more examples exist from the other type: hinterlands of exploitation. This generally took place in sparsely populated areas with unfamiliar conditions (Courtenay, 1980). There was readily available land in such areas, but in the 19th century European settlers and investors were not attracted by the harsh conditions of the tropics. Moreover the creation of plantations required a high initial investment per hectare, the benefits of which were uncertain as crop yields do not reach their full capacity until some years after planting. In the 19th century, colonial rulers increased their power and searched for an economic base for the acquired colonies (Courtenay, 1980). As a result, investors, adventure seekers and planters were attracted by exploitation prospects when plantation companies were given favourable investment conditions. With the spread of foreign planters and capital, colonies became integrated in the world economy (Loewenson, 1992).

The demand for land for plantation crops resulted in the loss of land for local farmers. In Africa this happened most heavily in the southern part but also in East and North Africa. In Zimbabwe, for example, 6 million ha or one-sixth of the country, passed into European hands (Blaikie and Brookfield, 1987). In the early 20th century, there were rapid developments in plantation agriculture in South America (banana, coffee), Asia (rubber, oil palm, cocoa) and Africa (oil palm, coffee, tea). Developments were market-driven, for example the demand for rubber used in the car industry resulted in large-scale rubber plantations in South-East Asia in the 1920s. Oil palm seeds from Sumatran plantations were the source for the large-scale planting of oil palm in Malaysia in 1917, whereas from about 1910 Europeans began to develop plantations in Nigeria, the Cameroons and the Congo (Henderson and Osborne, 2000). The plantation industry rapidly expanded after the Second World War in Africa and South-East Asia when new land was opened up for plantations, which sometimes followed large-scale logging activities.

Productivity on many of the older plantations was low because of low-yielding planting material, erosion resulting from clean weeding, little or no use of inorganic fertilizers, and overall poor crop husbandry practices. Europeans, who led in the developments of plantation crops in tropical regions, made some terrible mistakes, particularly in the earlier periods, when they attempted to raise rubber as an orchard crop and kept the land clean and well cultivated (Pendleton, 1954). This was not only expensive but also led to serious soil erosion. Productivity of the plantations increased with increasing knowledge, which mainly resulted from research.

Many tropical countries received independence in the 1960s and in some countries plantations were nationalized as they were considered synonymous with expropriation of land and exploitation of labour. Owing to mismanagement, fluctuating commodity prices and political turmoil, productivity of nationalized plantations often decreased. This happened, for example, to nationalized sisal and coffee plantations in Tanzania. In other countries, such as Papua New Guinea and Malaysia, the area under plantation crops increased after independence. Also in the 1960s, there was a tendency to integrate estate and smallholder production in nuclear estates. The idea was that commercial estates would be surrounded by a ring of smallholders, who would obtain advice and material inputs through the scheme management (Ruthenberg, 1972). It works well in a number of countries and for a number of crops, but has not been successful in the sisal industry of Tanzania. Overall the firm distinction between plantation and smallholder agriculture formed during the colonial period has disintegrated (Tiffen and Mortimore, 1988).

Importance and Spreading of Plantation Agriculture

The most important plantation crops are cocoa, coffee, tea, coconut, bananas, rubber, oil palm, jute, sisal and hemp (Burger, 1994). Other important plantation crops are sugarcane, tobacco, cinchona and pineapple. Oil palm is currently the most valuable plantation economy of the tropical world (Henderson and Osborne, 2000). The world plantation belt runs from Central and South America across the equatorial regions of Africa to Asia, to the Far East and Queensland in Australia (Fig. 3.1).

The exact extent of plantations in the tropics is not known, but several reports have indicated that the area under plantation agriculture has increased in the past decades. Estimates from the 1970s showed that about 4% of the farming areas in the tropics are plantations (Sanchez, 1976). Based on the cultivated area, the plantation

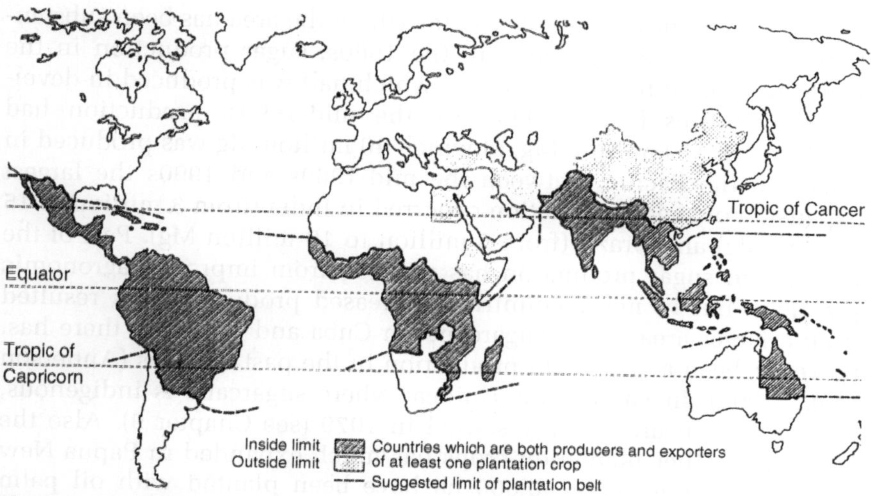

Fig. 3.1. The world plantation belt (Courtenay, 1980).

area in the mid-1970s was about 20 million ha. Nair (1984) estimated that perennial plantation crops in 1984 occupied about 8% of the total arable area in the tropics, which suggests an increase in area under plantations. Global statistics on the area under perennial cropping can be obtained from FAO databases but no distinction is made between smallholders and plantations.

Large plantation areas occur in Malaysia, where the extent of oil palm increased from about 150,000 ha in the early 1970s to 2.0 million ha in 1990 (Härdter *et al.*, 1997) to 2.6 million ha in 1997 (Jalani, 1998) and over 3 million ha at the end of 1998. The increase occurred at the expense of the rainforest, and rubber and cocoa (Pushparajah, 1998). The area under cocoa in Malaysia grew from 10,000 to 307,000 ha between 1970 and 1986 (Webster and Watson, 1988), but cocoa areas have declined to about 100,000 ha in 1998 because of labour shortages. It has been estimated that at least 0.8 million ha new oil palm plantings in Malaysia involved clearance of forest (PORIM, 1994).

The Malaysian government is actively promoting the cultivation of tree crops with a view to compensating for the loss of tree cover because of forest felling and to generating income for both the smallholder and the plantation section. Another country where tree crops are actively promoted is Côte d'Ivoire. Cocoa production in Côte d'Ivoire increased from 150 Mg in the 1960s to over 1100 Mg in the mid-1990s, which reflects a large increase in the area under cocoa, which occurred largely at the expense of the rainforest. In Côte d'Ivoire cocoa is mainly cultivated by smallholders.

An important plantation crop for which the area has been substantially increased is sugarcane. In the 1960s, sugar production in the world was about 64 million Mg, of which half was produced in developing countries (FAO, 1996). By the mid-1990s, production had increased to 119 million Mg, of which 40 million Mg was produced in Asia and the Pacific. Between the mid-1960s and 1990s the largest expansion of sugar production occurred in India (from 3 million to 15 million Mg) and Brazil (from 5 million to 10 million Mg). Part of the increase in sugar production has resulted from improved agronomic practices, but in many countries, increased production has resulted from a larger area under sugarcane. In Cuba and Barbados there has, however, been a decline in production in the past decades (Anderson *et al.*, 1995). In Papua New Guinea, where sugarcane is indigenous, commercial sugarcane only started in 1979 (see Chapter 9). Also the area under other plantation crops has much expanded in Papua New Guinea and more than 80,000 ha have been planted with oil palm since the mid-1960s. Overall, plantation crops cover less than 4% of the total land under agriculture in Papua New Guinea.

Very few plantation crops have been as drastically reduced as the sisal in East Africa, especially in Tanzania. In the 1960s, sisal production in Tanzania equalled 234,000 Mg or nearly one-third of the world's annual sisal fibre production, but in the mid-1980s, Tanzania's sisal production had declined to about 30,000 Mg of fibre per year (see Chapter 10). In the Tanga Region, there were more than 70 large sisal plantations in the 1960s, but in the late 1980s less than 20 were fully operational.

A growing proportion of perennial crop cultivation is in the hands of smallholders, such as the cultivation of cocoa and oil palm in West Africa, tea in Kenya, rubber in Indonesia, Malaysia, Thailand and Sri Lanka, and cashew in Tanzania (Ruthenberg, 1972; Webster and Wilson, 1980; Watson, 1990). Plantation agriculture has largely expanded in the past decades in many countries, although some plantation crops are also grown by smallholders either as monocrop or incorporated in mixed-cropped homegardens.

Research in Plantation Crops

Agricultural research was greatly expanded in the 1920s and 1930s and the research was mostly concentrated on the plantation industries and on crops that could be grown for sale (Hall, 1936). Cash crops brought in money, which could pay for the agricultural research. Of 42 technical agricultural officers in the Kenya service in 1929 only 12 were employed on native reserves. Much of the early work on African soils was pedological with land evaluations for plantation crops like coffee or tobacco (Hall, 1936). Until the outbreak of the Second World War, the Dutch in Indonesia were the leaders in research on most of the commodity crops, but Commonwealth countries became the major contributors after the Second World War. With considerable research effort in the 1950s and 1960s, plantation agriculture became more scientifically based and crop yields increased accordingly (Webster and Wilson, 1980).

Up to the 1950s, there was more scientific research on plantation crops than on food crops. Research was conducted by different groups who had little interaction, and this situation existed in Papua New Guinea, where systematic soil research only started in the 1950s. At the experimental stations for export tree crops (plantation crops), agronomists and soil fertility experts investigated optimum inorganic fertilizer rates and nutrient deficiencies using field trials, glasshouse trials and on-farm experiments. The second group were pedologists and soil surveyors, who mapped the soils of large parts of the country (Bleeker, 1983). Until the 1970s, agronomic and soil fertility research focused on plantation crops and no significant research on food crops took place (Hartemink and Bourke, 2000). As a result much more is known about nutrient deficiencies in plantation crops than in food crops in Papua New Guinea (Table 3.1). Rubber is a fairly marginal crop in Papua New Guinea but it has received more research than sago, which is the main staple for more than 10% of the people (Hartemink and Bourke, 2000).

Table 3.1. Nutrient deficiencies reported in plantation crops and food crops in Papua New Guinea; √ means that the deficiency has received serious research investigation. Table based on reconciliation of published literature over the period 1955–1998. Modified from Hartemink and Bourke (2000).

		Macronutrients					Micronutrients				
		N	P	K	Mg	S	B	Zn	Fe	Cu	Mn
Plantation crops	Arabica coffee	√	√	√	√	√	√	√	√	√	√
	Cocoa	√					√	√	√	√	√
	Coconuts	√	√	√		√	√	√			
	Rubber					√	√		√	√	√
Food crops	Sweet potato	√	√	√	√	√	√	√	√	√	
	Taro	√	√	√		√					
	Irish potato		√				√				
	Maize		√			√					
	Rice					√		√			
	Groundnuts					√	√	√			
	Citrus spp.							√			√

The situation in Papua New Guinea is probably no exception and throughout the tropical world, research into tree crops is fairly well advanced due to the long-term commitments of commercial research organizations. Examples of long-serving research institutes in the tropics are Rubber Research Institute of Malaysia (RRIM), West African Institute for Oil-palm Research (WAIFOR), Coffee Research Institute (CRI) and the Kericho Tea Research Foundation (TRFK) in Kenya. These research organizations usually receive their funds through a levy on the export of the commercial produce. Fruits, nuts and spices have received less research attention by commercial organizations.

Research investments in cash crops may be beneficial for smallholder farmers as well, although estimates of on-farm benefits from agricultural research are difficult and complex (Pannell, 1999). Research on tea plantations in Tanzania has been shown to be of great benefit to smallholders and the financial return on the research investment has been immense (Carr, 1999). Also Webster and Wilson (1980) noted that crop productivity of smallholders can be more easily raised when improved planting materials and technological advances have been developed and tested on plantations. In Papua New Guinea, for example, smallholder coffee farmers are benefiting greatly from the extensive research conducted at plantation research centres (Table 3.1).

It is difficult to distinguish between the real benefits of research on plantation crops and the fact that smallholders start to cultivate cash crops that generate financial returns. Studies in Indonesia have shown that the average net income of oil palm smallholders was seven times higher than that of their neighbours who are mainly involved in the subsistence production of food crops (Härdter *et al.*, 1997). Despite difficulties in the assessment, it seems likely that research in plantation crops is useful to smallholders provided sufficient interaction exists between plantation research centres and NGOs or government extension services.

International research organizations and NGOs are usually not interested in conducting research on plantation crops because their main focus is on resource-poor smallholders. Crops grown on plantations have been referred to as non-CGIAR crops (Smith, 2000). In the past decade, some CGIAR centres have been established to conduct research in tree crops (e.g. Center for International Forestry Research (CIFOR) and International Centre for Research in Agroforestry (ICRAF)). Coconuts and bananas have also been included in some genetic database networks (Smith, 2000). Little has been done on the major plantation crops through the CGIAR centres but much understanding on perennial crops has been gained by the agroforestry research conducted by ICRAF.

Pros and Cons of Plantation Agriculture

Agricultural plantations have a number of socio-economic, ecological and economic advantages and disadvantages when compared with smallholder agricultural systems. Disadvantages are the financial risks due to fluctuating world market prices, and the dependence on cheap labour and often advanced technology, which requires imports and foreign exchange. The drain of foreign exchange to pay for imports (machinery, equipment, fuel and inorganic fertilizer) may exceed the income through export of the plantation produce, because of low productivity or low world market prices. However, in most situations plantation agriculture is a profitable business and earns foreign exchange.

In a number of countries, the plantation sector contributes substantially to the Gross National Product (GNP) and the wealth of the nation. In the 1980s, rubber exports contributed more than 10% to the GNP of Malaysia whereas in Côte d'Ivoire a group of plantation crops produced 22% of GNP. As yields are usually higher on plantations, they may contribute more to GNP than the area they occupy, for example in Kenya tea plantations comprise 35% of the area under tea, but they produce more than 60% of the total output (Tiffen and Mortimore, 1988). In India, the major plantation crops tea, coffee, rubber and cardamom covered less than 0.5% of the total cropped area in the mid-1980s, but accounted for over 10% of the added value in agriculture (Giriappa, 1989).

From an economic viewpoint an important difference with smallholder agriculture is that plantation owners are not resource poor. They may not use proper soil conservation practices or apply inorganic fertilizers, but if there is a need to implement such practices, then the shareholders need to be convinced of the necessity and money may become available. Even when crop husbandry is poor and production declines or ceases, the plantation manager will probably lose his job, but not his life because of hunger.

There are many studies on the social impact of plantation agriculture, e.g. the study by Loewenson (1992), who emphasized the poor conditions at agricultural plantations in Zimbabwe. The political trend in Western Europe in the 1960s and 1970s was that overseas-owned plantations should be rejected on moral grounds as they are often situated on the best soils and exploit land and labour. In some cases, plantations might indeed be situated on the best soils, but many of the rubber and oil palm plantations in western and southwestern Nigeria are situated on poor soils because no soil surveys were conducted before the land was planted (Saylor and Eicher, 1970). There are several examples of historical failures in plantation development. The failure of Danish plantations on the former Gold Coast colony of Ghana was largely due to the unsuitable climate and soil conditions (Breuning-Madsen *et al.*, 2002).

Biodiversity and environmental effects

Crop monocultures are often regarded as unnatural, ecologically dysfunctional, and a threat to sustainable agriculture (Wood, 2000). In nature, however, there are also many productive monocultures, although they seem to occur in marginal or undisturbed conditions – an example is the large area of *Imperata* grasslands in South-East Asia. A common objection to plantations, which are mostly crop monocultures, is that they contain a low diversity of wildlife as compared to natural forests (Cannell, 1999). Biodiversity receives increasing public and political concern (Gowdy and McDaniel, 1995; Morin, 2000). In many cropping systems biodiversity has been reduced whereas productivity has remained high (Anderson, 1994).

Research conducted in Malaysia showed that the number of species of mammals found in primary forest was 75 but fewer than 20 species were found in oil palm or rubber plantations (PORIM, 1994). Monoculture systems severely restrict habitat and favour only a very restricted number of cohabiting species. Biodiversity conservation at the expense of agricultural development and poverty alleviation is a difficult discussion, to which soil scientists could contribute substantially. Soil biodiversity has been mainly the research terrain of the soil biologists and several recent papers have highlighted the relationship between biodiversity and ecosystems functioning (e.g. Lavelle *et al.*, 1994; Beare *et al.*, 1997; Fragoso *et al.*, 1997; Freckman *et al.*, 1997), but pedologists have also shown interest (Yan *et al.*, 2000).

The biggest threat to biodiversity is the loss of habitat through the expansion of the plantation area (Tinker, 1997). Currently, South-East Asian farmers and companies are exploiting the oil palm to the limit by converting forests, with all their biodiversity, to plantations for profit. The smoke clouds from the burning rainforests are the downside of a successful, or perhaps over-successful, industry (Henderson and Osborne, 2000). In many parts of Malaysia, Sabah and Kalimantan new oil palm plantings are increasingly being made on peatland dominated by Histosols (Fig. 3.2). Tropical peat lands are one of the largest near-surface reserves of terrestrial organic carbon (Page *et al.*, 2002). Clearing and draining peatland causes much subsidence (Wösten, 1997) and the decomposition of the peat emits large quantities of CO_2, which is a major greenhouse gas. Large areas of peatland are being cleared in South-East Asia and it has been estimated that up to 40% of the world annual carbon emission is due to peat decomposition and forest fires (Page *et al.*, 2002). Smoke and milling by-products (effluent) may cause pollution but in the oil palm industry this is increasingly controlled following treatment of the effluent (Basiron and Darus, 1996).

Fig. 3.2. Natural forest on peat soils (Histosols) in Malaysia being cleared for the planting of oil palm. Mature oil palm fallen over because of the subsidence following drainage of the peat. These soils would remain under forest if it were not economical to plant them with oil palm.

Few soil pollution and environmental impact studies have been conducted in developing countries, where both local industries and often foreign investment have shown a general lack of appreciation of the environment (Naidu, 1998). In developing countries environmental ethics and laws have not kept pace with the increase in pesticide use (Lal *et al.*, 1988). Moreover, there is shortage of data and the expected increase in agricultural production in developing countries may cause even more environmental damage (Tinker, 1997). In a recent review of environmental concerns about pesticides in soil and groundwater in Oceania, there were many examples from Australia and New Zealand, but hardly any from the tropical countries in Oceania – simply because they were not available (Theng *et al.*, 2000), but also because pesticides use in those countries is low compared with Australia and New Zealand.

It is likely that the appreciation of a healthy environment and the number of environmental studies are proportional to the wealth and inversely related to the poverty rate of a nation. In recent years, however, some soil pollution studies have been conducted in tropical countries (Matin *et al.*, 1998), and a considerable amount of work is done by the soil science department of Bayreuth University, Germany (Wilcke *et al.*, 1999a,b,c). Plantation crops are often grown with high levels of agrochemical inputs such as pesticides, herbicides and inorganic fertilizers. These inputs may pollute the environment when not used judiciously. Some studies have been conducted on plantations and these are discussed in the subsequent chapters. The environmental impact factor that has been most researched in the tropics is soil erosion, and there are excellent reviews available (Hudson, 1986; Lal, 1990; Morgan, 1995; El-Swaify, 1997).

Trends in Crop Yields

Global cereal yields increased on average from 1.1 to 2.8 Mg/ha (+155%) between 1950 and 1995 and broadly kept pace with population growth (Greenland, 1997). Yields of perennial crops also increased between the 1950s and 1990s and Table 3.2 shows 10-year averages for four major perennial crops. Yield increases were considerable for tea, coffee, cocoa and the oil palm in Africa, but the most spectacular yield increase occurred in the oil palm in Asia. Yield data were calculated as total annual production divided by the area under the crop from FAO databases.

In most countries, yields of the major plantation crops have increased in the past decades. In India, for example, coffee yield increased by almost 60% in 15 years (Table 3.3). The incidence of frost in Brazil in the mid-1970s and the resulting loss in coffee production and natural calamities in Colombia thereafter were great incentives to the Indian coffee export (Giriappa, 1989). The drought of 1980–1981

Table 3.2. Perennial crop yields (kg/ha/year) for different periods, based on FAO data provided by W. Stephens (Cranfield University, personal communication, 2000). Annual yield data were averaged for periods of 10 years.

				Oil palm	
Period	Tea	Coffee	Cocoa	Africa	Asia
1949–1959	640	384	229	932	236
1960–1969	769	451	307	1,135	391
1970–1979	768	493	343	1,245	1,718
1980–1989	867	524	373	1,466	5,519
1990–1996	1,072	549	439	1,782	11,095
% change 1950s–1990s	+68	+43	+92	+91	+4,601

Table 3.3. Crop yields (kg/ha/year) of major plantation crops in India over different periods. Modified from Giriappa (1989).

Crop	1971–1972	1975–1976	1980–1981	1985–1986	% change 1971–1986
Sugarcane	47,511	50,993	56,844	57,600	+21
Tea	1,271	1,405	1,477	1,777	+40
Coffee	498	490	702	787	+58
Rubber	678	772	771	821	+21
Coconuts	5,626	5,449	5,249	5,093	–9

affected the production of rubber and yields stagnated, but overall rubber yields increased by 21% between the early 1970s and the mid-1980s. Over the same period, yields of the food crops rice and ragi (finger millet: *Eleusine coracana*) increased by 235% and 41%, respectively.

Webster and Watson (1988) compiled plantation yields for perennial crops in the 1930s and 1980s (Table 3.4). Yields of coffee, cocoa and tea have tripled between the 1930s and 1980s and are much higher than those reported in Table 3.2. The main technical advances that have contributed to the higher yields are: improved planting material and nursery techniques; improved pest and disease control; soil conservation measures and leguminous ground cover; chemical weed control; improved diagnosis of crop nutrient requirements by leaf and soil analysis; and field experiments leading to better use of inorganic fertilizers (Webster and Watson, 1988).

The yield data in Table 3.4 are from plantations under good management and conditions. However, not all plantations, even if they are reasonably efficient, achieve the yields given in the table, partly because they are not fully replanted with modern planting material (Webster and Watson, 1988). Average yields on plantations are, as can be expected, lower than under good management and conditions (Table 3.5).

Yield differences with smallholders

For most crops and in most countries, yields on plantations are twice as high as on smallholder fields, and in some countries differences between plantation and smallholder yields are as high as 500%. They may become smaller if proper extension and training is given to smallholders in the cultivation of perennial (plantation) crops. Barlow and Tomich (1991) reported considerable yield increases for oil palm and cocoa grown by smallholders in Indonesia between 1980 and 1987.

Table 3.4. Perennial crop yields (kg/ha/year) for the 1930s and 1980s under good management and conditions on plantations. Data from Webster and Watson (1988).

	1930s	1980s	% change 1930s–1980s
Rubber (dry rubber)	500	2000	+300
Palm oil (oil)	2000	5500	+175
Coconut (copra)	1500	3000	+100
Cocoa (dry beans)	900	2200	+144
Coffee (dry, clean beans)	1000	2200	+120
Tea (dry leaf)	1000	3000	+200

Table 3.5. Plantation and smallholder yields (kg/ha/year) in various countries, compiled from data in Goldthorpe (1985), Barlow and Tomich (1991) and Stephens *et al.* (1998). Yield data were mostly from the 1980s and could be substantially higher nowadays.

Country	Crop	Plantation	Smallholding	% difference
Côte d'Ivoire	Oil palm (fresh fruit)	10,200	5,800	76
Kenya	Coffee	1,078	633	70
	Tea	2,000–3,500	1,000–1,400	43–250
Tanzania	Tea	3,000	500	500
Sri Lanka	Rubber	1,000	450	122
	Tea	900	700	29
Malaysia	Rubber	1,300–1,600	800–900	44–100
	Cocoa	1,080	850	27
	Copra	2,000	900	122
	Oil palm (fresh fruit)	24,100	15,700	54
Indonesia	Rubber	1,190	530	125
	Coconuts	960	960	0
	Cocoa	580	160	263
	Coffee	580	380	53
	Tea	1,410	480	194
	Oil palm	3,940	1,180	234
Papua New Guinea	Rubber	500–600	200–600	0–200
	Coconut	900	500	80
	Cocoa	440	330	33
	Coffee	2,000	700	186
	Oil palm (fresh fruit)	21,500	11,900	81

However, in many areas in the tropics, smallholders' stands of perennial tree crops consist of old stands, inherently low-yielding, unselected seedlings (Webster and Wilson, 1980). At establishment there was poor crop husbandry (no weeding, no fertilizer use, erosion) and after the trees reached maturity lack of equipment, input or skill was another important cause for low yields. Cash crops may be neglected in certain seasons when priority is given to food crops; and the young people may have migrated to towns leaving only those too old to work effectively (Webster and Watson, 1988). Another reason for the yield gap might be that plantations are situated on superior soils compared with the surrounding smallholders and that more inorganic fertilizers and other inputs are being used.

There are exceptions to this pattern, such as the smallholder sugar growers of Cuba, whose yields are reported to be higher than those of the government-owned estates. In Malaysia, high oil palm and rubber yields are obtained by trained and supervised smallholders comparable to the yields obtained at plantations. There has also been evidence that smallholders produce at least as much or more output per hectare than large plantation farms and that this lower productivity of plantations is due to the underuse of the land (Tiffen and Mortimore, 1988).

References

Anderson, D.L., de Boer, H.G. and Portier, K.M. (1995) Identification of nutritional and environmental factors affecting sugarcane production in Barbados. *Communications in Soil Science and Plant Analysis* 26, 2887–2901.

Anderson, J.M. (1994) Functional attributes of biodiversity in land use systems. In: Greenland, D.J. and Szabolcs, I. (eds) *Soil Resilience and Sustainable Land Use*. CAB International, Wallingford, pp. 267–290.

Barlow, C. and Tomich, T. (1991) Indonesia agricultural development: the awkward case of smallholder tree crops. *Bulletin of Indonesian Economic Studies* 27, 29–53.

Basiron, Y. and Darus, A. (1996) The oil palm industry – from pollution to zero waste. *The Planter* 72, 141–165.

Beare, M.H., Reddy, M.V., Tian, G. and Srivastava, S.C. (1997) Agricultural intensification, soil biodiversity and agroecosystem function in the tropics – the role of decomposer biota. *Applied Soil Ecology* 6, 87–108.

Blaikie, P. and Brookfield, H. (1987) Colonialism, development and degradation. In: Blaikie, P. and Brookfield, H. (eds) *Land Degradation and Society*. Methuen, London and New York, pp. 100–121.

Bleeker, P. (1983) *Soils of Papua New Guinea*. Australian National University Press, Canberra.

Breuning-Madsen, H., Awadzi, T.W., Bredwa-Mensah, Mount, H. and Hudson, B. (2002) The Danish plantations on the former Gold Coast colony of Ghana – why did they fail? *Soil Survey Horizons* 43, 1–8.

Burger, A. (1994) *The Agriculture of the World*. Avebury, Aldershot, UK.

Cannell, M.G.R. (1999) Environmental impacts of forest monocultures: water use, acidification, wildlife conservation, and carbon storage. *New Forests* 17, 239–262.

Carr, M.K.V. (1999) Evaluating the impact of research for development: tea in Tanzania. *Experimental Agriculture* 35, 247–264.

Courtenay, P. (1980) *Plantation Agriculture*, 2nd edn. Bell & Hyman, London.

El-Swaify, S.A. (1997) Factors affecting soil erosion hazards and conservation needs for tropical steeplands. *Soil Technology*, 11, 3–16.

Evans, J. (1986) Plantation forestry in the tropics – trends and prospects. *The International Tree Crops Journal* 4, 3–15.

Evans, J. (1992) *Plantation Forestry in the Tropics*, 2nd edn. Clarendon Press, Oxford.

FAO (1996) *The State of Food and Agriculture*. FAO, Rome.

Fragoso, C., Brown, G.G., Patron, J.C., Blanchart, E., Lavelle, P., Pashanasi, B., Senapati, B. and Kumar, T. (1997) Agricultural intensification, soil biodiversity and agroecosystem function in the tropics – the role of earthworms. *Applied Soil Ecology* 6, 17–35.

Freckman, D.W., Blackburn, T.H., Brussaard, L., Hutchings, P., Palmer, M.A. and Snelgrove, P.V.R. (1997) Linking biodiversity and ecosystem functioning of soils and sediments. *Ambio* 26, 556–562.

Giriappa, S. (1989) *The Role of Plantation Crops in Agricultural Development.* Daya Publishing House, Delhi.

Goldthorpe, C.C. (1985) *Plantation Agriculture in Papua New Guinea.* Institute of National Affairs Inc., Port Moresby.

Goldthorpe, C.C. (1987) A definition and typology of plantation agriculture. *Singapore Journal of Tropical Geography* 8, 26–43.

Goldthorpe, C.C. (1994) An organisational analysis of plantation agriculture. *The Planter (Kuala Lumpur)* 70, 5–18.

Gowdy, J.M. and McDaniel, C.N. (1995) One world, one experiment: addressing the biodiversity–economics conflict. *Ecological Economics* 15, 181–192.

Greenland, D.J. (1997) Inaugural Russell-memorial-lecture – soil conditions and plant growth. *Soil Use and Management* 13, 169–177.

Hall, A.D. (1936) *The Improvement of Native Agriculture in Relation to Population and Public Health.* Oxford University Press, London.

Härdter, R., Chow, W.Y. and Hock, O.S. (1997) Intensive plantation cropping, a source of sustainable food and energy production in the tropical rain forest areas in Southeast Asia. *Forest Ecology and Management* 91, 93–102.

Hartemink, A.E. and Bourke, R.M. (2000) Nutrient deficiencies of agricultural crops in Papua New Guinea. *Outlook on Agriculture* 29, 97–108.

Henderson, J. and Osborne, D.J. (2000) The oil palm in all our lives: how this came about. *Endeavour* 24, 63–68.

Hudson, N. (1986) *Soil Conservation.* B.T. Batsford, London.

Jalani, B.S. (1998) Oil palm research and development in Malaysia: progress and trends. *The Planter (Kuala Lumpur)* 74, 73–92.

Lal, R. (1990) *Soil Erosion in the Tropics – Principles and Management.* McGraw-Hill, New York.

Lal, R., Miller, F.P. and Logan, T.J. (1988) Are intensive agricultural practises environmentally and ethically sound? *Journal of Environmental Ethics* 1, 193–210.

Lavelle, P., Dangerfield, M., Fragoso, C., Eschenbrenner, V., Lopez-Hernandez, D., Pashanasi, B. and Brussaard, L. (1994) The relationship between soil macrofauna and tropical soil fertility. In: Woomer, P.L. and Swift, M.J. *The Biological Management of Tropical Soil Fertility.* John Wiley & Sons, Chichester, pp. 137–169.

Loewenson, R. (1992) *Modern Plantation Agriculture – Corporate Wealth and Labour Squalor.* Zed Books, London.

Matin, M.A., Malek, M.A., Amin, M.R., Rahman, S., Khatoon, J., Rahman, M., Aminuddin, M. and Mian, A.J. (1998) Organochlorine insecticide residues in surface and underground water from different regions of Bangladesh. *Agriculture Ecosystems and Environment* 69, 11–15.

Morgan, R.P.C. (1995) *Soil Erosion and Conservation.* Longman, Harlow.

Morin, P.J. (2000) Biodiversity's ups and downs. *Nature* 406, 463–464.

Naidu, R. (1998) Contaminants and the soil environment – preface. *Geoderma* 84, 1–2.

Nair, P.K.R. (1984) Agroforestry with coconuts and other tropical plantation crops. In: *Plant Research and Agroforestry.* ICRAF, Nairobi, pp. 79–102.
Page, S., Siegert, F., O'Rieley, J.O., Boehm, H.V., Jaya, A. and Limin, S. (2002) The amount of carbon released from peat and forest fires in Indonesia during 1997. *Nature* 420, 61–65.
Pannell, D.J. (1999) On the estimation of on-farm benefits of agricultural research. *Agricultural Systems* 61, 123–134.
Parrotta, J.A. (1992) The role of plantation forests in rehabilitating degraded tropical ecosystems. *Agriculture, Ecosystems and Environment* 41, 115–133.
Pendleton, R.L. (1954) The place of tropical soils in feeding the world. *Ceiba* 15, 201–222.
PORIM (1994) *Environmental Impacts of Oil Palm Plantations in Malaysia.* Palm Oil Research Institute of Malaysia, Malaysia.
Pushparajah, E. (1998) The oil palm – a very environmental friendly crop. *The Planter (Kuala Lumpur)* 74, 63–72.
Ruthenberg, H. (1972) *Farming Systems in the Tropics.* Clarendon Press, Oxford.
Sanchez, P.A. (1976) *Properties and Management of Soils in the Tropics.* John Wiley & Sons, New York.
Saylor, R.G. and Eicher, C.K. (1970) Plantations in Nigeria: lessons for West African development. In: Bunting, A.H. (ed.) *Change in Agriculture.* Gerald Duckworth, London, pp. 498–503.
Sedjo, R.A. and Botkin, D. (1997) Using forest plantations to spare natural forests. *Environment* 39, 14–20.
Smith, R.W. (2000) Tree crops: essential components of sustainable tropical agricultural systems. *TAA Newsletter* March, 17–19.
Stephens, W., Hamilton, A.P. and Carr, M.K.V. (1998) Plantation crops. In: Webster, C.C. and Williams, C.N. (eds) *Agriculture in the Tropics*, 3rd edn., Blackwell, Oxford, pp. 200–221.
Theng, B.K.G., Kookana, R.S. and Rahman, A. (2000) Environmental concerns of pesticides in soil and groundwater and management strategies in Oceania. In: Huang, P.M. and Iskandar, I.K. (eds) *Soils and Groundwater Pollution and Remediation – Asia, Africa and Oceania.* Lewis Publishers, Boca Raton, Florida, pp. 42–79.
Tiffen, M. and Mortimore, M. (1988) *The Role of Plantations in Development: a Literature Review.* Overseas Development Institute, London.
Tinker, P.B. (1997) The environmental implications of intensified land use in developing countries. In: Greenland, D.J., Gregory, P.J. and Nye, P.H. (eds) *Land Resources: On the Edge of the Malthusian Precipice?* The Royal Society, London, pp. 1023–1033.
Watson, G.A. (1990) Tree crops and farming systems development in the humid tropics. *Experimental Agriculture* 26, 143–159.
Webster, C.C. and Watson, G.A. (1988) Tree crops of the humid tropics. *TAA Newsletter* 8, 11–19.
Webster, C.C. and Wilson, P.N. (1980) *Agriculture in the Tropics*, 2nd edn. Longman, Harlow, UK.
Wilcke, W., Kretzschmar, S., Bundt, M. and Zech, W. (1999a) Metal concentrations in aggregate interiors, exteriors, whole aggregates, and bulk of Costa Rican soils. *Soil Science Society of America Journal* 63, 1244–1249.

Wilcke, W., Muller, S., Kanchanakool, N., Niamskul, C. and Zech, W. (1999b) Polycyclic aromatic hydrocarbons in hydromorphic soils of the tropical metropolis Bangkok. *Geoderma* 91, 297–309.

Wilcke, W., Muller, S., Kanchanakool, N., Niamskul, C. and Zech, W. (1999c) Urban soil contamination in Bangkok: concentrations and patterns of polychlorinated biphenyls (PCBs) in topsoils. *Australian Journal of Soil Research* 37, 245–254.

Wood, D. (2000) In defence of monocultures. *ILEIA Newsletter* 16, 14–15.

Wösten, J.H.M. (1997) Pedotransfer functions to evaluate soil quality. In: Gregorich, E.G. and Carter, M.R. (eds) *Soil Quality for Crop Production and Ecosystem Health*. Elsevier Science, Amsterdam, pp. 221–245.

Yan, F., McBratney, A.B. and Copeland, L. (2000) Functional substrate biodiversity of cultivated and uncultivated A horizons of vertisols in NW New South Wales. *Geoderma* 96, 321–343.

4

Soil Fertility Decline – Theoretical Considerations

Er is in de bodemkunde wellicht geen belangrijker, doch ook geen gecompliceerder probleem dan dat van de bodemvruchtbaarheid.[1]

F.A. van Baren (1934)

Soil fertility decline has been a matter of concern ever since sedentary agriculture commenced some 10,000 years ago. In most permanent agricultural systems soil fertility was maintained through applications of manure, other organic materials and the inclusion of legumes in the cropping systems. In the 19th century inorganic fertilizers became available, which transformed many farming systems and their traditional management. Through the application of inorganic fertilizers productivity could be enhanced and maintained under permanent cropping. In many parts of the world the availability, use and profitability of inorganic fertilizers has been very low, whereas there has been an intensification of land use and an expansion of crop cultivation on to marginal soils. As a result, soil fertility has declined and this decline is perceived to be widespread, particularly in some regions of the tropics. This chapter focuses on the theoretical considerations for the assessment of soil fertility decline, including boundary conditions, data types, errors and precision in assessment and interpretation of study results.

[1] 'Soil chemical fertility is the most important but also most complex problem in soil science.'
(First sentence of F.A. van Baren's PhD thesis, supervised by C.H. Edelman)

Introduction

Growing agricultural crops implies that nutrients are removed from the soil through the agricultural produce (food, fibre, wood) and crop residues. Nutrient removal may result in a decline of the soil fertility if replenishment with inorganic fertilizers or manure is inadequate. Soil fertility is defined as 'The quality of a soil that enables it to provide nutrients in adequate amounts and in proper balance for the growth of specified plants or crops' (SSSA, 1997). Although it omits the importance of soil physical and biological conditions for crop productivity, it is a useful simplification.

A decline in soil fertility implies a decline in the quality of the soil, and in this book, soil fertility decline is defined as the decline in chemical soil fertility, or a decrease in the levels of soil organic C, pH, CEC and plant nutrients. Soil fertility decline thus includes nutrient depletion or nutrient decline (larger removal than addition of nutrients), nutrient mining (large removal of nutrients and no inputs), acidification (decline in pH and/or an increase in exchangeable Al), the loss of organic matter, and an increase in toxic elements (e.g. Al, Mn). To assess soil fertility decline, it is necessary to define the spatial and temporal boundaries of the systems under study.

Spatial and Temporal Boundaries

The total amount of a nutrient in the soil declines when the output exceeds the input over a given period of time, soil depth and at a certain location. Spatial and temporal boundaries need to be chosen to ascertain whether the nutrient level declined. A spatial boundary is, for example, the plot or paddock, whereas a temporal boundary might be the period the plot was cultivated, or the number of growing seasons during which a crop is grown. When such boundaries are chosen it is possible, for example, to conclude that the soil fertility has declined in a paddock between 1970 and 1995. There are a number of difficulties and pitfalls before such conclusions can be drawn and these are discussed below.

The black box approach

In its simplest form the soil can be considered a black box from which nutrients are removed (output) and nutrients enter (input). The box is a pool of nutrients whereas input and output of nutrients are as fluxes or flows. Such an approach has been widely adopted in agricultural and ecological research since the seminal work of Nye and Greenland

(1960). When the content of the box is measured over a period of time, it may have changed in three ways. The first possible outcome is that no changes have occurred in the content of the black box, which implies that output matched input of nutrients over the period of observation (Fig. 4.1). The soil is apparently in a steady state condition in which the properties are in a sort of dynamic equilibrium (Yaalon, 1971). Such a condition is exceptional and there is increasing evidence that even under natural conditions nutrient losses occur (Poels, 1989; Stoorvogel *et al.*, 1997). The 'steady state condition' does not imply that losses are absent and it also provides no information on the environmental impact of a land-use system.

The second possibility is that the content of the box has increased over the period of observation. It could be due to nutrient inputs exceeding nutrient outputs, or because the output of nutrients is decreased whereas the inputs remained the same, or a combination of the two. Higher inputs may also mean higher outputs but as long as total input exceeds output, the content of the box increases and soil fertility builds up. This has occurred in many areas in the Pleistocene sandy areas of The Netherlands, Germany and Belgium, where plaggen soils, or Anthrosols, developed because of centuries of applications of a mixture of manure, sods, litter and sand (Pape, 1970). It continues to date through excessive applications of animal manure and inorganic fertilizers in agricultural soils of many Western European countries (de Walle and Sevenster, 1998). As a result, the box may overflow and nutrients may be leached into surface water and groundwater causing environmental problems. There are also examples in the tropics where soil fertility has built up as a result of long-term applications of organic materials (Sandor and Eash, 1995).

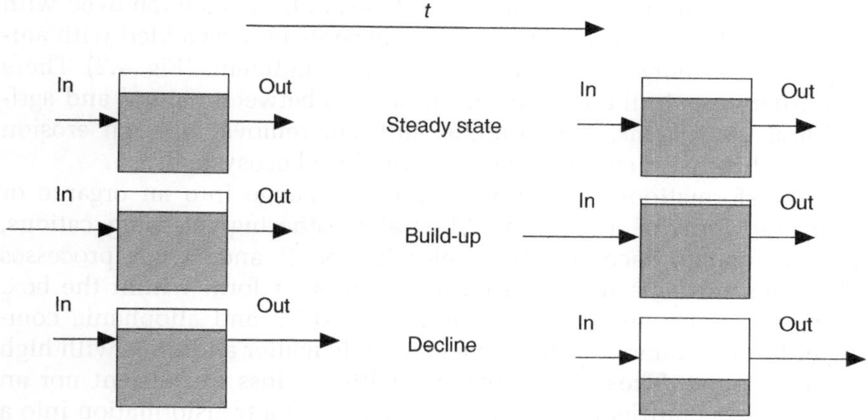

Fig. 4.1. Changes in the soil nutrient pool resulting from different ratios of input and outputs.

The third possibility is that the content of the box decreases and the soil fertility declines. It occurs when nutrient output exceeds input over a given period due to increasing outputs when inputs remain the same, for example when there is a sudden increase in the rate of erosion or leaching because of high rainfall events. It could also be caused by decreasing inputs when output levels remain unchanged, for example in alluvial soils where inputs through sedimentation stops after a river shifts its watercourse. Decreased inputs may also imply decreased outputs, but the net difference between nutrient input and output eventually determines whether there is a decline in soil fertility. The net difference is affected by a large number of soil processes and soil management factors and also has environmental implications, which are not directly visible.

Additions, removals, transformations and transfers

The arrows in Fig. 4.1 represent the sum of inputs and sum of outputs. Both the input and output of nutrients consist of various factors governed by complex and interacting soil processes. Factors affecting soil fertility decline are essentially the same as those in pedogenesis, as follows: additions, removals, transfers and transformations. Most of these have been studied at the pedon scale but mass balances and studies at smaller scales are scarce.

In natural ecosystems, additions of nutrients can occur through atmospheric deposition (nutrients in dust and precipitation) and biological nitrogen fixation (symbiotic and asymbiotic) whereas removals include leaching of N and cations, and gaseous losses (denitrification, volatilization). Removal of nutrients may also occur through soil erosion and runoff. In agricultural ecosystems, nutrients are removed with the yield and crop residues, whereas nutrients may be added with animal manure, inorganic fertilizers or other amendments (Fig. 4.2). There are differences in the rates of soil processes between natural and agricultural ecosystems. For example, nutrient removal through erosion and leaching is generally higher in agricultural ecosystems.

Transformations of nutrients imply a change into an organic or inorganic form of a nutrient. Mineral weathering releasing cations, organic matter decomposition releasing N, P and S are processes changing nutrients from one form into another form within the box. Other examples are P fixation by sesquioxides and allophanic compounds, or N immobilization due to organic matter additions with high lignin content. These processes are neither a loss of nutrient nor an output (these nutrients are still in the box), but a transformation into a less available form. The transformations may imply that the nutrient is not available for crop production for prolonged periods (>100 years).

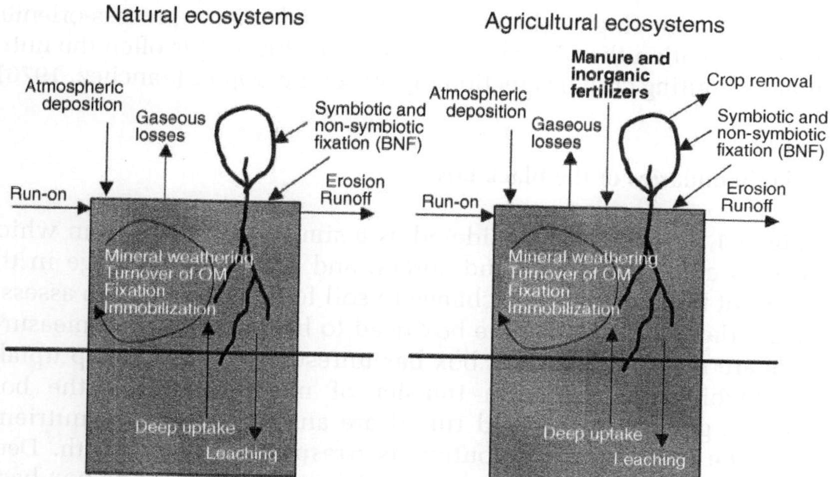

Fig. 4.2. The soil–plant black box under natural and agricultural conditions, with the main nutrient additions (run-on, atmospheric deposition, biological nitrogen fixation (BNF)), removals (runoff, leaching, gaseous losses), transformations (weathering, organic matter (OM) turnover, fixation and immobilization) and transfer (deep uptake, leaching). In agricultural ecosystems nutrients are removed with the crop and may be added through animal manure, inorganic fertilizers or other amendments. See text for explanation.

Transfer of a nutrient refers to replacement within a pedon, such as deep N uptake, or the eluviation of organic matter or K-rich clay minerals into deeper soil horizons. Deep uptake is not a nutrient input and eluviation is not a removal of nutrients but a transfer of nutrients to other soil horizons within the box.

For the assessment of soil fertility decline in this book, only those processes are considered which could be an input (addition) or an output (removal) of a nutrient in the black box, and these include leaching, denitrification, volatilization and biological nitrogen fixation. Each of these processes is affected by soil management and agroecological factors. Leaching is governed by intrinsic soil properties (porosity) in relation to climate (excess rainfall), cropping systems (rooting system) and soil management aspect (inorganic fertilizer applications, organic matter content). The fact that both physical and chemical soil properties as well as several other factors are involved explains why quantifying leaching losses is so difficult (Addiscott, 1996). Denitrification is also a difficult process to quantify, and it is generally assumed to be low in upland soils of the humid tropics (Grimme and Juo, 1985). Volatilization has been studied particularly in relation to the efficiency of N fertilizers in lowland rice (Raun and Johnson, 1999). Quantification of biological nitrogen fixation (BNF) has received much attention in tropical ecosystems as it is considered a viable way of increasing the N status of the soil (Giller *et al.*, 1997). No single element has received more process-oriented research attention than N, which is not surprising as it is often the nutrient most limiting crop production in soils of the tropics (Sanchez, 1976).

Spatial boundaries of the black box

In Fig. 4.1 the soil was considered as a simple black box from which nutrients can be removed and added, and in which a change in the content of the box reflects a change in soil fertility. In order to assess a change, the boundaries of the box need to be defined before measurements are made. If the black box has unrestricted depth, deep uptake and leaching are merely a transfer of nutrients within the box. However, gaseous losses and runoff are an output because nutrients depart the box – and this output is irrespective of its depth. Deep uptake can only be considered as a nutrient input when the box has a horizontal boundary at some depth. This is shown in Fig. 4.2, where the horizontal line is the border at which leaching is an output and deep uptake is an input to the soil–crop system. Replenishment of nutrients can only take place through inputs from outside the box, i.e. manure, inorganic fertilizers and BNF.

In addition to the depth of the box, the width and length are important boundaries. If the box is very large, then losses of nutrients by erosion and subsequent deposition in a lower position of the landscape should be regarded as a transfer of nutrients. The same applies for nutrients transported by subsurface flow. This is illustrated in Fig. 4.3. Nutrients are removed from box A whereas box B gains nutrients but no net changes occur within box C. Box A and B could be single

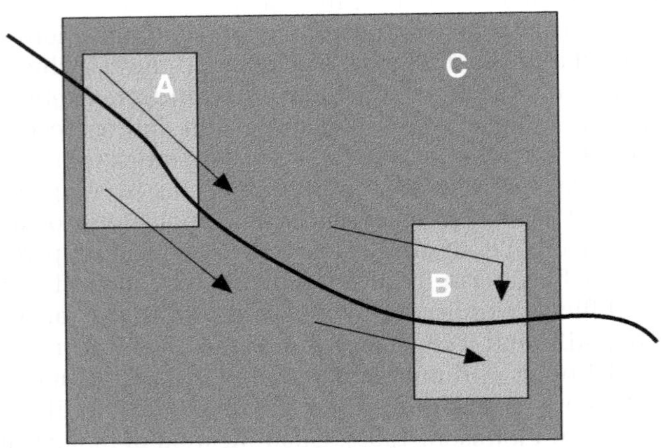

Fig. 4.3. Soil catena with flows of nutrient between boxes. Transfer of nutrients from A to B implies removal (Box A) and addition (Box B) but no net change within Box C.

pedons and box C could represent a catena. Likewise, box A and B may be a catena and box C a watershed, and at a smaller scale, box A and B are a region and box C represents a continent.

The transfer of nutrients from one area or spatial scale to another is natural, as for example the deposition of nutrient-rich sediments in alluvial plains and river deltas, or the Harmattan dust in West Africa, which reaches as far as Europe. The transfer of nutrients across spatial scales is not an easy topic and most mass-balance studies have focused on single or polypedons. This problem is not unique to soil fertility decline studies and also applies to soil erosion studies, where net soil loss within a watershed may be nil despite heavy soil loss in the upper part of the catchment.

Temporal boundaries

Monitoring soil properties implies that observations are made at different times. Observations are linked to the spatial scale chosen, and this was illustrated in Fig. 4.3: transfer of nutrients from box A to B results in no net changes in box C. So if box A and B were separately measured and evaluated, conclusions would be different compared with when only box C was considered. When the observational period is long enough, nutrients might also leave box C through rivers and streams but they may also enter box C. An example is the loss of N, first from pedon A and B, and then subsequently from the entire field (box C). In other words: whether a change occurs in the black box depends on the size of the box (depth, width and length) in relation to the period of observation.

The fixed-time-step approach, where the size of the box is monitored at the beginning and after some period, provides no information on what has happened during that period. It is possible that the input–output ratio of nutrients may have changed over time. For example, in shifting cultivation systems the level of exchangeable cations may be low before cutting fallow vegetation, but increases when the vegetation is burned and cations are released. The higher levels of exchangeable cations are usually followed by a rapid decrease due to losses and crop uptake. The size of the box depends on the time of observation and different conclusions might be reached when measurements and observations are made at large time steps (see also the section on 'Frequency, period and time of observation' on page 125).

The soil science literature is flooded with short-term observations, by which transient phenomena can be missed or misinterpreted (Pickett, 1991). In general, observations made over a long period allow more rigorous conclusions (see the section on 'Long-term Experiments' on page 92). If, however, long-term observations also imply large time steps, it may mask what has happened during the period of observation. The best approach is long-term monitoring with relatively short time steps.

There is another temporal factor affecting the assessment of soil fertility decline, and that is the soil property or indicator chosen and the techniques used for its analysis. Changes in the microbial population only need time scales of days or even shorter, whereas measuring changes in CEC may require years of cultivation between the measurements to detect a significant difference. A related problem is that some soil parameters fluctuate during the day or between seasons and detecting changes may therefore be confounded by the natural variability. Such variation in combination with results from different analytical techniques has been fairly well documented, and it is discussed in the section on 'Spatial and Temporal Variation' on page 103.

Data Types

Soil degradation features such as water erosion and salinization may be observed and assessed with remote sensing and LandSat™ images. Such techniques cannot be used to measure a decline in soil nutrient levels, and other types of observational methods are required. In the literature on soil fertility decline, three different data types are used to assess soil changes caused by agricultural production systems: expert knowledge, nutrient balances and the monitoring of soil chemical properties over time or at different sites. Some of these data can be relatively easily collected whereas other data require long-term commitment and are costly to collect. Each data type has specific advantages and disadvantages, and the type of data collected is determined by the research plan, and the financial conditions and objectives of the study.

Expert knowledge

Soil science has always used qualitative measures of soil properties, such as soil colour and field texture, but also soil mapping has many qualitative aspects including the delineation of mapping units. Qualitative approaches have greatly contributed to soil science and form the base for what can be called expert knowledge systems. Farmers and other users of the land have expert knowledge about their soils. The knowledge has been largely ignored by soil science but in the past decade there has been a tendency to value indigenous soil knowledge, or ethnopedology as it is sometimes called (Sillitoe, 1998; Warkentin, 1999; WinklerPrins, 1999).

Indigenous soil knowledge has different characteristics from knowledge gained by the scientific study of the soil. A farmer has empirical knowledge of his soils, which is not soil process or data oriented but yield or management oriented (Bouma, 1993). Yield decline as observed by a farmer could, however, be caused by a variety of factors including soil fertility decline, adverse weather conditions, invasion of weeds, soil physical deterioration or a combination of factors. Therefore it is difficult to distinguish soil fertility decline from other factors, and farmers' knowledge on soil fertility decline is difficult to interpret if not substantiated by other types of data, such as, for example, crop yield, weather conditions or pests and disease information.

In a recently published annotated bibliography on ethnopedology, more than 900 references and abstracts are listed and the bibliography provides a wealth of information on how farmers perceive soil fertility (Barrera-Bassols and Zinck, 2000). This perception is almost universally qualitative and could be affected by many biophysical and management factors including the memory or political motives of the farmer. This was the case in a land degradation study in Burkina Faso, where it was found that the perception of local farmers was frequently socially constructed and politically mediated (Gray, 1999). Perceptions of degradation were related to interethnic conflict over land between locals and migrants and each group blamed the other for the perceived degradation. Environmental development projects also offered tangible benefits for peasants who perceived their resource base to be degraded (Gray, 1999) and it was therefore beneficial to exaggerate the situation.

In conclusion: as expert knowledge may have political motives and is qualitative and in opposition to the study approach of this book (Chapter 1, 'Aims and Approach' on page 7), expert knowledge data are not further considered for the assessment of soil fertility decline. None the less, farmers' knowledge can be instrumental in selection of sampling sites or for additional information, and various examples exist in the literature where such information was used to investigate long-term changes in soil fertility (e.g. Sillitoe and Shiel, 1999; Peters, 2000).

Measured change in soil chemical properties – Type I data

Two different approaches have been used to monitor soil chemical properties. First, soil dynamics can be monitored over time at the same site, which is called chronosequential sampling (Tan, 1996) or Type I data (Sanchez *et al.*, 1985). Type I data show changes in a soil chemical property under a particular type of land use over time. Usually the original level is taken as the reference level to investigate the trend in such changes. It is most useful if trends are also followed under other land-use systems, for example under cultivation, secondary regrowth and natural forest over the same period. These data expose differences more readily and accurately because natural forest ecosystems are not *per se* stable, especially under marginal conditions (Poels, 1987).

Soil chemical data (Type I) can be derived from stored samples that are analysed at the same time as the newly collected soil samples but soil storage may affect some soil chemical properties. Alternatively, data from samples analysed in the past can be compared to newly collected and analysed soil samples. For this to be useful, analytical methods as well as systematic laboratory errors need to be the same (see the section on 'Soil Sampling, Soil Analysis and Errors' on page 97).

Type I data have been used for quantifying soil contamination by comparing soil samples collected before the intensive industrialization period with recent samples taken from the same locations (Lapenis *et al.*, 2000). Type I data are also useful in assessing the sustainability of land management practices in the tropics (Greenland, 1994b), but limited data-sets exist as they require long-term research commitment and detailed recordings of soil management and crop husbandry practices.

Measured change in soil chemical properties – Type II data

In the second approach, soils under adjacent different land-use systems are sampled at the same time and compared. This is called biosequential sampling (Tan, 1996), Type II data (Sanchez *et al.*, 1985), and 'sampling from paired sites' in the soil science literature from Australia (e.g. Bramley *et al.*, 1996; Garside *et al.*, 1997). It has also been named the 'space-for-time' method (Pickett, 1991), and the 'inferential method' in studies conducted in Nigeria (Ekanade, 1988).

The main underlying assumption is that the soils of the cultivated and uncultivated land are the same soil series, but that differences in soil properties can be attributed to the differences in land use. Obviously, this is not always the case and the uncultivated land may have been of inferior quality and therefore not planted. Also, spatial variability may be confused with changes over time when, for example, tree crops of different age are sampled at the same time, and soil properties are often confounded with genetic improvement and silvicultural practices (Sanchez *et al.*, 1985). Other confounding factors are differences in clay content, soil depth, or unknown history of land use, and these largely affect the usefulness of Type II data for the assessment of soil fertility decline. In ecology, Type II data studies have often proved to be misleading, as functional parameters such as nutrient availability and plant–animal interactions have been conspicuously under-represented (Pickett, 1991). When carefully taken, however, biosequential soil samples can provide useful information, and this sampling strategy has been followed in a considerable number of studies on soil changes under agricultural crops.

Data on changes in soil properties can also be obtained when the effects of a certain treatment or management practice on soil properties are measured. Examples of such studies are the measurements of soil pH after lime applications or the capacity for water infiltration after tractor traffic. These are referred to as Treatment–Effect (TE) studies and provide information on what happens to the soil when a certain treatment (management practice) is applied. In effect, Type I or chronosequential data are collected, although a simultaneous comparison of soil properties with different treatments yields Type II data. Most TE studies are short term (<2 years) and have been used to follow changes in soil physical properties (e.g. bulk density), because such changes can occur more rapidly than changes in soil chemical properties. There are, of course, soil physical properties that are also slow, such as clay eluviation or the formation of hardpans.

Monitoring of soil chemical properties gives information on how soils respond to agricultural activities and whether soil fertility decline takes place. Spatial and temporal variation requires a cautious selection of sampling sites, an appropriate number of replicated observations and a careful interpretation of the results (see sections on 'Soil Sampling, Soil Analysis and Errors' page 97, 'Spatial and Temporal Variation' page 103 and 'Soil Changes and Nutrient Removal' page 107). Unless these are met, it may be difficult to extrapolate the information or to derive maps of the patterns observed in single pedons. Coupling the soil information to soil maps in a geographical information system (GIS) provides opportunities to map soil fertility decline in different areas. A large amount of data of sufficient quality is needed before such maps can be derived and there are few examples where measured soil fertility decline was coupled to a GIS (e.g. Stoorvogel, 1993; Stoorvogel and Smaling, 1998). More studies exist in which semiquantitative approaches have prevailed to map soil fertility decline.

Semiquantitative

A third way of studying soil fertility decline embraces a semiquantitative approach, which operates at a much coarser (smaller) scale, namely national or supra-national scale. Existing soil data are combined with pedotransfer functions (regression analysis between some soil or other parameters) into a GIS to estimate the decline in soil fertility at a given location. This is essentially a mechanistic modelling approach in which expert knowledge is also important. It is generally perceived that such studies are not to replace soil-monitoring efforts but should be seen as the best possible way of getting the most out of available data. They may require considerable computer power but much depends on the complexity of the models involved. The outcome of such studies provides qualitative, comparative and spatial information on the decline in soil fertility. The maps created are appealing, and may result in more awareness among the public and policy makers than facts and figures collected by monitoring changes in soil chemical properties.

Table 4.1 summarizes the different data types including a short description and the characteristics of each type. Each data type has its own merits and drawbacks but the most comprehensive and effective characterization of soil behaviour is obtained when all three data types are available (Bouma, 1993).

Minimum data-sets

In the previous sections, the quality of the data as well the period and time of observation was discussed. What quantity of data is required and which soil attributes are needed to assess that at a given location the soil fertility has declined? In the literature, these questions have been discussed in relation to the assessment of sustainable land management (Smyth and Dumanski, 1995), and in particular in relation to such vague concepts as 'soil quality' and 'ecosystem health' (Doran and Parkin, 1996; Greer and Schoenau, 1997).

Table 4.1. Data types in soil monitoring studies and their advantages and disadvantages.

Data type	Short description	Advantages	Disadvantages	Typical example
Expert knowledge	Combination of field observations with general knowledge	Easy to obtain, rapid assessment, useful for small-scale studies	Subjective, not quantitative	Oldeman *et al.*, 1991
Type I Chronosequential	Monitoring soil properties over time	Accurate, hard data, using existing data	Slow, expensive, contamination of monitoring sites, spatial and temporal variability, sample storage	Billett *et al.*, 1990
Type II Biosequential	Comparing soil properties under different land use	Easy to obtain, rapid, relatively hard	Soils at sampling sites may differ, unknown land-use history of sites, spatial and temporal variability	Bramley *et al.*, 1996
Semiquantitative	Combination of existing data with pedotransfer functions or models	Using existing data, fairly rapid, indicative, appealing outcome	Not very hard, require computer power	Stoorvogel and Smaling, 1990

Determining what data are to be included depends on the type of study and its objectives. For example, if it is to be proved that continuous wheat cropping affects soil organic matter, then measurements of soil organic C, light fraction and particulate organic matter as well as mineralizable C, microbial biomass and soil carbohydrates and enzymes may be included. Also bulk density and texture may be included for such a study but it would make little sense to include DPTA-extractable Zn. Likewise, if it is to be determined whether leguminous crops enhance soil acidity, then the soil pH, buffering capacity, CEC and exchangeable cations and acidity are soil properties that should be measured. These are, however, specific types of studies. Most data in soil fertility decline studies were collected to supplement other agronomic investigations in long-term studies. A more general discussion on soil attributes to be included in soil fertility decline studies is given below.

Soil organic matter is a key component of soil fertility (Woomer *et al.*, 1994), and a decline in its content must be regarded as an important factor affecting the productivity of the soil. Therefore, soil organic matter contents should be regarded as an essential soil attribute in soil fertility decline studies. Gregorich *et al.* (1994) considered assessment of soil organic matter as a valuable step towards identifying the overall quality of a soil. Soil pH is easy to measure, and together with levels of plant nutrients (total N, inorganic N, available and total P, exchangeable Ca, Mg, K) these are important soil properties that should be included in a minimum data-set. In many studies, existing data have been used, which were sometimes supplemented with newly collected data. Soil survey data can be particularly useful to assess soil changes (Young, 1991; Hartemink, 1996).

In the semiquantitative study of Stoorvogel and Smaling (1990), the data-set was quite extensive (soil maps, agroecological maps, literature data). Several pedotransfer functions were used to arrive at a single parameter to assess nutrient depletion in sub-Saharan Africa: the difference between nutrient input and output at a national scale. The maximum use was made of existing data to arrive at an attractive presentation of the results: a map showing in different shades of red where soil nutrients were being mined. Reasoning the other way around (what is needed to assess the difference between nutrient input and output for different land-use systems in sub-Saharan Africa) would result in a long list of soil, climatic and agronomic attributes of which many would not be available. It is a clear example of where first the target was set (is there nutrient depletion?) and then the available data were compiled. In this example one cannot speak of a 'minimum data-set' but rather of the 'maximum use of available data'. In the case studies presented in Chapters 9 and 10, a similar approach was taken whereby existing data were used to select new sampling sites and to assess whether soil chemical properties had changed.

Long-term Experiments

Long-term experiments are important to obtain meaningful results on the decline in soil fertility. Such experiments not only yield valuable Type I data, but also provide hard data which can be used in semi-quantitative studies. Moreover, soil properties of long-term experimental plots can be compared with soils under other types of land use, giving Type II data. The principal benefits of long-term experiments are clear (shifting the noise from the trend) and a number of unforeseen uses and functions have been derived from them in recent years (Rasmussen *et al.*, 1998).

Definition

Long-term has been differently defined in different disciplines. For example, in the field of finance long-term capital will be 'generated by assets held for longer than six months', while long-term bonds will constitute a 'financial operation or obligation based on considerable term and especially one of more than 10 years' (Laryea *et al.*, 1995).

In agriculture and ecology, 'long-term' has been used for various periods. Lal and Stewart (1995a) considered long-term experiments those that are conducted on the same site for 10 years or more, whereas Kirkegaard (1995) considered agronomic trials exceeding 5 years as long-term experiments. Also Pickett (1991) stated that long-term studies in ecology are defined as those lasting more than 5 years. Rasmussen *et al.* (1998), Tinker (1994) and Poulton (1995) mentioned that at least 20 years are required for an agronomic experiment to be classified as long term. Steiner and Herdt (1993) made an inventory of non-European long-term agronomic experiments and found that there are many experiments varying in length from 5 to 150 years that call themselves long term. They made the following arbitrary division: (i) classical experiment (>50 years); (ii) medium-length experiments (20 to 50 years); and (iii) young long-term experiments (<20 years but with intention to continue). In this book long-term is defined as 10 years or more.

Long-term experiments in the tropics

There are a fair number of classical and long-term experiments in the tropics (Steiner and Herdt, 1993), and Table 4.2 lists some of these experiments and when they were started. Many more long-term experiments in the tropics are still in existence, but details could not be traced for inclusion in the list of Steiner and Herdt (1993).

Table 4.2. Selected existing long-term experiments conducted in the tropics. Based on experiments listed in Steiner and Herdt (1993).

	Country, Institute	Name of experiment	Primary crops	Initiation date
Africa	Egypt, Cairo	Bahtim plots	Cotton, wheat, maize	1912
	Burkina Faso, INERA	Intensive rotation exp.	Groundnuts, millet, maize	1960
	Ivory Coast, IDESSA	Nitrogen maintenance exp.	Maize, cotton	1969
Asia	India, Tamil Nadu	Old permanent manurial exp.	Sorghum, cotton	1909
	India, Tamil Nadu	New permanent manurial exp.	Sorghum, wheat	1925
	Sri Lanka	Effect of NPK on tea	Tea	1962
	India	Long-term effect of manure and N	Pearl millet, wheat	1967
S. America	Argentina, INTA	Breeding of improved potato cv.	Potatoes	1941
	Chile, La Platina Exp.	Monocultures in central Chile	Wheat, potatoes, maize	1963
	Uruguay, INIA	Crop–pasture rotations	Wheat, other crops	1963

In recent years, there have been several reviews on long-term soil experiments in the tropics. They are listed in Table 4.3 including experimental factors and period during which they were collected. Several other reviews are available but they contain much shorter periods (<10 years) (e.g. Harris, 1995). A number of reviews have focused on work conducted in Africa and only a few that include long-term studies in Asia and South America are available.

Although some experiments in the tropics have been conducted for prolonged periods, not many of them exceed those conducted in the USA (Sanborn fields, University of Missouri since 1888; long-term flax research at North Dakota State University since 1892), Canada (Historical Rotations 'B and C' at Scott since 1911), Australia (Permanent topdressing experiment at Rutherglen since 1914), and there are many in the UK (e.g. Palace Leas Meadow hay plots since 1896), and not to forget, the experiments at the Rothamsted Experimental Station.

The Rothamsted experience

The longest and best-known agricultural experiments are at Rothamsted, and were started between the 1840s and 1880s. These experiments have been crucial to unravelling the effects of soil amendments on soil chemical properties and have been instrumental in modelling exercises, and the development of statistical techniques for field experimentation. Moreover, they have yielded many unanticipated study possibilities such as the effects of atmospheric deposition or thermonuclear tests. Also sustainability, environmental quality, and

Table 4.3. Selection of recent reviews on long-term cropping experiments in the tropics.

Country	Crop	Experimental factors	Period	Reference
Burkina Faso	Sorghum	NPK, FYM	1960–1978	Greenland, 1994a
Burkina Faso	Groundnuts, millet	NPK, fallow	1962–1984	Pieri, 1995
Côte d'Ivoire	Maize, cotton	N, crop residues	1972–1990	Traore and Harris, 1995
Ghana	Cassava	NPK, lime mulch	1951–1979	Greenland, 1994a
Ghana	Groundnut	NPK, lime mulch	1954–1978	Greenland, 1994a
Ghana	Maize	NPK, lime mulch	1949–1980	Greenland, 1994a
Kenya	Maize	NPK, FYM, crop residues	1976–1994	Bekunda *et al.*, 1997
Niger	Sorghum	NPK, FYM	1960–1990	Bekunda *et al.*, 1997
Nigeria	Maize	NPK, tillage	12 years	Greenland, 1994a
Senegal	Groundnuts, millet	NPK	1957–1973	Pieri, 1995
Zambia	Maize	NPK	1965–1981	Singh and Goma, 1995
Zambia	Maize, groundnuts	Lime	1972–1985	Singh and Goma, 1995
India	Maize	Intercropping	1976–1989	Laryea *et al.*, 1995
India	Sorghum, chickpea, safflower	None	1976–1989	Laryea *et al.*, 1995
India	Maize, wheat, cowpea	NPK, FYM	1972–1986	Greenland, 1994a
Philippines	Rice	NPK	1965–1990	Greenland, 1994a
Peru	Rice, maize, soybeans	NPK	1972–1989	Smyth and Cassel, 1995

NPK, NPK inorganic fertilizers; FYM, farmyard manure.

species-adoption impacts can be studied but were never envisaged by the founders of the Rothamsted experiments (Powlson and Johnston, 1994; Rasmussen *et al.*, 1998).

Jenkinson (1991) listed the following principal advantages of the long-term experiments ('classicals' as he calls them) at Rothamsted:

- They have a continuous role as living demonstrations for farmers and students of the effects of organic and inorganic manures.
- They enable the monitoring of trends in slow-changing factors such as soil pH, soil organic matter and soil-borne diseases.
- They are a source of well-characterized experimental material for scientists, and have been used to study phosphate extractability, organic matter chemistry, soil microbial biomass, soil structure etc.
- They provide a test bed on which modern experiments can be superimposed.
- The stored crop and soil samples can be used to follow changes that could not have been measured (or envisaged) when the samples were taken.
- The experiments are of interest to ecologists because they show the effects of treatments on the flora and fauna over prolonged periods.

- They provide data for long-term studies of the relationship between crop yield and weather.
- They provide data on the effects of atmospheric pollution.
- Data can be used to validate computer simulations of field processes over the year-to-century time span.

Another reason for conducting long-term experiments is to document changing environmental influences and system states before they become lost to the historical record (Pickett, 1991). This is more relevant for observations in natural ecosystems such as the Amazon than to agricultural ecosystems.

Long-term agricultural experiments also have disadvantages. The main disadvantage is that agriculture changes in ways that cannot be foreseen, giving the following dilemma: leave the experiment unchanged and allow it to become irrelevant or change the experiment and lose continuity with the earlier years. Other disadvantages of long-term experiments are that their site may be representative of only a small area, that plot-to-plot soil movement may occur, and that data collection in the past has been imperfect (Steiner and Herdt, 1993). Moreover, many experiments have statistical design limitations, such as the lack of plot replication, which, for example, is found in the Broadbalk experiment at Rothamsted (Jenkinson, 1991), and the Permanent Rotation Trial at the Waite Agricultural Research Institute in South Australia (Grace *et al.*, 1995). Both experiments started, however, with large plots and have yielded enormous insight on the interaction between soils and crops and management practices.

Setting-up long-term experiments

Most agricultural problems do not need long-term experimentation for their solution and an experiment lasting a few years is usually enough (Jenkinson, 1991). A thorough analysis is required of whether a long-term experiment is needed and how it will solve a biophysical or socio-economic problem. Long-term experiments are also expensive and require financial and scientific commitment over prolonged periods. In conclusion: long-term experiments should not be started lightly, but, once started, they should not be abandoned lightly (Jenkinson, 1991).

In recent years, several authors have stated that it may be necessary to initiate new long-term experiments in developing countries as the increased need for food and fibre will require either more intensive use of existing cropland or greater conversion of forest to cropland (Greenland, 1994a; Lal and Stewart, 1995b; Rasmussen *et al.*,

1998). Starting new long-term studies requires careful planning and selection of treatments and experimental design. One of the most important considerations is the selection of the site, which needs to be representative for large parts of the surrounding farm area. Other considerations are that the site should be secure, easily accessible for routine observations and sampling, and if one is not to be installed at the site, be near a weather station (van Wambeke, 1988). To produce experimental results without excessive uncontrolled variability in the data, the site should be, as nearly as possible, on a single soil type and in the same erosion and textural class with a soil of uniform depth and drainage, and on an uniform slope within a single landscape position (Frye and Thomas, 1991). Site characterization is extremely important and the experiments should be preceded by uniformity tests (van Wambeke, 1988; IBSRAM, 1999).

There should be no shading of plots by trees because this could differentially affect yields on some plots (Leigh *et al.*, 1994). The layout should be such that the greatest variability occurs between replications. The experiment should be large in situations of high variability and the minimum size of an experiment is larger than one would expect for a well-controlled situation (Nelson, 1987). The plots should be big enough to allow for frequent soil sampling, which is always destructive. Furthermore, the experiment needs to be expandable in order to accommodate changes in the experimental design in the future without losing the continuity with the past. The design needs to minimize the risk for soil erosion/creep and soil deposition (Frye and Thomas, 1991), unless effects of erosion are to be part of the study.

In summary: long-term experiments in agriculture must be on sites with security of tenure; have acceptable levels of funding to achieve their aims; be multidisciplinary; have objectives that are well defined but not rigid; be designed so that the data can be analysed statistically but allow for acceptable change; have large plots, which can be subdivided to include additional tests; have clearly defined but flexible experimental protocols except for crop and soil sampling, which should remain unchanged; have sets of measurements that are agreed at the start; and provide a continuous output of good data with well-documented interpretations (Johnston and Powlson, 1994).

Finally, accidents do occur in long-term experiments, such as, for example, application of amendments to the wrong plot. In many cases, nothing can be done about accidents in long-term experiments, except to admit them and point out their effects on results. Researchers should thus keep accurate records of things that go wrong because knowledge of errors often helps to explain erratic results (Frye and Thomas, 1991). Accidents and errors are rarely admitted in scientific papers, so perhaps they hardly ever occur.

Soil Sampling, Soil Analysis and Errors

Soil chemical data, which are obtained by sampling and analysing soils, are required for the monitoring of soil properties (both Type I and II data). Soil sampling procedures and analytical techniques and methods have been a matter of continuing, careful improvement since soil science emerged some 150 years ago.

Soil analysis is undertaken in land capability assessments to assess the potential for a certain type of land use and to characterize mapping units in a soil survey. However, an increasing number of soil chemical measurements are undertaken to assess the risk for ecological and human health and for environmental regulations. To a lesser extent soil chemical analysis is undertaken for soil engineering properties for building developments or road construction. Most soil analysis is undertaken for diagnosing soil constraints for agriculture and usually the analysis is for chemical properties – biological and physical tests are more rare (McLaughin *et al.*, 1999). Three sources of errors can be distinguished: (i) during the sampling and handling of the soil samples; (ii) during the laboratory analysis; and (iii) in the interpretation of the results.

Errors in soil sampling

Errors in soil sampling methods have been well documented and are generally much greater than errors in the analysis (Cline, 1944; Tan, 1996). A key problem is that soil volumes, not areas, are sampled. All results are expressed as units per mass and not on a volume or area base, which would require bulk density data (see the section 'Effects of Bulk Density' on page 117). Sampling depth should be in line with the depth of soil horizons or the rooting depth of the crop. The soil sample needs to be representative for the section of the paddock or block and should adequately represent the soil type (Brown, 1999). The accuracy with which a soil sample represents the population sampled depends upon the soil variability, the number of sampling units contributing (i.e. the number of subsamples or cores), and the method of soil sampling (Cline, 1944). The problem is that soil variability is not known until soil samples have been taken and analysed, but much progress has been made in the quantification of soil variability and how many samples should be taken to characterize soil properties (see the section on 'Spatial and Temporal Variation' on page 103). Soil sampling equipment is also a potential source of error when it is difficult to clean or take cores or slices with different volumes.

The number of soil samples and the error are affected by the sampling design. Brown (1999) distinguished the following five sampling design strategies: (i) judgement or random sampling based on the selection of typical spots and avoiding unusual spots; (ii) proportional sampling, which uses a mathematically derived proportion of cores taken from unusual spots; (iii) simple random sampling, which includes sampling unusual spots; (iv) random grid sampling; and (v) systematic sampling, where samples are taken at regular distances along a transect. Each of these sampling strategies will affect the outcome of the study and are based on classical statistical techniques as discussed by Cline (1944, 1945). Classical or design-based sampling assumes the samples are random, whereas in more recent statistical techniques (geostatistics) the soil sampling is model-based (see 'Recent techniques – geostatistics' page 107).

In many soil fertility studies a random grid sampling method is used within each plot, whereby 10–20 subsamples are taken for each sample. Larger numbers of subsamples are often desirable but this means that the plots are thoroughly perforated, which may shorten their period use, and some soil properties are affected (i.e. increased water infiltration). The number of subsamples (or cores) is an important consideration and the minimum number to cope with soil variability differs per soil chemical property, soil type and cropping system. Based on a review of a large number of studies in Australia, Brown (1999) suggested a minimum of 5–10 cores to characterize organic C and total N, 10–20 cores for pH and exchangeable cations, 20–30 cores for extractable P and larger numbers when inorganic fertilizer or lime is applied (Table 4.4).

Errors in soil handling and storage

The next source of errors occurs in the handling, storage and preparation of the soil sample. Contamination of soil samples is possible in the field and properties may also change during transport to the laboratory. Except for N, little work has been carried out on the effects of temperature, moisture etc. on soil samples in transit from the field to the laboratory (Brown, 1999). Biological activity continues so there can be rapid changes in the NO_3-N content.

In most cases, soils are air dried at ambient laboratory temperature and humidity in the laboratory, which significantly affects some soil properties. Both temperature (Molloy and Lockman, 1979) and method of drying (Payne and Rechcigl, 1989) affect soil chemical properties. Tan (1996) and Landon (1991) listed the effects of air-drying based on earlier works. Air-drying will not affect total C or total N, but affects NO_3-N and NH_4-N. The pH may be slightly lowered by air-drying and is particularly altered in soils rich in S. Drying can cause changes in P-fixation, which is related to changes in Al and Fe chemistry. In soils

Table 4.4. Soil parameters, their variation and number of cores (subsamples) required. Modified from Brown (1999) and Landon (1991).

Property	Vertical changes	Variation	Recommended core numbers	Random error
pH	Often increases with depth, particularly where lime is present. In strongly leached soils, acidity increase with depth	SD = 0.03–0.90; 80% of studies SD <0.5	30–40 but 10–20 where site is uniform and lime never applied	0.2 units
Organic C	Highest near the surface sometimes accumulating as mat; may be high in subsoils of podzols and buried profiles	CV = 6–74%; 80% of studies CV <15%	5–10, but larger number of cores is required when surface organic matter varies	5%
Total N	Closely associated with organic C	CV = 1–85%; 80% of studies CV <20%	As for organic C	5%
Extractable P	Almost always concentrated in the topsoil, or where fertilizer P applied in bands or drill rows	CV = 8–126%; 80% of studies CV <90%	30–40, or 50–60 where fertilizer recently spread or banded or in paddocks with high stocking rates	10%
Exchangeable cations	Generally increases when clay content increases or where organic matter is high	CV = 3–224%; 80% of studies CV <70%	30–40, or 10–20 where site is uniform and no amendments are applied	5%

with a low pH, P soluble in diluted acid tends to increase, whereas P levels tend to decrease in high pH soils. Potassium may be fixed or released from the fixed form during drying, but it depends on the type of clay minerals present. During drying exchangeable K tends to increase in soils low in exchangeable K, whereas more K becomes fixed in soils with moderate or high K content.

Storage of samples is needed for Type I data when samples taken in the past are re-analysed jointly with newly collected soil samples. Except for air-dried soils, samples may also be stored under refrigeration but this cannot be recommended for long-term storage because of possible shifts in microbial community and the potential development of anaerobic conditions. Freezing at temperatures below 20°C can be suitable for long-term storage, given that microbial activity is effectively minimized, although it has some drawbacks (Boone *et al.*, 1999). Freezing promotes desiccation, lyses microbial cells and disrupts soil organic matter structure, and it may alter exchangeable ammonium and soluble P concentrations. Typically there is a flush of biological activity in thawed soils due to the decomposition of soil microbial cells lysed by the freezing (Boone *et al.*, 1999).

Air-dried soil samples, which are not kept in airtight containers, may absorb NH_3, SO_3 or SO_2 gases and therefore the container should be composed of materials that will not contaminate the soil samples (Tan, 1996). Even in airtight containers soil changes occur during prolonged storage. An important change that may occur in stored soil samples is an increase in surface acidity and increased solubility and oxidizability of soil organic matter (Bartlett and James, 1980). Research on Australian soils, however, showed that the pH (in water) of soils that were stored for 7 years was 0.55 units higher. There was no relationship between soil type and pH change due to storage of soils (Slattery and Burnett, 1992). Overall, the effects of soil storage are important for the long-term study of soil changes but available literature on this subject is limited. There seems to be no storage condition that is perfect and the absence of a storage effect on soil properties should be checked rather than assumed (Boone *et al.*, 1999). However, in reports on long-term agricultural and ecological studies storage methods and conditions of soil samples are seldom reported.

Storage is not relevant for soil samples that were analysed in the past and which plots were re-sampled. For such samples, however, it is necessary that the analytical laboratory has a consistent or systematic error (Kempthorne and Allmars, 1986), which means that it should not change over time, or the error should be quantified. With analytical apparatus and personnel changes, it may be difficult to keep the error constant in a soil laboratory.

Errors in soil analysis

The next step is the actual analysis of the soil and, as stated before, it is widely proclaimed that analysis yields fewer errors than soil sampling. Most soil tests have been calibrated for topsoil properties. After a series of pot trials and field trials the quantity of nutrients measured by soil tests can be expressed in terms of deficient, adequate or toxic for the crop considered. The quantity of nutrients extracted from the soil differs from that accumulated by the crop and is also different for different soil types. The large variation in soils and soil properties versus consistency of analytical methods has been a matter of concern since the beginning of the 20th century. Soil classification, which strongly emerged in the 1950s, required that soils were analysed by standard methods. As such it has greatly facilitated the standardization of soil analytical methods. It also means that some soil chemical analytical data are almost meaningless, like, for example, the CEC determined by NH_4OAc at pH 7.0 in highly weathered soils. In such acid soils there may be a considerable portion of pH-dependent charge, which results in a gross overestimation of the CEC. So the results of soil analysis can be soil-order specific.

The debate on soil analytical methods in relation to soil variation and nutrient availability is far from over, but in practice a standard set of soil testing procedures is now widely used. The selection of an inappropriate analytical method could be termed a fundamental error, and it is generally perceived that selecting the right procedure is more difficult than performing the actual analysis (Tan, 1996). Most errors, however, arise from the fact that variation is insufficiently dealt with (see the section on 'Spatial and Temporal Variation' on page 103) and errors could also be made in the actual analysis. In the past decades, several international programmes have been developed to check laboratory errors and precision through exchange of soil and plant samples (e.g. LABEX, WEPAL), which has greatly improved the accuracy and quality of many soil analytical laboratories. There are also many handbooks available on soil analytical techniques, and guidelines have been developed for quality management in soil laboratories (van Reeuwijk, 1998).

Errors – sensitivity analysis

A sensitivity analysis is a useful way to investigate the precision of soil analysis and the effects of errors in soil fertility studies. Such analyses are sometimes published in model studies (e.g. Janssen *et al.*, 1990) but they are rarely published in soil fertility research. In Fig. 4.4 an example is shown from inorganic N measurements on Oxisols in Western Kenya (Hartemink, 1994). Measurements involved moisture content (as a correction factor), bulk density (to calculate N-values into kg/ha) and laboratory analysis of NO_3-N and NH_4-N using the methods described in Anderson and Ingram (1993) and Dorich and Nelson (1984).

The relation between the error in a measurement and the effect on inorganic N content is linear for most parameters. Large errors (>20%) in the measurement of gravimetric water content at the time of subsampling for extraction and in the blank have little effect but errors in the bulk density and in the laboratory analysis have substantial impact (see also the section on 'Effects of Bulk Density' on page 117). Weighing and recording errors of soil for extraction have also considerable effects, and errors of 5–10% in a number of measurements substantially affect the outcome of the inorganic N values. A sensitivity analysis of errors in the field would be probably more difficult as some of them may be hard to quantify.

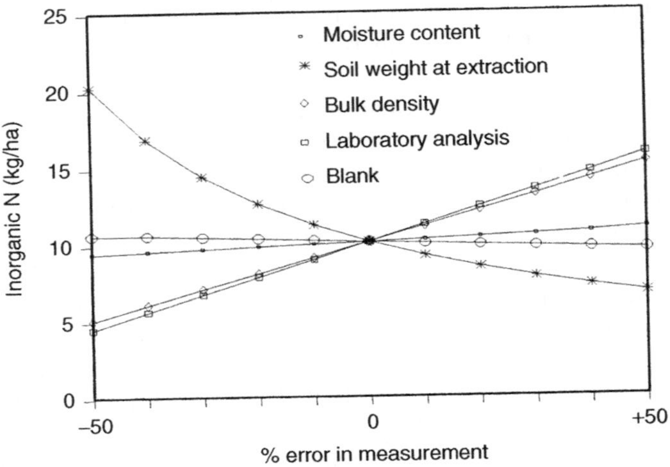

Fig. 4.4. Relation between errors in measurements and inorganic N in kg/ha. Errors range from –50 to +50% for different factors in the calculations. See text for explanation. Modified from Hartemink (1994).

Errors and the nutrient balance

Nutrient balances provide a convenient and biologically meaningful context within which to organize what is known about a system's biogeochemical cycles, put nutrient pools and fluxes into perspective, and can lead to considerable insight into processes that regulate nutrient cycling. Nutrient balances help to guide system management decisions and direct the course of future research (Robertson, 1982). In recent years, the use of nutrient balances has been encouraged through a systems approach to both improve food crop production and maintain the soil resource base (Johnston and Syers, 1998). Some of the factors in the nutrient balance depend on soil tests such as nutrient loss by erosion; but also the size of nutrient pools is determined by soil tests. In the previous sections, errors in soil analysis were discussed and it was shown that in the measurement of a single soil chemical property various errors can be made and accumulate.

When all factors in a nutrient balance are considered and accumulated errors are calculated, the difference between the nutrient outputs and inputs may show a wide range of values. For example, if the sum of nutrient inputs is on average 150 kg/ha (range: 125–175 kg/ha) and the sum of nutrient output is on average 200 kg/ha (range: 175–225 kg/ha), the difference between the averages is –50 kg/ha. If the range of values is considered, the difference could be nil or as large as –100 kg/ha. Range in values is rarely given in nutrient balance studies.

A recent overview of sources of biases and errors in nutrient budgets was prepared by Oenema and Heinen (1999). Possible biases in the budget could be personal (conceptual interpretation and simplification of the system and its flows), sampling biases of nutrient pools and flows, data manipulation biases through generalization, averaging and scaling up. Biases could also be due to fraud that could occur when nutrients budgets are being used as a policy instrument to enforce a nutrient management strategy with possible economic consequences for farmers and other stakeholders. Nutrient budgets may also suffer from sampling errors originating from within-plot heterogeneity and from spatial and temporal variation (see the section on 'Spatial and Temporal Variation' below). Also errors in measurements may occur due to variations in the determinations of the volume and composition of the sample but the error is usually smaller than the sampling error (Tan, 1996). In summary: biases are generally a more serious problem than errors because biases in the various inputs and outputs accumulate to a much larger extent in the surplus or deficit of the budget than errors, according to Oenema and Heinen (1999). When all measurement errors are taken into account, it may be that their sum may exceed the difference between nutrient inputs and outputs – as was shown in the example above. An accompanying problem with nutrient balances in the tropics is the very uneven availability of basic information regarding land use and nutrient cycling in important ecosystems, and the large ecosystem heterogeneity (Robertson, 1982).

Spatial and Temporal Variation

A complicating factor in soil fertility decline studies is the heterogeneity of the material in the black box – the soil. Soil varies in space (between two points) and in time (between two sampling times at the same site) and across a range of scales for both space and time. Variation in soil properties between two sampling points or with time can be natural or be induced by cultivation. Soil spatial variation has been a matter of concern ever since systematic soil inventories started. Soil fertility research has dealt with variation by taking a sufficiently large number of soil samples in order to differentiate treatment effects from random variance.

Variation by soil property

Variation in soil chemical properties is affected by many properties including the parent material from which the soil is derived, microrelief, soil fauna, texture, litter inputs and the effects of individual plants. In agricultural systems, amendments, tillage, cropping sequences, animal dung and manure as well as compaction from grazers or machinery also cause soil variation. The degree of variation differs per soil chemical property and some properties vary more than others, in both time and space. Table 4.4 summarizes the analysis of 44 studies on soil variation in Australia (Brown, 1999) supplemented with some general information of Landon (1991).

The data from Brown (1999) were from a large number of studies with a variety of sampling methodologies and agroecological conditions. Therefore, it is not surprising that the ranges in spatial variation are so large. Moreover, inorganic fertilizer banding explains some of the high variation. The errors quoted by Landon (1991) show that 5–10% is common for the major soil chemical properties, but these errors cannot be directly linked to the standard deviations and coefficients of variation found by Brown (1999).

Soil chemical properties may vary from year to year, between seasons in a year, or even between days depending on weather conditions and management factors. Several studies have been carried out in an attempt to find seasonal or climatic patterns in this variation, but many studies have failed because insufficient attention was given to spatial variation or laboratory variation. In soil science, spatial variation has been given more attention than temporal variation because more data are available to investigate spatial effects. Fewer data-sets are available to study temporal variation possibly because it requires observations over a period of time, which may be affected by weather, management and unknown factors. It has been suggested that seasonal variation on some soil properties may mask differences due to management or treatments. Therefore, characterization of some soil chemical properties requires more than one soil sample per year (Brown, 1999). For most standard soil chemical properties used in this book (pH, organic C, total N etc.) short-term temporal variation is relatively small.

The number of soil samples needed to characterize a soil chemical property is site-specific and also affected by land use. Prasolova *et al.* (2000) used a spatial analysis of soil chemical properties to calculate the number of samples required in *Araucaria* plantations (Table 4.5). The calculations were based on the obtained experimental estimates of the mean differences between the means for sampling dates and variance estimates of the soil properties. For total N, more samples need to be taken whereas soil pH can be assessed with fewer samples. There were considerable differences between the two sites in the number of samples required. No soil classification was given for the two sites.

Table 4.5. Sample size required for estimation of the pH, organic C, total N and CEC at different levels of error at two sites under *Araucaria cunninghamii* plantations in subtropical Australia. Modified from Prasolova *et al.* (2000).

	Site 1				Site 2			
	pH	Organic C	Total N	CEC	pH	Organic C	Total N	CEC
10% error	5	29	32	19	7	35	66	15
20% error	3	9	10	7	4	11	19	6

Variation due to cultivation

Natural soil variability is affected by cultivation and the cropping system. Most annual crops are sown by broadcasting over the field and usually no row effects exist, i.e. localized nutrient extraction or addition. Perennial crops are grown in rows (sisal, sugarcane, oil palm), which determines the rooting pattern and extraction of water and nutrients. This is influenced by the soil management of the crops such as the application of inorganic fertilizers in rings around trees, which induces spatial variability (Tinker, 1960). Soil variation under oil palm is illustrated in Table 4.6, which depicts the pH and exchangeable K values of a Typic Paleudult in an oil palm plantation in Malaysia.

The oil palm was fertilized with 1.5 kg N and 3.5 kg K/palm/year in the form of ammonium chloride and muriate of potash, respectively. Palm density was 140/ha, so inorganic fertilizer applications were 210 kg N and 520 kg K/ha/year. The fertilizers were applied in a ring around the palm, which caused significant acidification and an increase in the levels of exchangeable K as compared to the inter-row (between two rows of palms) and frond piles (area where pruned oil palm leaves are piled up).

Field-scale heterogeneity may be created when crop residues are piled up and burned creating 'hot spots' of soil fertility. Soil sampling should take into account the spatial arrangement of the crops, which

Table 4.6. Field-scale heterogeneity in pH and exchangeable K ($n = 4$) in a 20-year-old oil palm plantation in Malaysia. Only circles around the palm received N and K fertilizer. Modified from Kee *et al.* (1995).

	pH (1:2.5 w/v)			Exchangeable K ($mmol_c$/kg)		
Sampling depth (m)	Palm circles	Inter-rows	Frond piles	Palm circles	Inter-rows	Frond piles
0–0.15	3.4	4.4	4.3	8.4	3.1	2.9
0.15–0.30	3.5	4.2	4.4	8.8	2.8	3.4
0.30–0.45	3.5	4.1	4.2	8.5	2.3	3.1

might have created field-scale heterogeneity in soil properties. Although the cultivation-induced variation can be taken into account when the crops are still growing, it is virtually impossible to consider such variation when the previous crop has been slashed and a new crop is planted. For example, when oil palm plantations are replanted, the hot spots created by the inorganic fertilizer applications (Table 4.6) affect the soil sampling results. Old tree rooting patterns will affect the results of soil sampling replanted areas and may have caused variation that cannot be explained by the current cropping arrangement. The effect of roots is illustrated in Fig. 4.5, which shows P fractions at different distances from a tree in a tropical deciduous forest in Mexico.

There was a clear tree-related pattern in soil P levels, but this pattern was different for different P fractions. Both tree-related patterns and analytical technique used to investigate this pattern have an impact on the conclusions in such studies. In natural ecosystems, it

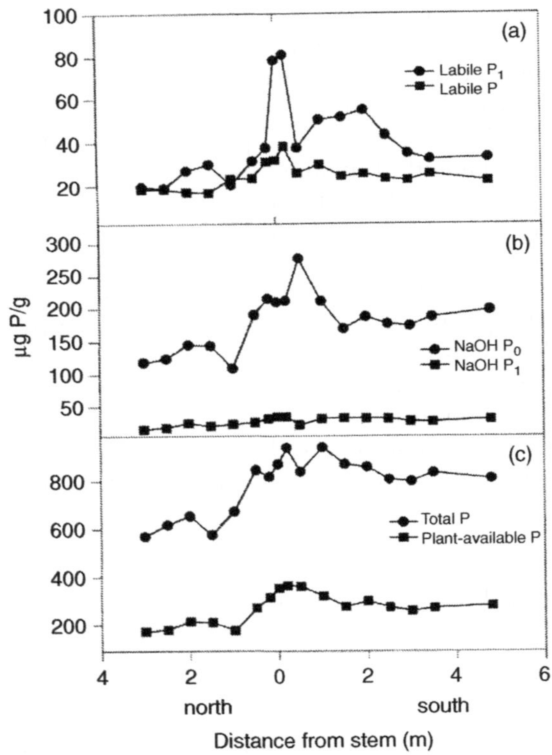

Fig. 4.5. Soil P fractions (0–0.05 m depth) with distance from a stem of a *Caesalpinia eriostachys* tree on Typic Ustorthents in Mexico. (a) Labile inorganic and organic P fraction; (b) sodium hydroxide extractable inorganic and organic P fraction; (c) total phosphorus and plant-available P (Dockersmith, 1999).

should be questioned what comes first: the effect of the tree on the soil, or the effect of the soil on tree dispersal and growth. In agricultural ecosystems where soils are cropped with crops grown in rows or other fixed spatial arrangements, that question is less relevant. It is, however, directly relevant when fallow or forest land is cleared and planted because decaying tree roots enhance the spatial variability in nutrient availability and extraction. In recent years, enormous progress has been made in quantifying and modelling the effects of roots (van Noordwijk *et al.*, 1991; Schroth, 1998).

Recent techniques – geostatistics

In 1980, new statistical methods were introduced in soil science by which the quantification and prediction of soil variability was greatly improved (Webster, 2000). These spatial statistics, which were subsequently named geostatistics, used functions describing the semi-variance of the distance between two observation points to characterize the degree of autocorrelation of a spatial attribute. With these functions (semi-variograms) interpolation of measured values became possible and such interpretation was named kriging. A series of advanced statistical techniques followed including stochastic modelling, Monte Carlo simulations, fuzzy set theory and the development of pedometrics as a subdiscipline in soil science (McBratney *et al.*, 2000).

Geostatistics have been used at many different scales and have found wide applications in environmental soil research. Only a few soil fertility studies have used geostatistics (Dobermann *et al.*, 1994; Goovaerts, 1998; Hoosbeek *et al.*, 1998). The limited use of geostatistics in soil fertility research might be because the data requirements in geostatistics are higher than requirements for conventional statistics. Although geostatistics have proved to be useful in soil science, they are not further considered in this book – partly because data were insufficient.

Soil Changes and Nutrient Removal

Soil fertility has to decline if nutrient output exceeds nutrient input over a given time span and for a given mass (area × depth × bulk density). In agroecosystems nutrient output occurs mainly by the removal of nutrients by the crop. Nutrient removal data are often the only quantified nutrient output in nutrient balance studies. It gives an insight into which nutrients are preferred by a crop, as different crops remove different quantities of nutrients in different ratios. In this section, nutrient removal by annual crops, perennial crops and forest trees, sisal and sugarcane is reviewed. Soil fertility decline under these cropping systems is presented in Chapters 5 to 10.

Annual crops

There is a wide range in nutrient removal for annual crops (Table 4.7) and this is related to the differences in cultivars, time of sampling and agroecologies, which affect yield and thus nutrient removal. Variation is also the result of the different crop parts that are measured. In the literature, it is not always indicated what was included in the measurements and husks and cobs are sometimes included whereas in other studies the nutrients in these plant parts were excluded. Also nutrients in below-ground biomass other than harvested parts are seldom reported.

Perennial crops

Nutrient removal data for some perennial crops are given in Table 4.8. There are quite a number of woody perennial crops that are heavy K-consumers (oil palm, coffee) whereas other crops remove mostly N. Bananas, sugarcane and sisal (Table 4.8) are heavy K-consumers.

Variation in the nutrient removal measurements can be large. As mentioned this is due to differences in agroecologies but it is also influenced by cultivar differences and soil amendments. Table 4.9 shows the range of values in nutrient removal per Mg sisal fibre, and

Table 4.7. Nutrient removal (kg/ha) of annual crops.

Crop	Yield (kg/ha)	Nutrients in kg/ha					Reference
		N	P	K	Ca	Mg	
Maize (grain)	1,000	18–77	2.2–9.7	8–72	5–14	3.3–10.7	Boxman and Janssen, 1990
	1,100	17	3	3	0.2	1	Cooke, 1982
	2,500	40	9	33	7.5	5.0	Sanchez, 1976
	12,500	298	55	247	nd	nd	IPI, 1995
Cassava	8,000	30	10	50	20	10	Sanchez, 1976
	11,000	25	3	65	6	nd	Cooke, 1982
	45,000	202	32	286	nd	nd	IPI, 1995
Yam	11,000	38	3	39	0.7	nd	Cooke, 1982
Sweet potato	16,500	72	8	88	nd	nd	Sanchez, 1976
	34,500	175	34	290	nd	nd	IPI, 1995
Groundnut	800	30	2.2	5	1	1	Cooke, 1982
	1,000	51–62	2.8–3.5	7–17	12–19	4.0–6.7	Boxman and Janssen, 1990
Soybean	1,000	49	7.2	21	nd	nd	Sanchez, 1976
	1,000	79–97	6.4–7.8	46–60	nd	4.7–5.4	Boxman and Janssen, 1990
	3,400	210	22	60	nd	nd	Cooke, 1982

nd, No data.

Table 4.8. Nutrient removal (kg/ha) of perennial crops.

Type	Crop	Yield (kg/ha)	Nutrients in kg/ha N	P	K	Ca	Mg	Reference
Woody perennial crops	Oil palm	2,500 (oil)	162	30	217	36	38	Cooke, 1982
		15,000	90	8.8	112	28	nd	Sanchez, 1976
		24,600	193	36	249	nd	nd	IPI, 1995
	Rubber	1,100	7	1	4	nd	nd	Cooke, 1982
	Cocoa	500	10	2.2	5	1	1	Sanchez, 1976
		1,000	19.3	4.6	10.9	1.3	3.4	Heuveldop *et al.*, 1988
		1,200	24	4	36	nd	nd	Cooke, 1982
	Coffee	2,000	253	19	232	nd	nd	IPI, 1995
	Tea	600	31	2.3	15	2	nd	Sanchez, 1976
		1,300	60	5	30	6	3	Cooke, 1982
	Coconut	1,400	62	17	56	6	12	Cooke, 1982
Herbaceous perennial crops	Sugarcane	88,000	45	25	121	nd	nd	Cooke, 1982
	Bananas	45,000	78	22	224	nd	nd	Cooke, 1982
		30,000	85	10	226	72	90	Sanchez, 1976
	Pineapple	12,500	9	2.3	29	3	nd	Sanchez, 1976

nd, No data.

Table 4.9. Range of values in nutrient removal of *Agave sisalana* and hybrid sisal (kg/ha per Mg fibre). From Hartemink (1995) based on 11 literature sources.

	Nutrients in kg/ha per Mg fibre				
	N	P	K	Ca	Mg
Agava sisalana	27–33	5–7	59–69	42–70	34
Hybrid 11648	22–26	3–4	30–44	79–83	nd
All data[a]	20–50	2–23	30–101	41–159	20–53

[a]This includes the removal data reported in the literature but in which it was not specified whether measurements were made on *Agave sisalana* or a hybrid.
nd, No data.

the range is considerable for both *Agave sisalana* and Hybrid 11648 sisal. It is likely that methodological differences and measurement errors also account for the range in nutrient-uptake values.

Forest plantations

Table 4.10 shows nutrient accumulation of forest plantation species in Tanzania, Sarawak, Indonesia and Nigeria. Large amounts of nutrients are accumulated in the above-ground biomass although there are differences between the species. *Cupressus* removes large amounts of Ca and *Gmelina* removes much K. *Pinus* species remove large amounts of N and Ca. The table also shows that above-ground biomass in forest plantations can be very high and reach up to 400–500 Mg/ha after 20 years. Such biomass stocks are comparable to primary forest (Whitmore, 1985).

Nutrients in the roots

In most field studies with annual crops, root biomass production and nutrient removal by the roots receive little attention. The reasons are obvious: the root system is hidden from direct observation and the quantification of roots is tedious and difficult because of problems in extracting roots from the soil. It is also complex because of the spatial and temporal variability of roots in the soil matrix (Fig. 4.6). Despite these problems various destructive and advanced non-destructive methods have been developed to study roots of field crops (Taylor *et al.*, 1991) in addition to sampling schemes for their quantification (van Noordwijk *et al.*, 1985). Much of the research on roots is conducted in the temperate regions and information on root biomass and its nutrient content in tropical crops is limited.

Table 4.10. Biomass and nutrient accumulation (kg/ha) of forest plantations in Tanzania, Sarawak, Indonesia and Nigeria. Based on data given in Evans (1992), Halenda (1993), Kadeba (1994) and Bruijnzeel (1983).

Species	Age of the plantation (years)	Soil order	Biomass (Mg/ha)	Nutrients (kg/ha)				
				N	P	K	Ca	Mg
Pinus patula	7	Not given	186	849	86	425	470	143
	22	Not given	506	1477	135	756	802	253
Pinus caribaea	14	Oxisol	125	397	31	217	191	74
	14	Inceptisol	231	622	44	357	294	117
	14	Inceptisol	156	450	36	269	288	90
Pinus merkusii	7	Andisols	182	nd	39	205	477	91
Tectona grandis	20	Andisols	107	nd	93	362	910	118
Cupressus lusitanica	7	Not given	133	587	63	457	1007	86
	22	Not given	398	1007	114	837	1588	145
Gmelina arborea	7	Ultisol	92	316	20	463	289	97
	6	Not given	122	352	63	185	79	nd

nd, No data.

Fig. 4.6. Rooting pattern of taro (*Colocasia esculenta*) on Typic Tropofluvents in the lowlands of Papua New Guinea. Large amounts of nutrients are taken up by the roots and recycled during the crop cycle. Quantification of nutrient uptake and recycling in the root system is difficult.

Nutrients in the below-ground biomass should be considered as a transformation of nutrients and not as a loss or removal. This applies to both annual and perennial crops, although the time scale at which the transformation should be considered is different. At the end of a crop cycle in a perennial crop system the trees are slashed and burned or left to decompose. The nutrients in the above- and below-ground biomass are returned to the soil. During the crop cycle the nutrients have been withdrawn from the soil solution. The withdrawal is temporary, i.e. 10 years for sisal, 20–30 years for forest plantations. Some of the nutrients taken up are recycled during the crop cycle, like litterfall and throughfall, which can be very high (particularly for K) in tree crop systems (Parker, 1983).

Hartemink and Johnston (1998) quantified nutrient uptake of fertilized and unfertilized taro (*Colocasia esculenta*) roots on a Typic Tropofluvents in the humid lowlands of Papua New Guinea. Fertilized (100–50–100 kg NPK/ha) and unfertilized plants were harvested at 126 DAP (days after planting) (mid-season) and 231 DAP (harvest). Root biomass at 126 DAP was 0.26 Mg/ha (15% of total biomass) in the unfertilized plots and 0.52 Mg/ha (13% of total) in the fertilized plots, but at 231 DAP root biomass was similar (0.50 Mg/ha). Nutrients in the root biomass as a fraction of the total nutrient uptake were similar at 126 DAP for both treatments (Table 4.11). At 231 DAP, however, the fraction of nutrients in the root biomass was considerable lower in the fertilized plots. The study has shown that the

Table 4.11. Nutrient removal (kg/ha) in roots, corms and leaves of unfertilized and fertilized taro, on Typic Tropofluvents in Papua New Guinea. Type I data, from Hartemink and Johnston (1998).

Sampling period	Plant part	Unfertilized taro					Fertilized taro				
		N	P	K	Ca	Mg	N	P	K	Ca	Mg
Mid-season (126 DAP)	Roots	3	<1	13	4	2	8**	1***	25**	8***	4***
	Corms	5	2	17	4	1	14**	2	22	6	2
	Leaves[a]	19	5	46	8	1	63	9	119	29*	5*
	Total	27	9	76	16	4	85*	13	166	42*	10*
At harvest (231 DAP)	Roots	6	1	25	5	3	5	1	23	5	3
	Corms	13	5	42	8	2	31*	12*	86*	23	5*
	Leaves[a]	34	10	80	21	3	55	18	106	46	6
	Total	53	16	147	35	8	91	31	215	74*	15*

*, **, *** Indicates significant difference between fertilized and unfertilized taro at $P<0.05$, 0.01 and 0.001 respectively.
[a] Leaf biomass includes petioles.
DAP, days after planting.
Values are rounded.

amount of nutrients taken up by roots of fertilized and unfertilized taro was similar at harvest, but that a much larger proportion of plant nutrients was allocated to the roots when no inorganic fertilizers were applied (Hartemink and Johnston, 1998).

The research on taro roots showed that nutrient uptake by roots can be considerable, but that the nutrient concentration (g/kg) varies during the growing season. Nutrient concentrations were also affected by inorganic fertilizer applications, which will affect uptake calculations. In the literature on nutrient removal, these factors are rarely considered.

Annual vs. perennial crops

In annual crops only part of the total amount of nutrients taken up is removed by the economic produce viz. the grain of maize, the tubers of sweet potato or the seeds of soybean. This is illustrated in Table 4.12, which gives the total nutrient uptake of sweet potato tubers and vines (= above-ground biomass). The majority of the nutrients taken up by sweet potato are found in the vines and for most nutrients less than one-third is found in the marketable tubers (= economic produce). Farmers only remove the tubers from the field and the vines remain behind. As vines decompose, nutrients become available for the subsequent sweet potato crop. Similar to what was found in the research with taro (Table 4.11), less than 25% of the total N and K uptake was found in the corms (the economic produce) – the remaining nutrients were in the leaves and roots and are returned to the soil when the crop is harvested. Although both examples are from root and tuber crops, it is generally recognized that crop residues are

Table 4.12. Nutrient uptake (kg/ha ± 1SD) of sweet potato in the humid lowlands of Papua New Guinea. Hobu soils were classified as Typic Eutropepts and the soils at Unitech were Typic Tropofluvents (Hartemink *et al.*, 2000).

Site	Plant part	Fresh yield (Mg/ha)	Nutrients in kg/ha N	P	K	Ca	Mg
Hobu	Marketable tubers	18.2 ± 3.7	30 ± 6	12 ± 2	93 ± 20	5 ± 1	5 ± 1
	Non-marketable tubers	4.0 ± 1.0	8 ± 2	3 ± 1	25 ± 6	1 ± 0.5	1 ± 0.5
	Vines	26.2 ± 4.8	80 ± 8	18 ± 2	180 ± 30	61 ± 13	20 ± 2
	Total		118 ± 10	33 ± 3	298 ± 46	67 ± 12	26 ± 2
Unitech	Marketable tubers	9.0 ± 3.8	15 ± 17	7 ± 3	39 ± 19	4 ± 2	2 ± 1
	Non-marketable tubers	2.9 ± 1.3	5 ± 5	2 ± 1	12 ± 5	1 ± 0.5	1 ± 0.5
	Vines	30.1 ± 8.2	59 ± 21	22 ± 2	189 ± 15	37 ± 8	10 ± 2
	Total		79 ± 40	31 ± 5	241 ± 23	42 ± 10	13 ± 3

Values are rounded up.

extremely important for recycling of nutrients in many cropping systems in the tropics (Giller *et al.*, 1997; Kumar and Goh, 2000).

The concept of nutrient uptake, removal, recycling and crop residues is no different in perennial crop systems but important differences are the time scale or the length of the crop cycle and the much greater biomass in perennial crops. Nutrients in the yield are a fraction of the nutrients immobilized in the above- and below-ground biomass. This is illustrated in Table 4.13, which shows nutrients in various pools of cocoa ecosystems of Costa Rica. Large amounts of nutrients are found in the topsoil and above-ground biomass. For N and to a lesser extent for K, the amounts of nutrients removed with the cocoa beans are much smaller than the amounts in the vegetation. The difference in soil N contents of about 1000 kg is caused by the fact that *Erythrina* is a N_2 fixer.

Nutrient removal data of annual and perennial crops cannot be easily compared, because: (i) data from annual crops usually consist of the economic produce per growing season; (ii) nutrients in forest tree crops usually consist of total uptake whereas only a part is removed; or (iii) nutrients in perennial tree crops do not take into account the uptake required for the total biomass. In order to obtain a more comparable picture of the nutrient removal capacity of different crops, an attempt has been made to quantify the annual removal data for different crops. Average data are presented for annual and perennial crops whereas for the forest tree species total nutrient accumulation was divided by the number of years (Table 4.14). Annual nutrient removal varies within crops and between cropping systems. Perennial crops remove on average more nutrients per year than forest trees, although there are differences between species. Some of these data

Table 4.13. Nutrients in cocoa with *Cordia alliodora* and *Erythrina poeppigiana* as shade trees in Costa Rica. Modified from Hartemink (1993) based on CATIE work (Alpízar *et al.*, 1986; Fassbender *et al.*, 1988, 1991; Heuveldop *et al.*, 1988).

	Compartment		Nutrients in kg/ha		
			N	P	K
Cocoa with *Cordia alliodora*	Soil (0–0.30 m depth)		5327	nd	385
	Vegetation	Cocoa	110	14	105
		Cordia	260	32	258
	Litterfall		115	14	66
	Removed with cocoa beans		19	4	28
Cocoa with *Erythrina poeppigiana*	Soil (0–0.30 m depth)		6370	nd	475
	Vegetation	Cocoa	109	10	52
		Erythrina	279	29	252
	Litterfall		175	9	54
	Removed with cocoa beans		26	4	27

nd, No data.

Table 4.14. Approximate annual nutrient removal of the harvested parts of selected annual crops, perennial crops and forest tree species (kg/ha/year). Based on data in Tables 4.7 to 4.10.

	Crop	Nutrients in kg/ha				
		N	P	K	Ca	Mg
Annual crops	Maize	40	9	33	7.5	5.0
	Cassava	30	10	50	20	10
	Sweet potato	72	8	88	nd	nd
	Groundnuts	30	2.2	5	1	1
Woody perennial crops	Oil palm	162	30	217	36	38
	Cocoa	24	4	36	nd	nd
	Coffee	253	19	232	nd	nd
	Tea	31	2.3	15	2	nd
	Coconuts	62	17	56	6	12
Herbaceous perennial crops	Sugarcane	45	25	121	nd	nd
	Bananas	85	10	226	72	90
	Pineapple	9	2.3	29	3	nd
	Sisal	45	10	60	90	35
Forest plantation trees	*Pinus patula*	67	6	34	36	12
	Pinus caribaea	28	2	16	14	5
	Pinus merkusii	nd	11	55	123	22
	Tectona grandis	nd	5	23	45	8
	Cupressus lusitanica	46	5	38	72	7
	Gmelina arborea	45	3	66	41	14

nd, No data.

can be useful to compare the rates of change in soil chemical properties and it is further explored in Chapter 11.

Conclusions

Nutrient removal data are determined by multiplying dry matter yield by nutrient concentration. Nutrient removal differs between crops and for most crops a range of values is reported, which is related to time of sampling, crop part sampled, soil type and fertilizer applications, and crop cultivars. Also variation in plant part analysis and dry matter determination contribute to the variability of nutrient removal data. The variation will affect the nutrient balance as nutrient output by the yield is often the best quantified component in a nutrient balance study. The range in nutrient removal rates within a crop is rarely taken into account and reported in nutrient balance studies.

Effects of Bulk Density

Why bulk density is useful

An important aim of this book is the quantification of the effect of cropping on soil chemical properties. Cropping brings about changes in soil physical and soil biological properties and these also influence soil chemical properties. For example, changes in the soil moisture or temperature regime affect soil microbial biomass, which influences mineralization of organic matter and many other processes. Measured changes in soil chemical properties are the net effect of these processes but quantification of such changes is, amongst others, dependent on the bulk density of the soil, which may change under cropping. In this book, changes in soil chemical properties are mostly expressed as concentrations, e.g. $mmol_c$/kg or g/kg. Nutrient concentration can be expressed as nutrient content (kg/ha), which is used in nutrient balance studies and more easily translated in nutrient replacement by inorganic fertilizers.

Suppose an Alfisol permanently cropped with maize contained 1.5 g N/kg in the topsoil (0–0.20 m) in 1990, and 1.2 g N/kg in 2000. The rate of change in total N content is 0.03 g N/kg/year. If the topsoil has a bulk density of 1.3 Mg/m^3, the decrease of 0.03 g N/kg/year is equivalent to a loss of 78 kg N/ha/year. This figure is appealing, particularly when it is expressed as inorganic fertilizer: the loss of N from the topsoil is equivalent to 170 kg urea or 390 kg sulphate of ammonia. The next step is to translate nutrient losses into economic terms and progress is being made in the economic quantification of soil nutrient losses (Alfsen *et al.*, 1997; Drechsel and Gyiele, 1999; FAO, 2001). It should be realized that the slightest deviation in nutrient concentration or bulk density values results in large differences when nutrients are expressed in kilograms per hectare, and the effects of bulk density variation on nutrient content was illustrated in Fig. 4.4.

Expressing soil chemical properties in kg nutrient/ha requires soil bulk density values (Fig. 4.7). Often these are not measured in soil fertility studies. Moreover, many soil chemical properties are determined by an extraction method and the values obtained are expressed in terms of availability. Available means that the nutrient is susceptible to absorption by plants whereas availability means effective quantity (Black, 1993). The amount of available nutrients extracted may hold little relation with the total amount of the nutrient in the soil and its availability over a given time span. The availability aspect is irrelevant for C and N because total pools are measured. Bulk density measurements improve the quantification of nutrient depletion as it would be possible to relate N loss to the total N pools. For P, K or Ca that is not possible unless the total element concentrations were determined.

Fig. 4.7. Bulk density measurements in a Kandic Paleustalf soil pit of a maize field near Machakos (Kenya). Three 100-ml cores were used per horizon. Soil moisture content at the time of sampling is an important factor for correct measurements of the bulk density.

Changes in bulk density

Bulk density is likely to change under cropping but much depends on the cropping system. In annual cropping systems where no mechanization is used, increases in bulk density are not so likely to occur. Increases may be caused by human traffic or occur naturally but generally these increases are not spectacular. Table 4.15 presents bulk densities of Typic Eutropepts before and after 2 years of cropping with sweet potato and maize. Bulk densities were low in these soils and this was possibly related to the high organic C levels of the soil (55 g C/kg). Very few changes occurred in the bulk density of the sweet potato plots because the harvesting of sweet potato involves topsoil digging with a fork to about 0.2 m. Maize was grown without tillage and the bulk density after five consecutive seasons was slightly increased, but remained very low.

In mechanized annual cropping systems, where tractor traffic is common, compaction may occur (Soane, 1990) and it may severely reduce nutrient availability (Lipiec and Stepniewski, 1995; Arvidsson, 1999). On sugarcane plantations where heavy machinery is used for land preparation, harvesting, and husbandry practices like the applications of inorganic fertilizers and the spraying of pesticides, soil compaction is fairly common (Yates, 1978; Blackburn, 1984). Sugarcane is usually cultivated on low ridges (intra-rows) with tractors and harvesters passing through the inter-row. As a result, the variation in soil

Table 4.15. Bulk density (Mg/m^3 ±1 SD) of Typic Eutropepts before planting and 2 years after cropping with sweet potato and maize in the humid lowlands of Papua New Guinea. The soils had been fallow for 7 years before planting. Type I data from Hartemink (unpublished).

	Sweet potato		Maize	
	0–0.05 m	0.10–0.15 m	0–0.05 m	0.10–0.15 m
Before first planting	0.67 ± 0.10	0.62 ± 0.06	0.66 ± 0.08	0.78 ± 0.11
After 2 years	0.64 ± 0.06	0.75 ± 0.08	0.74 ± 0.06	0.73 ± 0.07

physical properties, which naturally can be large within a field (Cassel and Lal, 1992), is enhanced. On Spodosols in Australia, McGarry *et al.* (1996) found a topsoil bulk density of 1.55 Mg/m^3 in sugarcane intra-rows compared with 1.85 Mg/m^3 in the inter-row. An adjoining uncultivated site had a topsoil bulk density of 1.40 Mg/m^3. Soil compaction under sugarcane has also been reported from India (Srivastava, 1984; Rao and Narasimham, 1988), Hawaii (Trouse and Humbert, 1961), South Africa (Swinford and Boevey, 1984) and in various other sugarcane-producing countries (Hartemink and Wood, 1998).

Bulk density in Fluvents and Vertisols under sugarcane in Papua New Guinea was increased up to 0.30–0.50 m depth (Table 4.16). A negative exponential relationship was observed between topsoil bulk density and water intake, and bulk densities causing slow water

Table 4.16. Bulk density (Mg/m^3) of Fluvents and Vertisols under sugarcane and natural grassland.[a] Type II data from Hartemink (1998).

		Land use			
Soil order	Sampling depth (m)	Sugarcane Within the rows	Sugarcane Inter-row	Natural grassland	SED[c]
Fluvents[b]	0–0.15	1.10	1.29	1.07	0.04
	0.15–0.30	1.18	1.34	1.17	0.06
	0.30–0.50	1.35	1.39	1.26	0.05
Vertisols	0–0.15	0.98	1.18	1.00	0.03
	0.15–0.30	1.08	1.19	1.02	0.05
	0.30–0.50	1.14	1.21	1.12	0.06
	0.50–0.70	1.13	1.22	1.17	0.06

[a] Values reported are the arithmetic mean of six core samples of 100 ml taken in two soil pits.
[b] The 0.50–0.70 m soil horizons could not be sampled accurately with 100-ml cores because of abundant gravel.
[c] Standard error of the difference in means (10 d.f.).

intake (<50 mm/h) were only 1.20 Mg/m^3 in Fluvents and 1.16 Mg/m^3 in Vertisols. For both soil orders, an increase of about 0.2 Mg/m^3 drastically reduced the water intake and it will affect nutrient availability.

In tree crop plantations, bulk density may increase following (human) traffic and when the plantation ages, but few data are available. Sanchez *et al.* (1985) summarized bulk density data from various forest plantations at Jarí, Pará, Brazil based on the PhD work of C.E. Russell (Table 4.17). Although differences appear in the bulk density of these forests, statistical analysis revealed no significant differences in bulk density for any of the depths. However, nutrient stocks (i.e. multiplying nutrient concentrations with bulk density corrected for change in soil depth due to compaction) may have been significantly different but no data were available to have this verified.

Effects on nutrient content

An increase in the soil bulk density reduces the nutrient availability for crops because rooting is restricted, which limits the volume of soil from which nutrients can be extracted. When compaction occurs resulting in a shallow rooting pattern, the crop becomes susceptible to water stress, which may have a larger impact than the reduced nutrient availability. It is difficult to distinguish these factors and their effects on crop productivity.

A problem arising in relating soil compaction to nutrient availability is that the thickness of the layer decreases when the soil is more compacted. This means that if the sampling depth remains the same, part of the subsoil is being sampled which affects calculations on nutrient contents. So in essence sampling should be corrected for decrease in the thickness of the compacted layer (Dias and Nortcliff, 1985).

An increase in bulk density does not necessarily mean that nutrient availability is reduced. Table 4.18 shows the nutrient concentration and nutrient content of an Oxisol cropped with sugarcane. The nutrient content was calculated for three depths using bulk densities determined in 1978 and 1983. Absolute and relative differences in the

Table 4.17. Bulk density (Mg/m^3) of Ultisols under native forest and forest plantations of different ages at Jarí, Pará, Brazil. Type II data, modified from Sanchez *et al.* (1985).

Sampling depth (m)	Native forest	*Pinus caribaea* 0.5 year	*Pinus caribaea* 9.5 years	*Gmelina arborea* 8.5 years
0–0.01	1.22	0.98	1.23	1.33
0.01–0.30	1.62	1.61	1.58	1.54
0.30–1.00	1.70	1.76	1.60	1.73

Table 4.18. Nutrient concentration and nutrient content of Oxisols under sugarcane in 1978 and in 1983. Calculated from data in Masilaca *et al.* (1985).

Sampling depth (m)	Soil property	Nutrient concentration		Difference		Nutrient content (kg/ha)		Difference	
		1978	1983	Absolute	In %	1978	1983	Absolute	In %
0–0.12	BD (Mg/m^3)	0.76	1.02	+0.26	+34				
	Organic C (g/kg)	68.2	41.3	–26.9	–39	62,198	50,551	–11,647	–19
	Total N (g/kg)	4.0	1.9	–2.1	–53	3,648	2,326	–1,322	–36
	Total P (g/kg)	1.1	0.9	–0.2	–18	1,003	1,102	+99	+10
	Ca ($mmol_c$/kg)	29.0	9.1	–19.9	–69	530	223	–307	–58
	Mg ($mmol_c$/kg)	2.9	1.6	–1.3	–45	32	24	–8	–26
	K ($mmol_c$/kg)	3.0	1.2	–1.8	–60	107	57	–50	–46
0.30–0.40	BD (Mg/m^3)	0.86	1.06	+0.2	+23				
	Organic C (g/kg)	7.1	10.9	+3.8	+54	6,106	11,554	+5,448	+89
	Total N (g/kg)	1.2	1.0	–0.2	–17	1,032	1,060	+28	+3
	Total P (g/kg)	0.9	1.1	+0.2	+22	774	1,166	+392	+51
	Ca ($mmol_c$/kg)	1.6	4.0	+2.4	+150	28	85	+57	+208
	Mg ($mmol_c$/kg)	0.5	0.4	–0.1	–20	5	5	0	0
	K ($mmol_c$/kg)	0.6	0.5	–0.1	–17	20	21	+1	+3
0.70–0.80	BD (Mg/m^3)	1.01	1.10	+0.09	+9				
	Organic C (g/kg)	3.6	3.0	–0.6	–17	3,636	3,300	–336	–9
	Total N (g/kg)	0.5	0.5	0	0	505	550	+45	+9
	Total P (g/kg)	1.1	1.1	0	0	1,111	1,210	+99	+9
	Ca ($mmol_c$/kg)	2.0	2.2	+0.2	+10	40	48	+8	+20
	Mg ($mmol_c$/kg)	0.4	0.2	–0.2	–50	5	3	–2	–46
	K ($mmol_c$/kg)	0.3	0.5	+0.2	+67	12	22	+10	+82

nutrient concentration and nutrient content were calculated for both periods. No correction was made for the decrease of soil layer thickness due to the increased bulk density.

In the topsoils (0–0.12 m) bulk density increased from 0.76 to 1.02 Mg/m^3 and although this is a considerable increase, the absolute value is still low (Yates, 1978; Landon, 1991). Between 1978 and 1983, there was a relatively lower decrease in nutrient content than in nutrient concentration. As a result of the increase in bulk density, different conclusions would be reached with regard to total P change in the topsoils: total P decreased from 1.1 to 0.9 g/kg whereas the P content of the topsoil increased by 99 kg/ha due to the 34% increase in topsoil bulk density. Similar discrepancies can be found in some other soil chemical properties. Annual losses of total N from the topsoil exceed 200 kg/ha but a slight increase in N contents of the subsoils was found, which may have been caused by leaching (see Chapter 9 for similar observations).

A second example on the effects of bulk density on soil nutrient contents comes from Nigeria, where Aina (1979) sampled Alfisols that had been cropped for 10 years and Alfisols that had been fallowed for 20–25 years (Table 4.19). Differences in the thickness of the soil layer due to compaction were not taken into account. The relative decrease in soil nutrient contents is lower than the decrease in nutrient concentration with the exception of NO_3-N, which varies greatly with time. Nutrient concentration and content drastically decreased in permanently cropped soils, but the relative decrease in nutrient contents was lower. One could also conclude that soil nutrient contents increase in soils under fallow as compared to cropping. From this example, it is not clear whether the increase in bulk density of cropped soils is favourable (increasing the availability or accessibility of nutrients) or unfavourable (decreasing the rootability).

Table 4.19. Nutrient concentration and nutrient content (0–0.15 m depth) of Alfisols under fallow and 10 years of cropping. Calculated from data in Aina (1979).

	Nutrient concentration		Difference		Nutrient content (kg/ha)		Difference	
Soil property	Fallow	Cropped	Absolute	In %	Fallow	Cropped	Absolute	In %
BD (Mg/m^3)	1.24	1.58	+0.34	+27				
NO_3-N (g/kg)	19.3	2.3	–17.0	–88	36	5	–30	–85
Available P (g/kg)	15.4	6.0	–9.4	–61	29	14	–14	–50
Ca ($mmol_c$/kg)	45.1	15.0	–30.1	–67	1681	712	–969	–58
K ($mmol_c$/kg)	2.4	0.9	–1.5	–63	174	83	–91	–52

Values are rounded up.

Conclusions

This section has illustrated the importance of bulk density in soil fertility studies. Bulk density affects the nutrient availability and content. It is likely to change under cropping, and its measurement is prone to spatial and temporal variation. In most studies on soil fertility decline, bulk density values are not reported despite its relatively easy measurement and rapid changes under cropping. As there are few bulk density data in soil fertility studies, it is not further considered. Also other soil physical attributes such as texture or water-holding capacity, which change less quickly than bulk density and are more complicated to measure, affect nutrient availability but are not taken into account in the subsequent chapters.

Interpretation of the Results

Provided spatial and temporal boundaries are correctly set, sampling and analysis have been done appropriately, and variation is dealt with, how should the results of soil fertility decline studies be interpreted? Is the difference between input and output in a nutrient balance the only criterion or does size of the flux also matter? Should the size of the flux be linked to the size of the nutrient pool or stock? For example, are K losses from Oxisols with few weatherable minerals more serious than from young alluvial soils or those developed in recent volcanic material? How is the size of the pool measured? Is it possible to differentiate between factors and soil processes governing soil fertility decline? How should the issue of scale be taken into account when evaluating the results of soil fertility decline studies? Is the decline in organic matter more serious than a decline in exchangeable Ca? Could thresholds be established for different soils in different agroecological conditions and is there a universal key to the interpretation of soil fertility decline? What is a minimum data-set before it can be firmly stated that 'the soil fertility has significantly declined'? Is a decline in exchangeable K more serious for annual crops than for perennials or pastures?

These questions form important considerations for the interpretation of research results on soil fertility decline. As can be expected, answers are complex and unique to each situation, which is determined by the agroecological conditions, the spatial and temporal boundaries of the study and other factors including the type of data and how they were collected. Soil fertility decline must be differently appraised for soils in different agroecologies but some common rules apply and these are briefly discussed here.

Resilience and reversibility

The decline in a soil fertility parameter in relation to the pool is closely related to the resilience of the soil. Resilience is the ability of the soil to recover from a period of stress as for example the cultivation of agricultural crops (Greenland and Szabolcs, 1994; Lal, 1997). Some soils withstand cultivation more easily and quickly recover after a period of intense cultivation. Resilience is an intrinsic property of the soil and shows that the assessment of soil fertility decline has temporary boundaries. The net removal of nutrients in relation to the size of the pool could be an indicator used to evaluate the seriousness of soil fertility decline. This approach is also taken in soil erosion studies whereby net soil loss is sometimes coupled to net soil formation and the effective rooting depth of the soil. Much depends, of course, on how the size of the pool is measured, i.e. the bioavailability concept in soil fertility (see previous sections) but also on the bulk density and its changes.

Not only different soils require different appraisal, also individual soil chemical properties require a different appraisal. For example, a slight increase in exchangeable acidity may have less effect on crop production than a considerable decrease in total N. Likewise, the decrease in soil organic C may have no direct yield effect, but could drastically reduce the resistance of the soil to physical deterioration or to supply N or P to the crop.

Since von Liebig, it has been generally assumed that output of nutrients needs to match the input in order to sustain crop production (van Noordwijk, 1999), or in other words: replace what was lost. However, the time frame at which the replacement is required is different for different soils. Inherently fertile soils might resist the drain of nutrients and remain productive for a considerable period of time (i.e. the resilience concept), but at some stage these soils require replenishment for what has been taken and lost. Inherently poor soils might need nutrient replenishment before a second crop is grown and their fertility may swiftly decline when permanently cultivated. The average annual nutrient balance may be negative but it may differ between years or seasons, which should be taken into account in the replenishment concept.

Another aspect that affects the interpretation of results is the degree of reversibility of a change in a soil property. A decreasing level of exchangeable K may be less of a problem than a large decrease in soil organic C, because K may be replaced by weathering minerals or inorganic fertilizers whereas a doubling of the soil organic C contents to the original level is very difficult. A strongly acidified topsoil may be easy to correct by the judicious application of lime, but it may be much harder to raise the pH of a strongly acidified subsoil. So the reversibility is different for the various soil attributes and it is also important to consider the depth to which the soil chemical changes have occurred.

The time-lag effect

To assess whether soil fertility decline has occurred depends on what properties are measured and the rates at which the properties change. This is somewhat similar to pedogenesis, where slowly and rapidly changing soil properties are distinguished (Yaalon, 1971). Rapidly changing properties include organic C, N and pH. These properties usually reach dynamic equilibrium within 100 to 1000 years. They are affected by changes in the environment but if these changes are relatively constant they will reach equilibrium. The second group of features changes slowly and appear to be at equilibrium mainly because their rate of change is so slow (Yaalon, 1971).

Some soil processes, once established, continue for some time despite changes in the environment, and the resistance to change may be related to what has been termed 'pedogenic inertia' (Bryan and Teakle, 1949; Chadwick and Chorover, 2001). The simplest example of a lag is the soil temperature, both diurnal and annual, which invariably lags behind the atmospheric temperature wave (Yaalon, 1971). In essence, this means that soil fertility may continue to decline for an indefinite period even if the cause of the decline (permanent cropping without nutrient inputs) has been removed and the soil has been left fallow. Not all soil properties would show this effect and at the same pace. Although this has been receiving attention in pedogenesis, it has received no attention in soil fertility decline studies under agroecosystems. In agroecosystems, the rates and direction of soil processes are different compared with natural ecosystems because of periodic disturbance, including tillage, weeding, and the application of soil amendments.

Frequency, period and time of observation

The frequency with which phenomena that affect soil properties occur is important. For example, a single and destructive soil erosion event may take place once every 10 years and could have substantial impact on the soil chemical fertility. On the other hand, there are very gradual processes, such as soil acidification, and these have been named subtle or slow phenomena (Pickett, 1991). For both rare events and slow phenomena to be recorded, long-term observations are needed.

Besides spatial and temporal variation in soil chemical properties and the pace of soil change, another important factor is the period during which the observations are made. Whether a consistent trend in a soil chemical property can be observed and quantified depends on the property itself and the period and time of observation. This is illustrated in Fig. 4.8, in which trends in a fictitious soil chemical property are depicted over time. In Fig. 4.8a the soil property shows some noise,

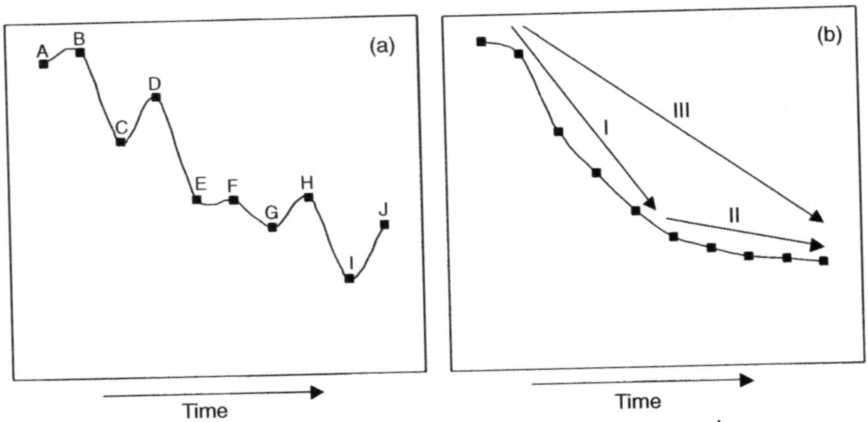

Fig. 4.8. Theoretical changes in soil chemical properties over time when no amendments are made and the soils are permanently cropped: (a) noise and trend, (b) exponential decline. See text for explanation.

which may have been the result of weather conditions or management factors, or a combination of the two. This could be the variation in soil pH over the years. On a different time scale it could be the variation in a soil property during a single day following the warming of the soil by the sun, or directly after rain or inorganic fertilizer applications. Soil chemical properties show variation at different time scales, but for most of the standard soil tests, long-term variation is of greater importance than the diurnal or short-term variation. The decline of the soil property in Fig. 4.8a (i.e. an interpolated line) is more or less linear.

A gradual decline in a soil chemical property is shown in Fig. 4.8b. This may represent a decline in exchangeable K in the soils of an unfertilized oil palm plantation. Quantification of the rate of decline depends on the period and time of observation. In the beginning the decline is fast (arrow I) but since the decline is non-linear, the rate of decline (Δ/t) is decreasing with time (arrow II). If observations would be made over time I, a different rate of decline would be observed compared with time II – even if the length of observational period is the same. Rates of decline over the whole period (arrow III) would again give a different conclusion and this would largely ignore the non-linearity of the relationship. It is not necessarily the case that the decline based on III is half the sum of I and II. To assess whether a soil chemical property declines exponentially, measurements at relatively short time steps are required. If time steps are larger, it should be known whether period I, II or III is evaluated.

The pattern in Fig. 4.8a may result in different conclusions when two points in the curves are compared. This is exemplified in Table 4.20, where long-term, medium-term and short-term comparisons are

Table 4.20. Changes in a soil property between different sampling times (A, B, C, D...etc.) – see Fig. 4.8a.

	B	C	D	E	F	G	H	I	J	
A	+/–	–	+/–	–	–	–	–	– –	–	
B				–	–	–	–	– –	–	Long-term comparison
C				–	–	–	+/–	–	+/–	
D				–	–	–	–	– –	–	
E					+/–	+/–	+/–	–	+/–	Medium-term comparison
F						+/–	+/–	–	+/–	
G							+/–	–	+/–	
H								–	+/–	Short-term comparison
I									+	

– – Large decrease +/– No change
– Moderate decrease + Moderate increase

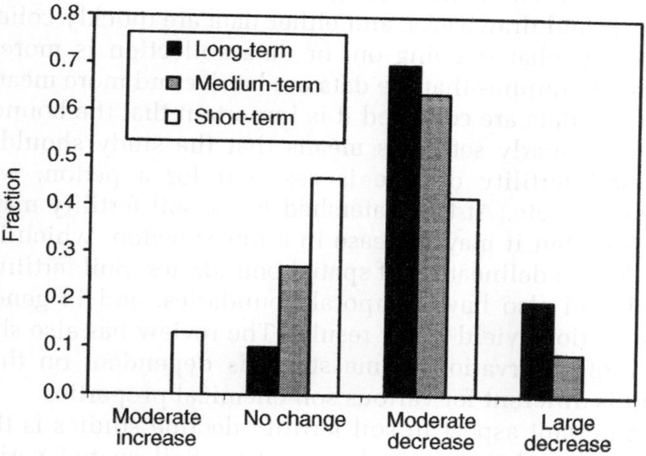

Fig. 4.9. Effect of length of comparison on soil changes. Data from Fig. 4.8a and Table 4.20. See text for explanation.

grouped. Comparisons were termed long term when they exceeded five data points of the *x*-axis (time), medium term when there were three to four data points, and short term when there were fewer than two periods between two data points.

The general pattern emerging is that long-term observations yield a stronger decline in soil fertility whereas short-term observations yield no clear pattern. Owing to the noise there is also a difference within the periods of comparison. A large decrease in the soil property was found in 20% of the long-term comparisons, whereas 70% of the comparisons yielded a moderate decrease (Fig. 4.9). In 10% of the long-term comparisons, no change was apparent. Medium-term comparisons yielded a large decrease in 9% of the cases, whereas in more than 25% of the comparisons no change was found. Short-term comparisons yielded no change in soil properties in almost half of the cases and a moderate increase in 15% of the comparisons.

Discussion and Conclusions

In this chapter, theoretical aspects for the assessment of soil fertility decline were critically reviewed. Evaluating soil fertility decline can be done with different types of data. There are data from measured soil chemical properties, and such data can be from the same plot that is permanently cultivated (Type I data), or from plots under different land use (Type II data). Soil fertility decline can also be assessed in a more semiquantitative way using nutrient balances. Each data type has it merits and drawbacks, and either data are quickly collected and indicative for what is going on, or data collection is more tedious, which usually implies that the data are harder and more meaningful.

Whatever data are collected it is important that the boundary conditions are properly set. This means that the study should indicate whether soil fertility decline is assessed for a pedon, watershed, region, country etc. At the watershed level, soil fertility may decline in one pedon but it may increase in a lower pedon, which illustrates the need for the delineation of spatial boundaries. Soil fertility decline studies should also have temporal boundaries, and in general long-term observations yield better results. The review has also shown that frequency of observations (time steps) is dependent on the type of study and is different for various soil chemical properties.

An important aspect in soil fertility decline studies is the spatial and temporal variation in soil properties. Soil spatial variation has been sufficiently tackled by research and various methods exist to quantify the variation. Temporal variation is a more difficult issue and fewer studies are available. As with spatial variation, it requires a sufficient number of subsamples and samples before rigid conclusions can be drawn. Temporal variation may also be confused with trends in the data but some soil chemical properties are more vulnerable to temporal variation than others.

Soil fertility decline studies are largely dependent on soil chemical analysis, which includes soil sampling, soil analysis and interpretation of the results. Errors are possible in all three steps although most errors are generally made during soil sampling because soil variation is insufficiently dealt with and too few samples are taken. The choice of the analytical technique in relation to the soil property or soil type is another potential source of errors. The effects of soil sample storage and a constant laboratory error are relevant for long-term studies on soil change, but data on storage effects and laboratory errors are scarce.

Bulk density is an important factor to consider in soil fertility studies. It is needed to convert nutrient concentrations into nutrient contents that can be used in nutrient balance studies. The decrease in soil depth should be considered when soils have been compacted. Slight deviations in bulk densities have a significant effect on the outcome of the nutrient content calculations. Nutrient removal by the economic produce is also an important component in nutrient balance studies. Published values on nutrient removal vary greatly, which, among other factors, is related to differences in cultivars, measured plant portion and age of the crop, soil type and the soil nutrient status.

For the interpretation of studies on soil fertility decline, resilience and reversibility are important concepts that reflect the ability of the soil to withstand stress and the ability to reverse changes brought about by cropping. The frequency at which observations are made also determines the interpretation of the results since some phenomena rarely occur whereas others take place gradually. The period of observation should be long enough to accommodate slow phenomena and rare events, but also to deal with temporal variation. Owing to noise in the data caused by temporal or other sources of unknown variation, different conclusions can be reached – even if the period of observation is substantial. The pattern of the decline, the time of observation and the size of the time steps should be known, and are important for accurately quantifying soil fertility decline.

This chapter has reviewed the main conditions and pitfalls in soil fertility decline studies with a focus on permanent cropping in the tropics. In the next five chapters evidence for soil fertility decline in the tropics is reviewed.

References

Addiscott, T.M. (1996) Measuring and modelling nitrogen leaching: parallel problems. *Plant and Soil* 181, 1–6.

Aina, P.O. (1979) Soil changes resulting from long-term management practises in Western Nigeria. *Soil Science Society of America Journal* 43, 173–177.

Alfsen, K.H., Bye, T.A., Glomsrod, S. and Wiig, H. (1997) Theory and applications. Soil degradation and economic development in Ghana. *Environment and Development Economics* 2, 119–143.

Alpízar, L., Fassbender, H.W., Heuveldop, J., Fölster, H. and Enríquez, G. (1986) Modelling agroforestry systems of cacao (*Theobroma cacao*) with laurel (*Cordia alliodora*) and poro (*Erythrina poeppigiana*) in Costa Rica.

I. Inventory of organic matter and nutrients. *Agroforestry Systems* 4, 175–189.

Anderson, J.M. and Ingram, J.S.I. (eds) (1993) *Tropical Soil Biology and Fertility, a Handbook of Methods*. CAB International, Wallingford.

Arvidsson, J. (1999) Nutrient uptake and growth of barley as affected by soil compaction. *Plant and Soil* 208, 9–19.

Barrera-Bassols, N. and Zinck, J.A. (2000) *Ethnopedology in a Worldwide Perspective – an Annotated Bibliography*. International Institute for Aerospace Survey and Earth Sciences (ITC), Enschede, The Netherlands.

Bartlett, R. and James, B. (1980) Studying dried, stored soil samples – some pitfalls. *Soil Science Society of America Journal* 44, 721–724.

Bekunda, M.A., Bationo, A. and Ssali, H. (1997) Soil fertility management in Africa: a review of selected research trials. In: Buresh, R.J., Sanchez, P.A. and Calhoun, F. (eds) *Replenishing Soil Fertility in Africa*. SSSA Special Publication Number 51. SSSA, Madison, Wisconsin, pp. 63–79.

Billett, M.F., Parker-Jervis, F., Fitzpatrick, E.A. and Cresser, M.S. (1990) Forest soil chemical changes between 1949/50 and 1987. *Journal of Soil Science* 41, 133–145.

Black, C.A. (1993) *Soil Fertility Evaluation and Control*. Lewis Publishers, Boca Raton, Florida.

Blackburn, F. (1984) *Sugar-cane*. Longman, Harlow, UK.

Boone, R.D., Grigal, D.F., Sollins, P., Ahrens, R.J. and Armstrong, D.E. (1999) Soil sampling, preparation, archiving, and quality control. In: Robertson, G.P., Coleman, D.C., Bledsoe, C.S. and Sollins, P. (eds) *Standard Methods for Long-term Ecological Research*. Oxford University Press, pp. 3–28.

Bouma, J. (1993) Soil behavior under field conditions – differences in perception and their effects on research. *Geoderma* 60, 1–14.

Boxman, O. and Janssen, B.H. (1990) Availability of nutrients and fertilizer use. In: Janssen, B.H. and Wienk, J.F. (eds) *Mechanized Annual Cropping on Low Fertility Acid Soils in the Humid Tropics. Wageningen Agricultural University Papers 90–5*. Wageningen Agricultural University, Wageningen, pp. 73–99.

Bramley, R.G.V., Ellis, N., Nable, R.O. and Garside, A.L. (1996) Changes in soil chemical properties under long-term sugar cane monoculture and their possible role in sugar yield decline. *Australian Journal of Soil Research* 34, 967–984.

Brown, A.J. (1999) Soil sampling and sample handling for chemical analysis. In: Peverill, K.I., Sparrow, L.A. and Reuter, D.J. (eds) *Soil Analysis – an Interpretation Manual*. CSIRO Publishing, Collingwood, Australia, pp. 35–53.

Bruijnzeel, L.A. (1983) Hydrological and biogeochemical aspects of man-made forests in south-central Java, Indonesia. PhD dissertation, Free University, Amsterdam, The Netherlands.

Bryan, W.H. and Teakle, L.J.H. (1949) Pedogenic inertia – a concept in soil science. *Nature* 164, 969.

Cassel, D.K. and Lal, R. (1992) Soil physical properties of the tropics: common beliefs and management restraints. In: Sanchez, P.A. and Lal, R. (eds) *Myths and Science of Soils in the Tropics*. SSSA, Madison, Wisconsin, pp. 61–89.

Chadwick, O.A. and Chorover, J. (2001) The chemistry of pedogenic thresholds. *Geoderma* 100, 321–353.

Cline, M.G. (1944) Principles of soil sampling. *Soil Science* 58, 275–288.
Cline, M.G. (1945) Methods and collecting and preparing soil samples. *Soil Science* 59, 3–5.
Cooke, G.W. (1982) *Fertilizing for Maximum Yield*, 3rd edn. Collins, London.
de Walle, F.B. and Sevenster, J. (1998) *Agriculture and the Environment: Minerals, Manure and Measures.* Kluwer Academic Publishers, Dordrecht.
Dias, A.C.C.P. and Nortcliff, S. (1985) Effects of two land clearing methods on the physical properties of an Oxisol in the Brazilian Amazon. *Tropical Agriculture* 62, 207–212.
Dobermann, A., Gaunt, J.L., Neue, H.U., Grant, I.F., Adviento, M.A. and Pampolino, M.F. (1994) Spatial and temporal variability of ammonium in flooded rice fields. *Soil Science Society of America Journal* 58, 1708–1717.
Doran, J.W. and Parkin, T.B. (1996) Quantitative indicators of soil quality: a minimum data set. In: Doran, J.W. and Jones, A.J. (eds) *Methods for Assessing Soil Quality.* SSSA, Madison, Wisconsin, pp. 25–37.
Dorich, R.A. and Nelson, D.W. (1984) Evaluation of manual cadmium reduction methods for determination of nitrate in potassium chloride extracts of soil. *Soil Science Society of America Journal* 48, 72–75.
Drechsel, P. and Gyiele, L.A. (1999) *The Economic Assessment of Soil Nutrient Depletion. Analytical Issues for Framework Development.* IBSRAM, Bangkok.
Ekanade, O. (1988) The nutrient status of soils under peasant cocoa farms of varying ages in southwestern Nigeria. *Biological Agriculture and Horticulture* 5, 155–167.
Evans, J. (1992) *Plantation Forestry in the Tropics*, 2nd edn. Clarendon Press, Oxford.
FAO (2001) *The Economics of Soil Productivity in Sub-Saharan Africa.* FAO, Rome.
Fassbender, H.W., Alpízar, L., Heuveldop, J., Fölster, H. and Enríquez, G. (1988) Modelling agroforestry systems of cacao (*Theobroma cacao*) with laurel (*Cordia alliodora*) and poro (*Erythrina poeppigiana*) in Costa Rica. III. Cycles of organic matter and nutrients. *Agroforestry Systems* 6, 49–62.
Fassbender, H.W., Beer, J., Heuveldop, J., Imbach, A., Enríquez, G. and Bonneman, A. (1991) The year balances of organic matter and nutrients in agroforestry systems at CATIE, Costa Rica. *Forest Ecology and Management* 45, 173–183.
Frye, W.W. and Thomas, G.W. (1991) Management of long-term field experiments. *Agronomy Journal* 83, 38–44.
Garside, A.L., Bramley, R.G.V., Bristow, K.L., Holt, J.A., Magarey, R.G., Nable, R.O., Pankhurst, C.E. and Skjemstad, J.O. (1997) Comparisons between paired old and new land sites for sugarcane growth and yield and soil chemical, physical, and biological properties. *Proceedings of Australian Society of Sugarcane Technologists* 60–66.
Giller, K.E., Cadish, G., Ehaliotis, C., Adams, E., Sakala, W.D. and Mafongoya, P.L. (1997) Building soil nitrogen capital in Africa. In: Buresh, R.J., Sanchez, P.A. and Calhoun, F. (eds) *Replenishing Soil Fertility in Africa.* SSSA Special Publication Number 51. SSSA, Madison, Wisconsin, pp. 151–192.

Goovaerts, P. (1998) Geostatistical tools for characterizing the spatial variability of microbiological and physico-chemical soil properties. *Biology and Fertility of Soils* 27, 315–334.

Grace, P.R., Oades, J.M., Keith, H. and Hancock, T.W. (1995) Trends in wheat yields and soil organic carbon in the permanent rotation trial at the Waite agricultural research institute, South Australia. *Australian Journal of Experimental Agriculture* 35, 857–864.

Gray, L.C. (1999) Is land being degraded? A multi-scale investigation of landscape change in southwestern Burkina Faso. *Land Degradation and Development* 10, 329–343.

Greenland, D.J. (1994a) Long-term cropping experiments in developing countries: the need, the history and the future. In: Leigh, R.A. and Johnston, A.E. (eds) *Long-term Experiments in Agricultural and Ecological Sciences*. CAB International, Wallingford, UK, pp. 187–209.

Greenland, D.J. (1994b) Soil science and sustainable land management. In: Syers, J.K. and Rimmer, D.L. (eds) *Soil Science and Sustainable Land Management in the Tropics*. CAB International, Wallingford, UK, pp. 1–15.

Greenland, D.J. and Szabolcs, I. (eds) (1994) *Soil Resilience and Sustainable Land Use*. CAB International, Wallingford, UK.

Greer, K.J. and Schoenau, J.J. (1997) Toward a framework for soil quality assessment and prediction. In: Gregorich, E.G. and Carter, M.R. (eds) *Soil Quality for Crop Production and Ecosystem Health*. Elsevier Science, Amsterdam, pp. 313–321.

Gregorich, E.G., Carter, M.R., Angers, D.A., Monreal, C.M. and Ellert, B.H. (1994) Towards a minimum data set to assess soil organic matter quality in agricultural soils. *Canadian Journal of Soil Science* 74, 367–385.

Grimme, H. and Juo, A.S.R. (1985) Inorganic nitrogen loss through leaching and denitrification in soils of the humid tropics. In: Kang, B.T. and van der Heide, J. (eds) *Nitrogen Management in Farming Systems in Humid and Subhumid Tropics*. IB, Haren and IITA, Ibadan, pp. 57–71.

Halenda, C.J. (1993) Aboveground biomass production and nutrient accumulation of *Gmelina arborea* plantation in Sarawak, Malaysia. *Journal of Tropical Forest Science* 5, 429–439.

Harris, H.C. (1995) Long-term trials on soil and crop management at ICARDA. In: Lal, R. and Stewart, B.A. (eds) *Soil Management – Experimental Basis for Sustainability and Environmental Quality*. CRC Press, Boca Raton, Florida, pp. 447–469.

Hartemink, A.E. (1993) *Nutrients in Cocoa Ecosystems – a Desk Study*. Department of Agronomy, Agricultural University, Wageningen, 31 pp.

Hartemink, A.E. (1994) Inorganic nitrogen dynamics under different land-use systems on an Oxisol in Western Kenya. MSc Thesis, Wageningen Agricultural University, The Netherlands.

Hartemink, A.E. (1995) *Soil Fertility Decline under Sisal Cultivation in Tanzania*. ISRIC, Wageningen, The Netheralnds.

Hartemink, A.E. (1996) Using soil survey data for the assessment of nutrient depletion. In: *Australian and New Zealand National Soils Conference*. ASSSI, Melbourne, pp. 103–104.

Hartemink, A.E. (1998) Soil chemical and physical properties as indicators of sustainable land management under sugar cane in Papua New Guinea. *Geoderma* 85, 283–306.

Hartemink, A.E. and Johnston, M. (1998) Root biomass and nutrient uptake of taro in the lowlands of Papua New Guinea. *Tropical Agriculture* 75, 1–5.

Hartemink, A.E. and Wood, A.W. (1998) Sustainable land management in the tropics: the case of sugarcane plantations. In: *16th World Congress of Soil Science.* ISSS, Montpellier, 7 pp.

Hartemink, A.E., Poloma, S., Maino, M., Powell, K.S., Egenae, J. and O'Sullivan, J.N. (2000) Yield decline of sweet potato in the humid lowlands of Papua New Guinea. *Agriculture Ecosystems and Environment* 79, 259–269.

Heuveldop, J., Fassbender, H.W., Alpízar, L., Enríquez, G. and Fölster, H. (1988) Modelling agroforestry systems of cacao (*Theobroma cacao*) with laurel (*Cordia alliodora*) and poro (*Erythrina poeppigiana*) in Costa Rica. II. Cacao and wood production, litter production and decomposition. *Agroforestry Systems* 6, 37–48.

Hoosbeek, M.R., Stein, A., van Reuler, H. and Janssen, B.H. (1998) Interpolation of agronomic data from plot to field scale – using a clustered versus a spatially randomized block design. *Geoderma* 81, 265–280.

IBSRAM (1999) Site selection and characterization: focus on biophysical and socioeconomic inventory. *Issues in Sustainable Land Management No. 6.* IBSRAM, Bangkok.

IPI (1995) *Potassium in Asia – Balanced Fertilization to Increase and Sustain Agricultural Production. 24th Colloquium of the International Potash Institute.* IPI, Basel, Switzerland.

Janssen, B.H., Guiking, F.C.T., van der Eijk, D., Smaling, E.M.A., Wolf, J. and van Reuler, H. (1990) A system for quantitative evaluation of the fertility of tropical soils (QUEFTS). *Geoderma* 46, 299–318.

Jenkinson, D.S. (1991) The Rothamsted long-term experiments: are they still of use? *Agronomy Journal* 83, 2–10.

Johnston, A.E. and Powlson, D.S. (1994) The setting-up, conduct and applicability of long-term, continuing field experiments in agricultural research. In: Greenland, D.J. and Szabolcs, I. (eds) *Soil Resilience and Sustainable Land Use.* CAB International, Wallingford, pp. 395–422.

Johnston, A.E. and Syers, J.K. (1998) Working group reports and conference summary. In: Johnston, A.E. and Syers, J.K. (eds) *Nutrient Management for Sustainable Crop Production in Asia.* CAB International, Wallingford, pp. 375–382.

Kadeba, O. (1994) Growth and nutrient accumulation by *Pinus caribaea* on three savanna sites in Northern Nigeria. *Agriculture Ecosystems and Environment* 49, 139–147.

Kee, K.K., Goh, K.J. and Chew, P.S. (1995) Effects of NK fertiliser on soil pH and exchangeable K status on acid soils in an oilpalm plantation in Malaysia. In: Date, R.A., Grundon, R.J., Rayment, G.E. and Probert, M.E. (eds) *Plant–Soil Interactions at Low pH.* Kluwer Academic Publishers, Dordrecht, pp. 809–815.

Kempthorne, C. and Allmars, R.R. (1986) Errors and variability of observations. In: Klute, A. (ed.) *Methods of Soil Analysis. Part 1 Physical and Mineralogical Methods*, 2nd edn. ASA-SSSA, Madison, Wisconsin, pp. 1–31.

Kirkegaard, J.A. (1995) A review of trends in wheat yield responses to conservation cropping in Australia. *Australian Journal of Experimental Agriculture* 35, 835–848.

Kumar, K. and Goh, K.M. (2000) Crop residues and management practices: effects on soil quality, soil nitrogen dynamics, crop yield, and nitrogen recovery. *Advances in Agronomy* 68, 197–319.

Lal, R. (1997) Degradation and resilience of soils. *Philosophical Transactions of the Royal Society of London – Series B: Biological Sciences*, 997–1008.

Lal, R. and Stewart, B.A. (1995a) Foreword. In: *Soil Management – Experimental Basis for Sustainability and Environmental Quality*. CRC Lewis Publishers, Boca Raton, pp. i–ii.

Lal, R. and Stewart, B.A. (1995b) Need for long-term experiments in sustainable use of soil resources. In: Lal, R. and Stewart, B.A. (eds) *Soil Management – Experimental Basis for Sustainability and Environmental Quality*. CRC Press, Boca Raton, pp. 537–545.

Landon, J.R. (ed.) (1991) *Booker Tropical Soil Manual*. Longman, Harlow, UK.

Lapenis, A.G., Torn, M.S., Harden, J.W., Hollocker, K., Babikov, B.V., Timofeev, A.I., Hornberger, M.I. and Nattis, R. (2000) Scientists unearth clues to soil contamination by comparing old and new soil samples. *EOS – Transactions American Geophysical Union* 81.

Laryea, K.B., Anders, M.M. and Pathak, P. (1995) Long-term experiments on Alfisols and Vertisols in the semiarid tropics. In: Lal, R. and Stewart, B.A. (eds) *Soil Management – Experimental Basis for Sustainability and Environmental Quality*, CRC Press, Boca Raton, pp. 267–292.

Leigh, R.A., Prew, R.D. and Johnston, A.E. (1994) The management of long-term agricultural field experiments: procedures and policies evolved from the Rothamsted classical experiments. In: Leigh, R.A. and Johnston, A.E. (eds) *Long-term Experiments in Agricultural and Ecological Sciences*. CAB International, Wallingford, UK, pp. 253–268.

Lipiec, J. and Stepniewski, W. (1995) Effects of soil compaction and tillage systems on uptake and losses of nutrients. *Soil and Tillage Research* 35, 37–52.

Masilaca, A.S., Prasad, R.A. and Morrison, R.J. (1985) The impact of sugarcane cultivation on three Oxisols from Vanua Levu, Fiji. *Tropical Agriculture* 63, 325–330.

McBratney, A.B., Odeh, I.O.A., Bishop, T.F.A., Dunbar, M.S. and Shatar, T.M. (2000) An overview of pedometric techniques for use in soil survey. *Geoderma* 97, 293–327.

McGarry, D., Pillai-McGarry, U.P., Bray, S.G., Bigwood, R.C. and Moody, P.W. (1996) The effect of 10 years of sugar production on soil physical and chemical properties of a Gleyed Podzolic soil, Bundaberg, Queensland. In: *Australian and New Zealand National Soils Conference*. ASSSI, Melbourne, pp. 163–164.

McLaughin, M.J., Reuter, D.J. and Rayment, G.E. (1999) Soil testing – principles and concepts. In: Peverill, K.I., Sparrow, L.A. and Reuter, D.J. (eds) *Soil Analysis – an Interpretation Manual*. CSIRO Publishing, Collingwood, Victoria, Australia, pp. 1–21.

Molloy, M.G. and Lockman, R.B. (1979) Soil analysis as affected by drying temperatures. *Communications in Soil Science and Plant Analysis* 10, 545–550.

Nelson, L.A. (1987) Design of experiments for land development and management of acid soils in Africa. In: Latham, M. and Ahn, P. (eds) *Land*

Development and Management of Acid Soils in Africa II. IBSRAM, Bangkok, pp. 209–216.

Nye, P.H. and Greenland, D.J. (1960) *The Soil under Shifting Cultivation.* Commonwealth Bureau of Soils, Harpenden, UK.

Oenema, O. and Heinen, M. (1999) Uncertainties in nutrient budgets due to biases and errors. In: Smaling, E.M.A., Oenema, O. and Fresco, L.O. (eds) *Nutrient Disequilibria in Agroecosystems: Concepts and Case Studies.* CAB International, Wallingford, UK, pp. 75–97.

Oldeman, L.R., Hakkeling, R.T.A. and Sombroek, W.G. (1991) *World Map of the Status of Human-induced Soil Degradation: an Explanatory Note*, 2nd revised edn. ISRIC, Wageningen.

Pape, J.C. (1970) Plaggen soils in the Netherlands. *Geoderma* 4, 229–255.

Parker, G.G. (1983) Throughfall and stemflow in the forest nutrient cycle. *Advances in Ecological Research* 13, 57–133.

Payne, G.G. and Rechcigl, J.E. (1989) Influence of various drying techniques on the extractability of plant nutrients from selected Florida soils. *Soil Science* 148, 275–283.

Peters, J.B. (2000) Gambian soil fertility trends, 1991–1998. *Communications in Soil Science and Plant Analysis* 31, 2201–2210.

Pickett, S.T.A. (1991) Long-term studies: past experience and recommendations for the future. In: Risser, P.G. (ed.) *Long-term Ecological Research – an International Perspective. SCOPE 47.* John Wiley & Sons, Chichester, UK, pp. 71–88.

Pieri, C. (1995) Long-term soil management experiments in semiarid Francophone Africa. In: Lal, R. and Stewart, B.A. (eds) *Soil Management – Experimental Basis for Sustainability and Environmental Quality.* CRC Press, Boca Raton, Florida, pp. 225–266.

Poels, R.L.H. (1987) Soils, water and nutrients in a forest ecosystem in Suriname. Doctoral thesis, Wageningen Agricultural University.

Poels, R.L.H. (1989) Nutrient input and output in undisturbed and silviculturally treated tropical rain forest in Suriname. *Netherlands Journal of Agricultural Science* 37, 383–386.

Poulton, P.R. (1995) The importance of long-term trials in understanding sustainable farming systems – the Rothamsted experience. *Australian Journal of Experimental Agriculture* 35, 825–834.

Powlson, D.S. and Johnston, A.E. (1994) Long-term field experiments: their importance in understanding sustainable land use. In: Greenland, D.J. and Szabolcs, I. (eds) *Soil Resilience and Sustainable Land Use.* CAB International, Wallingford, pp. 367–394.

Prasolova, N.V., Xu, Z.H., Saffigna, P.G. and Dieters, M.J. (2000) Spatial-temporal variability of soil moisture, nitrogen availability indices and other chemical properties in hoop pine (*Araucaria cunninghamii*) plantations of subtropical Australia. *Forest Ecology and Management* 136, 1–10.

Rao, S.R. and Narasimham, R.L. (1988) Effect of soil compaction in sugarcane: part I. On soil moisture, porosity and nitrification. *International Journal of Tropical Agriculture* 6, 232–236.

Rasmussen, P.E., Goulding, K.W.T., Brown, J.R., Grace, P.R., Janzen, H.H. and Korschens, M. (1998) Long-term agroecosystem experiments: assessing agricultural sustainability and global change. *Science* 282, 893–896.

Raun, W.R. and Johnson, G.V. (1999) Improving nitrogen use efficiency for cereal production. *Agronomy Journal* 91, 357–363.

Robertson, G.P. (1982) Regional nitrogen budgets: approaches and problems. *Plant and Soil* 67, 73–79.

Sanchez, P.A. (1976) *Properties and Management of Soils in the Tropics.* John Wiley & Sons, New York.

Sanchez, P.A., Palm, C.A., Davey, C.B., Szott, L.T. and Russel, C.E. (1985) Tree crops as soil improvers in the humid tropics? In: Cannell, M.G.R. and Jackson, J.E. (eds) *Attributes of Trees as Crop Plants.* Institute of Terrestrial Ecology, Huntingdon, pp. 79–124.

Sandor, J.A. and Eash, N.S. (1995) Ancient agricultural soils in the Andes of Southern Peru. *Soil Science Society of America Journal* 59, 170–179.

Schroth, G. (1998) A review of belowground interactions in agroforestry, focussing on mechanisms and management options. *Agroforestry Systems* 43, 5–34.

Sillitoe, P. (1998) Knowing the land: soil and land resource evaluation and indigenous knowledge. *Soil Use and Management* 14, 188–193.

Sillitoe, P. and Shiel, R.S. (1999) Soil fertility under shifting and semi-continuous cultivation in the Southern Highlands of Papua New Guinea. *Soil Use and Management* 15, 49–55.

Singh, B.R. and Goma, H.C. (1995) Long-term soil fertility management experiments in Eastern Africa. In: Lal, R. and Stewart, B.A. (eds) *Soil Management – Experimental Basis for Sustainability and Environmental Quality.* CRC Press, Boca Raton, Florida, pp. 347–382.

Slattery, W.J. and Burnett, V.F. (1992) Changes in soil pH due to long term soil storage. *Australian Journal of Soil Research* 30, 169–175.

Smyth, A.J. and Dumanski, J. (1995) A framework for evaluating sustainable land management. *Canadian Journal of Soil Science* 75, 401–406.

Smyth, T.J. and Cassel, D.K. (1995) Synthesis of long-term soil management research on Ultisols and Oxisols in the Amazon. In: Lal, R. and Stewart, B.A. (eds) *Soil Management – Experimental Basis for Sustainability and Environmental Quality.* CRC Press, Boca Raton, Florida, pp. 13–60.

Soane, B.D. (1990) The role of organic matter in soil compactibility: a review of some practical aspects. *Soil and Tillage Research* 16, 179–201.

Srivastava, A.K. (1984) Soil compaction: a problem in sugarcane culture. *Indian Sugar* 34, 503–504.

SSSA (1997) *Glossary of Soil Science Terms.* SSSA, Madison, Wisconsin.

Steiner, R.A. and Herdt, R.W. (eds) (1993) *A Global Directory of Long-term Agronomic Experiments.* Vol. 1: *Non-European Experiments.* Rockefeller Foundation, New York.

Stoorvogel, J.J. (1993) Optimizing land use distribution to minimize nutrient depletion: a case study for the Atlantic zone of Costa Rica. *Geoderma* 60, 277–292.

Stoorvogel, J.J. and Smaling, E.M.A. (1990) Assessment of soil nutrient decline in Sub-Saharan Africa, 1983–2000. Winand Staring Centre–DLO, Wageningen.

Stoorvogel, J.J. and Smaling, E.M.A. (1998) Research on soil fertility decline in tropical environments – integration of spatial scales. *Nutrient Cycling in Agroecosystems* 50, 151–158.

Stoorvogel, J.J., Janssen, B.H. and van Breemen, N. (1997) The nutrient budgets of a watershed and its forest ecosystems in the Taq National Park in Côte d'Ivoire. *Biogeochemistry* 37, 159–172.

Swinford, J.M. and Boevey, T.M.C. (1984) The effects of soil compaction due to infield transport on ratoon cane yields and soil physical characteristics. In: *Proceedings of the South African Sugar Technologists' Association*, pp. 198–203.

Tan, K.H. (1996) *Soil Sampling, Preparation, and Analysis.* Marcel Dekker, New York.

Taylor, H.M., Upchurch, D.R., Brown, J.M. and Rogers, H.H. (1991) Some methods of root investigations. In: McMichael, B.L. and Persson, H. (eds) *Plant Roots and their Environment.* Elsevier Science, Amsterdam, pp. 553–564.

Tinker, P.B. (1960) Soil heterogeneity and sampling procedures under oil palms. *Journal of West African Institute for Oil Palm Research* III, 9–15.

Tinker, P.B. (1994) Monitoring environmental change through networks. In: Leigh, R.A. and Johnston, A.E. (eds) *Long-term Experiments in Agricultural and Ecological Sciences.* CAB International, Wallingford, UK, pp. 407–421.

Traore, S. and Harris, P.J. (1995) Long-term fertilizer and crop residual effects on soil and crop yields in the savanna region of Cote d'Ivoire. In: Lal, R. and Stewart, B.A. (eds) *Soil Management – Experimental Basis for Sustainability and Environmental Quality.* CRC Press, Boca Raton, Florida, pp. 141–180.

Trouse, A.C. and Humbert, R.P. (1961) Some effects of soil compaction on the development of sugar cane roots. *Soil Science* 91, 208–217.

van Baren, F.A. (1934) Het voorkomen en de betekenis van kali-houdende mineralen in Nederlandse gronden. Doctoral thesis, Landbouwhogeschool, Wageningen, The Netherlands.

van Noordwijk, M. (1999) Nutrient cycling in ecosystems versus nutrient budgets of agricultural systems. In: Smaling, E.M.A., Oenema, O. and Fresco, L.O. *Nutrient Disequilibria in Agroecosystems: Concepts and Case Studies.* CAB International, Wallingford, UK, pp. 1–26.

van Noordwijk, M., Floris, J. and de Jager, A. (1985) Sampling schemes for estimating root density distribution in cropped fields. *Netherlands Journal of Agricultural Science* 33, 241–262.

van Noordwijk, M., Hairiah, K.M., Syekhfani, M.S. and Flach, E.N. (1991) *Peltophorum pterocarpa* (Dc.) Back (*Caesalpiniaceae*). A tree with a root distribution suitable for alley cropping on acid soils in the humid tropics. In: McMichael, B.L. and Persson, H. (eds) *Plant Roots and their Environment.* Elsevier Science, Amsterdam, pp. 526–532.

van Reeuwijk, L.P. (1998) *Guidelines for Quality Management in Soil and Plant Laboratories.* FAO, Rome and ISRIC, Wageningen.

van Wambeke, A. (1988) Site selection and soil variability. In: *First Training Workshop on Site Selection and Characterization.* IBSRAM Technical Notes No. 1. IBSRAM, Bangkok, pp. 43–51.

Warkentin, B.P. (1999) The return of the "other" soil scientists. *Canadian Journal of Soil Science* 79, 1–4.

Webster, R. (2000) Is soil variation random? *Geoderma* 97, 149–163.

Whitmore, T.C. (1985) *Tropical Rain Forests of the Far East*. Oxford University Press.

WinklerPrins, A. (1999) Local soil knowledge: a tool for sustainable land management. *Society and Natural Resources* 12, 151–161.

Woomer, P.L., Martin, A., Albrecht, A., Resck, D.V.S. and Scharpenseel, H.W. (1994) The importance and management of soil organic matter in the tropics. In: Woomer, P.L. and Swift, M.J. (eds) *The Biological Management of Tropical Soil Fertility*. John Wiley & Sons, Chichester, pp. 47–80.

Yaalon, D.H. (1971) Soil-forming processes in time and space. In: Yaalon, D.H. (ed.) *Paleopedology – Origin, Nature and Dating of Paleosols*. Israel Universities Press, Jerusalem, pp. 29–39.

Yates, R.A. (1978) The environment for sugarcane. In: *Land Evaluation for Standards for Rainfed Agriculture*. World Soil Resources Report No 49. FAO, Rome, pp. 58–72.

Young, A. (1991) Soil monitoring: a basic task for soil survey organizations. *Soil Use and Management* 7, 126–130.

Annual Crops

5

How much our country has been impoverished during the last fifty years, cannot be determined by a satisfactory testimony. But however we may differ on this head, there are but few who will not concur in the opinion, that our system of cultivation has been lessening the productive power of our lands.

E. Ruffin (1832)

Most food crops are annual crops and they usually complete their life cycle within 12 months. Annual crops are widely grown in the tropics and there is a vast body of literature on how the productivity of these crops could be maintained and improved, particularly in relation to more permanent cropping systems. This chapter focuses on soil changes under annual cropping. In the literature, evidence for soil fertility decline under annual crops has come from measured change in soil chemical properties and semiquantitative studies using pedotransfer functions. The evidence presented in the first sections of this chapter comes from soil changes measured at subsistence farmers' fields or experimental stations. It is followed by calculations on the rates of change in soil chemical properties in different soils and an overview of some of the major semiquantitative studies. Soil fertility changes under plantation crops are reviewed in the next chapters.

Selection Criteria and Soil Chemical Properties

There are few soil fertility studies involving long-term (>10 years) measurements in soil chemical fertility in the same field. Most studies measured changes over shorter periods and were focused on the effects of organic and inorganic nutrient inputs on soil properties and crop production. These studies often included a control treatment in which no nutrient inputs were made. They are considered here as baseline data showing what happens to the soil chemical fertility if no organic and inorganic fertilizers are used. As a considerable number of farmers in the tropics use little or no nutrient inputs, particularly in sub-Saharan Africa, such data are useful and could be treated as the point of departure for subsequent studies on integrated nutrient management. The data also provide evidence for the widespread notion that soil fertility is declining under permanent cropping in the tropics. Few studies have been conducted with the sole aim of investigating changes in soil fertility under permanent cropping.

Studies from the humid and subhumid tropical regions in Africa, Asia, and Central America are reviewed. Occasionally data from tropical Australia and China, and from New Zealand are included, for example when they were of high quality and to supplement data for some soil orders and cropping systems. Such data are also included to enhance the understanding between soil change and soil management under the given set of agroecological conditions.

In this book only data from upland, well-drained soils are included. The difference between sampling times should have been at least 2 years for both Type I and Type II data. Soil classification of the experimental site had to be provided in all the research papers reviewed. For convenience, all data in this book are presented by the soil orders of Soil Taxonomy. Although the World Reference Base is the international classification system endorsed by the International Union of Soil Sciences (IUSS), it contains 30 reference soil groups and was therefore found not workable. If other classifications were given (i.e. FAO–UNESCO, Canadian, French, or one of the Australian systems) reference tables were used as given in Sanchez (1976) or Isbell (1996).

Only studies in which replicated data are given were selected. The data are grouped based on the mode of collection (Type I and Type II – see Chapter 4, 'Data Types', page 86). Problems with Type II data arise when: (i) no soil classification is given; (ii) soils under cultivation differ from uncultivated soils; (iii) the period of cultivation is unknown; (iv) the land-use history of the cultivated soils is unknown, i.e. whether inorganic fertilizers were used. Studies in which one or more of these factors were uncertain are not included in this book.

In summary, the principal selection criteria for data to be included were: subsistence farmer or experimental plots that were unfertilized and permanently cultivated with annual crops in the humid or subhumid tropics, replicated data, soil classification provided, and the period of observation had to exceed 2 years.

Soil chemical properties – selection and standardization

A wide range of analytical procedures is available to measure soil chemical properties. Soil chemical data presented throughout this book were restricted to certain methods in order to make comparison between sites, crops and soils possible; these methods were:

- Soil pH usually measured in a soil–water suspension in the ratio 1:2.5 or 1:5. If the pH was determined in 1 M KCl, as is sometimes done in strongly acid soils to measure the zero point of charge, or in NaF (in allophane-rich soils) the data were not used. If the pH was determined in 0.01 M $CaCl_2$ (common in Australia), this has been indicated and the data have been included.
- Soil organic C is mostly determined by $K_2Cr_2O_7$ and H_2SO_4 oxidation (Walkley & Black) or dry combustion, and total N by Kjeldahl (concentrated H_2SO_4) or dry combustion. Although both methods have their drawbacks, they are very widely used, which makes exchange and comparison of soil C and N data possible. As will be shown later, total soil organic C is not a very sensitive parameter for short-term comparisons.
- Various methods exist for the determination of available P. The most common methods are Bray I and II, which use NH_4F and HCl as extractant, Olsen (uses $NaHCO_3$ extractant), Truog (extraction by H_2SO_4 and $(NH_4)_2SO_4$) and Mehlich, which uses H_2SO_4 and HCl as extractant.
- Exchangeable cations Ca, Mg, K, Na and CEC, percolation by 1 M NH_4OAc followed by spectrophotometry (K, Na), AAS (Ca, Mg) and titration (CEC).

All units for soil fertility properties are standardized as follows: organic matter (in %) has been multiplied by 5.8 to obtain organic C in g/kg; organic C and total N in % has been multiplied by 10 to obtain C and N in g/kg; exchangeable cations in ppm or mg/kg have been multiplied by 0.0499 (Ca), 0.0822 (Mg), and 0.0255 (K) to obtain levels in $mmol_c$/kg soil; exchangeable cations (Ca, Mg, K) in meq 100/g have been multiplied by 10 to obtain levels in $mmol_c$/kg soil.

Calculations and aberrant results

For convenience, all soil chemical data were entered in spreadsheets for storage and in order to facilitate calculations on the data. For each soil chemical property (χ) measured over a given time span (t), the following were calculated:

- the absolute difference: $\chi_1 - \chi_2$
- the change per year: $(\chi_1 - \chi_2) / (t_1 - t_2)$
- and the rates of change in soil chemical properties: $((\chi_1 - \chi_2) / \chi_1) / (t_1 - t_2) \times 100$ which gives the change in percentage per year of the initial level t_1, and this value has been used in many parts of this book.

Very few studies have been conducted in which rates of change in soil chemical properties were calculated. Much more needs to be known of the rate at which soil productivity declines under stress due to cultivation, and the reversibility of this degradation, which differs between soils because of the differences in their resilience to different kinds and intensities of stress (Greenland *et al.*, 1997).

In some of the published data-sets, it was found that, for example, exchangeable K slightly increased under cropping whereas no amendments had been given. That is very unlikely and they must be regarded as an erroneous result. Since the data were well replicated and the other selection criteria were met, there was no reason to exclude such data in the analysis. It illustrates the difficulties and variation that can be expected when using soil chemical data to assess soil fertility decline. Some of the aberrant data result from the fact that standard analytical methods are used for different soils. For example, recovery of C by oxidization (Walkley & Black) depends on the soil type and the C in the organic matter. Also CEC determinations are dependent on the pH and mineralogy of the soil type.

Measured Change in Soil Properties – Type I Data

Various soil fertility studies have been conducted in the humid lowlands of West Africa. Changes were mostly researched in Alfisols, and these soils cover extensive areas in West Africa, and cover about 10% in the whole of Africa (Eswaran *et al.*, 1997). On Alfisols at the International Institute of Tropical Agriculture (IITA) in Nigeria, Lal (1998) reported changes in soil fertility under ploughed and no-till plots cropped with maize. If crop residues were removed and no tillage was practised, soil organic C in the 0.05 m soil horizon decreased from 22.4 to 16.9 g/kg in 5 years when no inorganic fertilizers were applied. Total N levels decreased from 2.2 to 1.5 g/kg and the pH declined from 6.9 to 6.1 in 5 years. Levels of exchangeable Ca and the CEC decreased, but no changes were observed in exchangeable Mg or K. Changes in the 0.05 to 0.15 m soil horizon were similar. The decline in soil chemical fertility was lower when crop residues were returned to the field and inorganic fertilizers were applied. In another experiment on Alfisols in Nigeria, a decrease in levels of exchangeable K and available P was found under permanent maize cropping when no fertilizers were applied (Sobulo and Osiname, 1986).

Gray (1999) sampled farmers' fields ($n = 34$) on Alfisols in south-western Burkina Faso in 1988 and 1996. The most dramatic change was in the levels of available P, which declined from 15 to 8 mg/kg. The soils were already under 8 years of cultivation when they were sampled in 1988 and it was concluded that most changes had probably occurred in the first years after they had been taken into cultivation.

Lal (1997a) measured a ploughed and uncropped Alfisol for 4 years to monitor the effects of soil erosion on soil chemical fertility. In these plots there was no nutrient removal by crops but there was a reduction in pH from 6.4 to 6.0 and a reduction in soil organic C from 18.7 to 6.2 g C/kg. Also the levels of exchangeable cations declined, but little change was observed in available P between 1985 and 1989. This decline was mainly due to soil erosion, which varied from 30 to 50 Mg soil/ha/year (Lal, 1997b).

A serious decline in soil fertility has also been reported from other soils in the tropics where erosion is considerable (e.g. Moberg, 1972; Zöbisch *et al.*, 1995; Ruppenthal *et al.*, 1997). However, the decline in soil fertility in erosion studies could be confused with the effect that the subsoil fertility is measured because the more fertile topsoil has been washed away. This is likely to occur when sampling depth remains constant whereas topsoil thickness decreases because of soil erosion. A similar problem exists because of soil compaction (see Chapter 4. 'Effects of Bulk Density', page 117).

Soil changes were followed in a shifting cultivation system on Oxisols near Aiyinasi in Ghana (Nye, 1955). The site was cleared from old secondary forest, and maize and cassava were grown for 8 years. Topsoil pH declined from 6.0 to 5.0 and this was accompanied by a considerable decrease in CEC and exchangeable Ca and K. Levels of organic C decreased in the topsoil from 21.9 to 12.8 g/kg and total N levels decreased from 1.6 to 1.0 g/kg. Soil chemical properties also decreased in the subsoil (0.15–0.30 m), although the absolute decrease was lower than in the topsoil.

On Ultisols at Samaru in Nigeria, experiments were laid out in 1949 to measure the effects of farmyard manure and inorganic fertilizers on levels of soil organic C and total N (Jones, 1971). When no nutrient inputs were made, soil organic C decreased in 8 years from 2.2 g/kg to 1.5 g/kg. A similar degree of decline was observed in the total N content of the topsoil. In Oxisols in Congo, cultivation also led to considerable decline in soil organic C and within 5 months the soils had lost on average 13% of their organic C levels (Barthes *et al.*, 1996). Losses were higher under mechanized cultivation than under manual cultivation. Five years of cropping of cleared Oxisols in Sierra Leone reduced the organic C content from 58 to 29 g/kg at one site and from 20 to 12 g C/kg at another site (Brams, 1971). However, the rate of organic C loss was sharply reduced after the third year and organic C contents were more or less stabilized by the fifth year of cultivation. The decline in organic C content caused a 30% reduction in the CEC (Brams, 1971).

Fewer studies on soil fertility decline have been conducted in East Africa. Soil chemical data are available from FAO's Fertilizer Utilization Project (FURP), which was conducted in various parts of Kenya in the 1980s and early 1990s. Smaling and Braun (1996) listed soil chemical properties from several unfertilized soils, which were sampled at intervals of 3–4 years. Variation is considerable but the values of most soil chemical properties declined and there was no clear difference between soil orders (Alfisols, Oxisols, Psamments, Ultisols).

A summary of the main findings on soil fertility decline in annual cropping systems using Type I data is given in Table 5.1. All studies are from sub-Saharan Africa. Although the data are few and variation within and between soil orders is considerable, the picture emerging is that most soil fertility properties declined. There seems no clear relationship between the decline in a soil property and the period between two soil samplings, but this is examined in the section on 'Rates of Change' (page 151) and Chapter 11.

Other studies containing Type I data

In Gambia soil fertility properties were monitored at a national scale. The data were presented by crop and no information was presented on the soil types. An initial soil fertility survey of the country was conducted during the 1991 and 1992 cropping seasons. In 1998, a follow-up survey was conducted to determine if soil fertility had changed (Peters, 2000). In the 1998 survey no attempt was made to return to the same field originally sampled in the 1991–1992 survey and sample locations were selected at random. The study does not contain Type I data (as defined the section on 'Data Types' page 86 in Chapter 4), but as the sample size was large (and the country small) the data are of interest. Soil samples were taken from fields planted to groundnut, maize and millet, and some soil chemical properties increased or remain unchanged between 1992 and 1998 (Table 5.2). No correlations were found between tissue concentrations in groundnut, maize or millet and the soil chemical properties. The results sug-

Table 5.1. Changes in soil chemical properties under permanent cultivation with annual crops in the absence of nutrient inputs. Type I data.

Soil order	Period[a] (years)	Sampling depth (m)	pH	C (g/kg)	N (g/kg)	P (mg/kg)	CEC and exchangeable cations ($mmol_c/kg$): CEC	Ca	Mg	K	Ref.[b]
Alfisols	3	0–0.20	–0.2	–0.8	nd	–0.6	nd	nd	nd	+0.9	(1)
Alfisols	3	0–0.15	nd	nd	nd	–0.5	nd	nd	nd	–1.2	(2)
Alfisols	3	0–0.05	–0.4	–12.5	–1.5	+1.6	–38.1	–30.1	–8.8	–0.2	(3)
Alfisols	3	0.05–0.10	–0.1	–12.9	–1.2	–0.6	–16.1	–12.0	–5.7	–0.2	
Alfisols	5	0–0.05	–0.8	–5.5	–0.7	–23.9	–21.0	–30.0	0	nd	(4)
Alfisols	5	0.05–0.10	+0.7	–3.7	–0.4	–3.0	–16.0	–20.0	+1.0	nd	
Alfisols	8	0–0.20	0	–0.1	–0.2	–6.4	–2.6	–3.5	–0.5	–0.4	(5)
Oxisols	4	0–0.20	–0.4	–0.4	nd	–3.8	nd	nd	nd	–0.4	(1)
Oxisols	5	0.03–0.18	nd	–29.0	nd	nd	–60.0	nd	nd	nd	(6)
Oxisols	5	0.03–0.18	nd	–8.4	nd	nd	–10.0	nd	nd	nd	(6)
Oxisols	8	0–0.15	–1.0	–9.1	–0.6	–10.0	–28.0	–32.0	nd	–0.6	(7)
Oxisols	8	0.05–0.30	–0.2	–4.8	–0.4	–1.0	–7.0	–9.0	nd	–0.1	
Psamments	3	0–0.20	–0.7	–1.2	nd	–1.9	nd	nd	nd	–2.0	(1)
Ultisols	4	0–0.20	–0.3	–0.9	nd	–1.8	nd	nd	nd	–1.7	(1)
Ultisols	4	0–0.20	–0.5	–1.0	nd	–0.3	nd	nd	nd	–0.2	(1)
Ultisols	8	0–0.15	nd	–0.7	–0.09	nd	nd	nd	nd	nd	(8)

[a] Period between two soil samplings in years.
[b] Calculated from data given in: (1) Smaling and Braun (1996); (2) Sobulo and Osiname (1986); (3) Lal (1997a); (4) Lal (1998); (5) Gray (1999); (6) Brams (1971); (7) Nye (1955); (8) Jones (1971).
nd, No data.

Table 5.2. Soil chemical properties (0–0.15 m depth) under groundnut, millet and maize in 1992 and 1998 in Gambia. Type I data, modified from Peters (2000).

Sampling time	Number of samples	Crop	pH	Organic C (g/kg)	Available P (mg/kg)	Exchangeable cations ($mmol_c$/kg) Ca	Mg	K
1992	472	Groundnut	6.0	3.7	4	12	5	0.7
1998	171		6.1	3.5	7	19	7	0.8
1992	408	Millet	6.1	3.8	5	13	6	0.8
1998	120		6.1	4.1	8	20	7	0.7
1992	279	Maize	6.9	4.5	15	18	9	1.8
1998	40		6.5	4.9	10	23	10	1.6
1992	1159	All crops	6.3	4.0	8	14	7	1.1
1998	331		6.2	4.2	8	21	8	1.0

gest that soil fertility decline is very low or absent in the soils of subsistence farmers in Gambia. Although a large number of soil samples were taken, no *t*-test was conducted because the number of samples in 1992 and 1998 was different. There are various techniques to statistically compare paired data with uneven numbers (Snedecor and Cochrane, 1989), which would have been useful in this study to separate the random variation from the time effect.

A nationwide study of soil fertility change was also conducted in Bangladesh. Changes in C and N storage in the soil from different physiographic units were evaluated using 460 samples from 43 profiles collected in 1967 and 1995 (Ali *et al.*, 1997). Most sites had been cropped since 1967 but some of the original sites could not be re-sampled in 1995. Table 5.3 presents the results in total C and N for different depths and the two different sampling times. These are mean data from Inceptisols, and show that both total C and total N decreased, but total N losses of the top-

Table 5.3. Total C and N contents (Mg/ha) in the 0–0.15 and 0–1.00 m soil horizon of 28 soil profiles in Bangladesh sampled in 1967 and 1995. Calculated from Ali *et al.* (1997). Type I data.

	Sampling depth (m)	Sampling time 1967	1995	Change (in %) 1967–1995	Per year
Total C	0–0.15	22.4	21.2	–5	–0.2
	0–1.00	72.3	65.2	–10	–0.4
Total N	0–0.15	2.4	2.2	–8	–0.3
	0–1.00	7.4	6.9	–7	–0.2

soil (0–0.15 m) were relatively higher than C losses. No statistical analysis was attempted to differentiate the random soil variation from the long-term cropping effects on the C and N storage in the soil.

There are various Type I data studies in which soil fertility changes are monitored over time but in which the values are only presented in graphs and not in tables. This is, for example, the case in the paper by Daliparthy *et al.* (1992), which contained data under different management systems on an Entisol in West Bengal, India, and in the paper by McAlister *et al.* (1998), who studied soils under forest and cultivation in the São Francisco area of Niterói, Brazil. Both studies showed that the soil chemical fertility declines when no nutrient inputs are being made, but the data could not be extracted and were not used in the calculations for the rates of change in soil chemical properties.

Measured Change in Soil Properties – Type II Data

Ayanaba *et al.* (1976) summarized soil chemical properties from various trials at IITA in Nigeria. Soils that were cultivated for 2 years were compared with soils under bush regrowth. Differences in soil chemical properties were small and soil organic C had declined from 11.2 to 10.0 g/kg during the 2 years of cropping whereas the pH differed only by 0.1 unit (Ayanaba *et al.*, 1976). Juo and Lal (1977) measured soil chemical properties in bush fallow and after 3 years of cultivation with maize including annual applications of NPKZn on an Alfisol at IITA. Organic C in the topsoil under bush was 14.4 g/kg but 11.0 g C/kg under maize with removal of crop residues. Also the pH and levels of exchangeable cations were much lower under maize than under bush regrowth. The differences with soils under bush regrowth were small when maize crop residues were returned to the field (Juo and Lal, 1977). On Alfisols at the University of Ife in Nigeria, Aina (1979) compared soil chemical properties under long-term fallows with those of fields that were cropped for 10 years with maize but also with yam, cassava and sorghum. The pH after 10 years of cropping was 0.5 units lower than under fallow and levels of exchangeable Ca and K were reduced. Under 20–25 years fallow, levels of available P were 15 mg/kg compared with 6 mg P/kg after 10 years of cropping (Aina, 1979).

In Belize, Rendols developed over limestone, which had been under different periods of cultivation, were sampled. The soils were fertile, and a decline in pH, total N, available P and exchangeable K was observed, whereas no changes were found in the levels of exchangeable K (Arnason *et al.*, 1982). Various studies have been conducted in the Amazon region to investigate the effects of forest conversion to pastures (e.g. Piccolo *et al.*, 1994; Feigl *et al.*, 1995; de Moraes *et al.*, 1996; Fujisaka *et al.*, 1996). In the northeast of the state of Parà, Koutika *et al.* (1999) measured soil C and N contents in forests and under pastures of different age (7, 12 and 17 years). Soil organic C content under forest was 25 g/kg and levels under 7- and 12-year-old pastures were 23 g C/kg, whereas the C content of topsoils of 17-year-old pastures was 27 g/kg. Levels of total N followed a similar trend. Also in an experiment in Rondônia, soil C and N contents were slightly higher in 4- and 20-year-old pastures with *Brachiaria* sp. than under forest. This was partly explained because of an improved stability of C associated with clay and silt fractions under pasture vegetation, which retarded decomposition (Koutika *et al.*, 2000). The C and N associated with clay fractions were higher with increasing age of the pastures (Koutika *et al.*, 1999).

In the Gazipur District, Dhaka division of Bangladesh, Islam and Weil (2000) measured soil chemical properties in a Typic Paleudult under natural forest and in soils cropped for 7 years with mustard, rice, sugarcane and cotton. Compared with the soils under forest, the pH had been significantly increased, which was explained as a residual effect of the biomass burning when the site was cultivated. This is, however, unlikely as such effects do not often last for 7 years. Organic C levels were not significantly different between soils under natural forest and cultivation, but total N was lower in soils under cultivation (Islam and Weil, 2000).

An interesting data-set has come from southern Queensland (Australia), where Dalal and Mayer (1986) investigated changes in Alfisols and Vertisols under winter cereals. The crops received no inorganic fertilizers and the plots had an average cropping intensity of 0.7 crops per year. Organic C and total N significantly decreased in the topsoils whereas pH changes were minimal. Also inorganic P levels were not affected by cropping in both soil orders. Total C in the light fraction as well as mineralizable N declined in both soils at a higher rate than levels of organic C or total N. Such rapid changes in soil organic C fractions have also been reported for clayey Oxisols in the Congo (Barthes *et al.*, 1996) but have only recently received attention in soil fertility decline studies conducted in other tropical regions.

A summary of the main findings on soil fertility changes in the tropics using Type II data is given in Table 5.4. Although the data are few, a declining trend occurred in many soil chemical properties under permanent cultivation with annual crops.

Other studies containing Type II data

There are studies in which soil samples have been taken in cultivated soils and adjacent forest soils, but the period of cultivation is unknown. An example is given in Table 5.5 where soil samples were

Table 5.4. Changes in soil chemical properties under permanent cultivation with annual crops in the absence of nutrient inputs. Type II data.

Soil order	Period[a] (years)	Sampling depth (m)	pH	C (g/kg)	N (g/kg)	P (mg/kg)	CEC and cations ($mmol_c$/kg) CEC	Ca	Mg	K	Ref.[b]
Alfisols	2	0–0.15	–0.1	–1.1	0	–9.4	nd	nd	nd	nd	(1)
Alfisols	3	0–0.15	–1.2	–3.4	–0.2	nd	nd	–3.3	–4.3	–2.9	(2)
Alfisols	7	0–0.15	0	–4.4	–0.3	0	nd	nd	–2.0	nd	(3)
Alfisols	10	0–0.15	–0.5	nd	nd	–9.4	–74	–30.1	nd	–1.5	(4)
Oxisols	4	0–0.10	nd	+2.7	+0.1	nd	nd	nd	nd	nd	(5)
Oxisols	7	0–0.10	nd	–2.0	–0.2	nd	nd	nd	nd	nd	(6)
Oxisols	12	0–0.10	nd	–2.0	0	nd	nd	nd	nd	nd	(6)
Oxisols	17	0–0.10	nd	+2.0	+0.2	nd	nd	nd	nd	nd	(6)
Oxisols	20	0–0.10	nd	+1.7	+0.1	nd	nd	nd	nd	nd	(5)
Ultisols	7	0–0.15	+0.4	–1.0	–0.2	nd	nd	nd	nd	nd	(7)
Vertisols	12	0–0.10	+0.5	–1.5	–0.2	0	nd	nd	0	nd	(3)

[a] Period of cultivation, i.e. years difference between cropped and uncultivated soils.
[b] Calculated from: (1) Ayanaba *et al.* (1976); (2) Juo and Lal (1977); (3) Dalal and Mayer (1986); (4) Aina (1979); (5) Koutika *et al.* (2000); (6) Koutika *et al.* (1999); (7) Islam and Weil (2000).
nd, No data.

Table 5.5. Soil chemical properties of Inceptisols under primary forest and cultivated land in West Lampung, South Sumatra. Modified from Lumbanraja *et al.* (1998).

Land use	Sampling depth (m)	pH	C (g/kg)	N (g/kg)	P (mg/kg)	CEC and cations ($mmol_c$/kg) CEC	Ca	Mg	K
Primary forest	0–0.20	4.4	60.4	5.5	4.0	432	97.5	12.2	5.2
	0.20–0.40	4.8	25.0	2.3	1.5	182	36.7	14.2	1.4
Cultivated land	0–0.20	4.4	15.0	1.7	1.6	124	14.2	10.2	1.4
	0.20–0.40	4.3	7.5	0.8	0.7	122	16.8	9.4	1.2

taken in primary forest and in cultivated land cropped with tomato and beans (Lumbanraja *et al.*, 1998). There is considerable difference in soil chemical properties between the two types of land use, and the study provides some indication of what can be expected when forests are converted to crop land.

Moberg (1972) sampled different soil orders near Lake Victoria in Tanzania (Table 5.6). In both Oxisols and Psamments (very sandy Entisols) soil chemical properties were much higher in uncultivated (virgin) soils than under grazing or cotton. It was concluded that when Oxisols are used for shifting cultivation and grazing, they become very acid, the organic C and total N declines and the base saturation becomes extremely low (Moberg, 1972).

Although the studies of Moberg (1972) and Lumbanraja *et al.* (1998) provide evidence for soil fertility decline under cropping, the length of the period of cultivation is unknown. The data are illustrative but have limited use because it is not possible to calculate rates of change in soil

Table 5.6. Soil chemical properties in virgin land and soils under grazing and cotton near Lake Victoria, Tanzania. Modified from Moberg (1972).

Soil order	Land use	Sampling depth (m)	pH	C (g/kg)	P (mg/kg)	CEC and cations ($mmol_c$/kg) CEC	Ca	Mg	K
Oxisols	Virgin	0–0.15	6.1	4.3	28	279	119	55	13.4
		0.15–0.30	5.9	1.6	3	180	56	30	11.6
	Grazing	0–0.15	5.2	2.2	3	191	60	21	1.7
		0.15–0.30	4.7	1.4	5	286	34	10	1.3
Psamments	Virgin	0–0.15	6.2	8.5	7	58	22	13	1.5
		0.15–0.30	4.8	4.0	4	32	3	5	0.5
	Cotton	0–0.15	4.6	5.8	16	50	10	7	2.7
		0.15–0.30	4.0	5.1	12	50	4	3	2.4

properties. There are also Type II data studies in which the results are only presented in graphs and not in tables (e.g. de Moraes *et al.*, 1996), and these data could not be used in further calculations.

Rates of Change

In the previous sections, studies were reviewed in which soil chemical properties were given for individual soil orders. In this section, rates of change are calculated for soil properties of the different soil orders. Both absolute and relative changes in soil chemical properties are presented.

Absolute change

The absolute change in $mmol_c$/kg or g/kg/per year is presented in Table 5.7 (data from Tables 5.1 and 5.4). The absolute change per year was negative for most soil properties and in nearly all of the soil orders. There was considerable variation, and pH changes in the topsoils of Alfisols ranged from −0.40 to 0 units per year. The decline in exchangeable Ca ranged from −0.4 to −10.0 $mmol_c$/kg/year. There seems to be little relation between the rate of change and the period of cultivation, and plotting the relationship showed no obvious pattern (graph not shown). The data in Table 5.7 show what rates of change may be expected in different soils under annual cropping.

Relative change

Relative changes may provide better insight into the relationship between soil fertility changes and cropping. Most data presented in the previous sections had single time steps varying from 2 to 20 years. The rate of change in a soil property can be calculated as the percentage change per year compared with the initial value. This is a straightforward calculation when a study contains Type I data. If we assume t_1 is the total N in 1990 (lets say 1.6 g N/kg) and t_2 is the total N in 1998 (1.0 g N/kg). The rate of change is: (1.6 – 1.0)/1.6 * 100 = 38%. So the total N content decreased by 38% over the period of observation ($t_2 - t_1$ = 8 years), or on average total N decreased by 4.7% per year. For pH, which is expressed on a logarithmic scale, the change in per cent per year means a different amount of protons when the decline is from 4.5 to 4.0 or when it is from 6.5 to 6.0. For the soil nutrients and soil organic C that does not apply.

The relative change approach can also be used in Type II data studies. Juo and Lal (1977) calculated change in soil chemical proper-

Table 5.7. Absolute change (unit/year) in soil fertility (topsoils only) in the tropics under permanent cultivation with annual crops in the absence of nutrient inputs. Type I and II data.[a]

Soil type	Data type	Period[b] (years)	pH	C (g/kg)	N (g/kg)	P (mg/kg)	CEC and cations (mmol$_c$/kg)			
							CEC	Ca	Mg	K
Alfisols	II	2	−0.1	−0.6	<−0.1	−4.7	nd	nd	nd	nd
Alfisols	I	3	−0.1	−0.3	nd	−0.2	nd	nd	nd	nd
Alfisols	I	3	−0.1	−4.2	−0.5	nd	−12.7	−10.0	−2.9	−0.1
Alfisols	I	3	nd	nd	+0.2	−0.2	nd	nd	nd	−0.4
Alfisols	II	3	−0.4	−1.1	−0.1	nd	nd	−1.1	−1.4	−0.9
Alfisols	I	5	−0.2	−1.1	−0.1	−4.8	−4.2	−6.0	0	+0.2
Alfisols	II	7	0	−0.6	−0.1	0	nd	nd	−0.3	nd
Alfisols		8	0	<−0.1	<−0.1	−0.8	−0.3	−0.4	−0.1	−0.1
Alfisols	II	10	−0.1	nd	nd	−0.9	−7.4	−3.0	nd	−0.2
Oxisols	II	4	nd	+0.7	<+0.1	nd	nd	nd	nd	nd
Oxisols	I	4	−0.1	−0.1	nd	−0.9	nd	nd	nd	−0.1
Oxisols	I	5	nd	−5.8	nd	nd	−12.0	nd	nd	nd
Oxisols	I	5	nd	−1.7	nd	nd	−2.0	nd	nd	nd
Oxisols	II	7	nd	−0.3	<−0.1	nd	nd	nd	nd	nd
Oxisols	I	8	−0.1	−1.1	−0.1	−1.3	−3.5	−4.0	nd	−0.1
Oxisols	II	12	nd	−0.2	0	nd	nd	nd	nd	nd
Oxisols	II	17	nd	+0.1	<+0.1	nd	nd	nd	nd	nd
Oxisols	II	20	nd	+0.1	0	nd	nd	nd	nd	nd
Psamments	I	3	−0.2	−0.4	nd	−0.6	nd	nd	nd	−0.7
Ultisols	I	4	−0.1	−0.2	nd	+0.5	nd	nd	nd	−0.4
Ultisols	I	4	−0.1	−0.2	nd	−0.1	nd	nd	nd	−0.1
Ultisols	II	7	+0.1	−0.1	<−0.1	nd	nd	nd	nd	nd
Ultisols	I	8	nd	−0.1	−0.0	nd	nd	nd	nd	nd
Vertisols	II	12	+0.1	−0.1	<−0.1	0	nd	nd	0	nd

[a] Calculated from the data in Tables 5.1 and 5.4.
[b] Period of cultivation (Type I); or years difference between cropped and uncultivated soils (Type II).
nd, No data.

ties relative to the values under bush growth. In this Type II data study, soils under maize cropping were compared with soils under bush regrowth. So the change under cropping relative to non-cropped soils was compared. This approach was also used in Nigeria by Adejowun and Ekanade (1988), who named it the 'soil deterioration index' and it was also applied by Islam and Weil (2000) in a study in Bangladesh.

The relative change in soil chemical properties presented in the previous sections was calculated for the different soil orders (Table 5.8). Soil organic C declined in most soils but there were differences within and between soil orders. Soil pH in Alfisols declines from 0% to 6% per year, whereas in Oxisols the decline in pH was as high as 10% per year. It could have been expected that the relative rate of change would decrease with increasing period of cultivation, i.e. that the rate should be higher when soils are taken in production (see Chapter 4, 'Interpretation of the Results', page 123). This is further investigated in Chapter 11.

First-order kinetics

Calculating the rate of change in percentage per year using two data points assumes a linear change in a soil property. However, many soil chemical processes are non-linear and the rate of change therefore differs at different periods (Jenny, 1980). For example, an average decline in organic C at a rate of −0.3 g/kg/year observed between 1980 and 2000 may have been −0.8 g C/kg/year in the 1980s, but less than −0.2 g C/kg/year in the 1990s.

A different method is to assume that loss of a nutrient, χ, is a first-order kinetic process that can be fitted to single exponential model. The first-order process is: $-d\chi/dt = kt$, where the rate factor k can be calculated from plotting $\ln\chi/\chi_0$ versus t where k represents the slope of the line. Calculating the k-factor thus provides insight into the rates of change in a property over the period of observation. This was suggested by Nye and Greenland (1960) and first-order kinetics have been widely used in crop residue and organic matter decomposition studies.

First-order kinetics were used by Arnason *et al.* (1982) in a soil fertility decline study in Belize. Table 5.9 lists the results and shows the k-factor and the relative change in soil chemical properties of Rendols under permanent cropping in Belize. For the use of the single exponential model several data points and relatively short time steps are needed, and since most soil fertility studies have only single time steps (t_1 and t_2), it may explain why the model has not been used more widely. Moreover, it may be that such a model fits well for C and N but less well for exchangeable cations or the pH.

Table 5.8. Relative change (% per year) in soil chemical properties (topsoils only) in the tropics under permanent cultivation with annual crops in the absence of nutrient inputs.[a] Type I and II data.

							CEC and cations ($mmol_c$/kg)			
Soil type	Data type	Period[b] (years)	pH	C (g/kg)	N (g/kg)	P (mg/kg)	CEC	Ca	Mg	K
Alfisols	II	2	–1	–5	–1	–31	nd	nd	nd	nd
Alfisols	I	3	–1	–1	nd	–1	nd	nd	nd	+4
Alfisols	I	3	–2	–2	–24	+4	–14	–16	–15	–2
Alfisols	I	3	nd	nd	nd	–4	nd	nd	nd	–24
Alfisols	II	3	–6	–8	–6	nd	nd	–3	–16	–23
Alfisols	I	5	–2	–5	–6	–7	–4	–8	0	nd
Alfisols	II	7	0	–5	–4	0	nd	nd	–2	nd
Alfisols	I	8	0	–0.5	–0.5	–5	–1	–1	–1	–2
Alfisols	II	10	–1	nd	nd	–6	–6	–7	nd	–6
Mollisols	II	3	–1	nd	–5	–10	nd	–16	nd	–4
Oxisols	II	4	nd	+5	+2	nd	nd	nd	nd	nd
Oxisols	I	4	–2	–0.5	nd	–6	nd	nd	nd	–2
Oxisols	I	5	nd	–10	nd	nd	–7	nd	nd	nd
Oxisols	I	5	nd	–8	nd	nd	–3	nd	nd	nd
Oxisols	II	7	nd	–1	–1	nd	nd	nd	nd	nd
Oxisols	I	8	2	–5	–5	–8	–4	–9	nd	–3
Oxisols	II	12	nd	–1	0	nd	nd	nd	nd	nd
Oxisols	II	17	nd	+1	+1	nd	nd	nd	nd	nd
Oxisols	II	20	nd	+1	0	nd	nd	nd	nd	nd
Psamments	I	3	–3	–5	nd	–2	nd	nd	nd	–14
Ultisols	I	4	–1	–2	nd	+3	nd	nd	nd	–6
Ultisols	I	4	–2	–5	nd	–1	nd	nd	nd	–3
Ultisols	II	7	+1	–2	–3	nd	nd	nd	nd	nd
Ultisols	I	8	nd	–4	–4	nd	nd	nd	nd	nd
Vertisols	II	12	+1	–2	–2	0	nd	nd	0	nd

[a] Calculated from the references listed in Tables 5.1 and 5.4.
[b] Period of cultivation (Type I) or years difference between cropped and uncultivated soils (Type II).
nd, No data.

Table 5.9. Decline of soil fertility in relative values (% per year) and calculated *k*-factor based on first-order kinetics. Modified from Arnason *et al.* (1982).

Soil chemical property	Relative rate of decline (% per year)	*k*-factor (per year)
pH	1.2	0.013
Available P	10	0.11
Total N	4.8	0.05
Exchangeable Ca	16	0.19
Exchangeable K	3.9	0.035

Paired sequential samples

In some studies, several paired samples are available but all of them with different single time steps. This is the case when various fields are being sampled at different times, for example some fields may have been sampled in 1985 and again in 1997 whereas other were sampled in 1988 and again in 1994. The data-set from such a sampling scheme has several values of a soil property but with different time steps. It is possible to calculate from such data the rate of change, whereby the difference in years between the initial sample (t_1) and the second sample at (t_2) is plotted against the difference in the measured soil property values. This relationship does not take into account the time elapsed between the measurements. Based on a large number of sample pairs, the decline in a soil chemical property can be calculated, whereby t_1 is the initial value and t_2 the value of the second sampling. Thus it can be calculated whether a soil property had increased or not changed (i.e. value at t_2 minus value at $t_1 \geq 0$) or whether there has been a decline (i.e. value at t_2 minus value at $t_1 < 0$). Despite the fact that there are only two data points for each property, the *k*-factor can be calculated, as there are many data pairs with different time steps. This method has been used in Chapter 9, page 273.

Semiquantitative Studies

In the 1990s, several studies were conducted which indicated that soil fertility decline is a problem in many tropical countries and particularly in sub-Saharan Africa (Pieri, 1989; Probert, 1992; van der Pol and Traore, 1993; Dev, 1994; Prasad *et al.*, 1994; Rhodes, 1995; Folmer *et al.*, 1998; Henao and Baanante, 1999). Most of these studies were based on nutrient budgets in which fluxes and pools were estimated from published data or data derived from pedotransfer functions or some other method. The studies were mostly conducted on a large scale and not on a plot or field scale.

Supra-national scale

One of the most influential studies on soil nutrient depletion was conducted by Stoorvogel and Smaling (1990). It was initiated by the FAO, and in the study budgets for N, P and K were calculated for the arable soils of 38 countries in sub-Saharan Africa for the years 1983 and 2000. The area concerned was 201 million ha and this was divided into agroecological zones. A further stratification into land-use systems was made using the 1:5 million FAO soil maps and based on cropping patterns and soil management practices. Nutrient budgets were estimated as the difference between the inputs with: (i) the application of mineral fertilizers, (ii) organic manure, (iii) atmospheric deposition, (iv) biological nitrogen fixation, (v) sedimentation; and nutrient outputs: (i) removal with yield, (ii) removal of crop residues, (iii) leaching, (iv) denitrification and volatilization, and (v) water erosion. The difference between inputs and outputs was found to be negative in most countries in sub-Saharan Africa (Table 5.10).

The sub-Saharan Africa nutrient depletion study had methodological problems. Some of the input and output data were easily obtained

Table 5.10. Average nutrient balances of nitrogen, phosphorus and potassium (kg/ha) of the arable land for some selected sub-Saharan countries in 1983 and 2000. Modified from Stoorvogel *et al.* (1993).

	N		P		K	
	1983	2000	1983	2000	1983	2000
Benin	−14	−16	−1	−2	−9	−11
Cameroon	−20	−21	−2	−2	−12	−13
Gambia	−14	−17	−3	−3	−16	−24
Ghana	−30	−35	−3	−4	−17	−20
Kenya	−42	−46	−3	−1	−29	−36
Malawi	−68	−67	−10	−10	−44	−48
Mali	−8	−11	−1	−2	−7	−10
Nigeria	−34	−37	−4	−4	−24	−31
Senegal	−12	−16	−2	−2	−10	−14
Tanzania	−27	−32	−4	−5	−18	−21
Zimbabwe	−31	−27	−2	−2	−22	−26

(i.e. nutrient removal with the yield) whereas others were hard to quantify or relied heavily on FAO soil and crop databases, which are not always accurate. Denitrification was estimated from studies conducted in Puerto Rico (Dubey and Fox, 1974) whereas nutrients inputs by BNF and wet deposition were estimated from average annual rainfall. The simplifications were necessary because of the lack of data. In the supranational study of Stoorvogel and Smaling (1990), nutrient depletion in Gambia for the year 2000 was estimated to be –17 kg N, –3 kg P and –24 kg K/ha. Henao and Baanante (1999), who used a similar approach to Stoorvogel and Smaling, estimated the nutrient balance for Gambia over the years 1993–1995 to be: –30 kg N, –5 kg P and –18 kg K/ha/year. A nationwide soil fertility evaluation in 1992 and 1998 showed, however, that available P and exchangeable K levels had actually increased in soils under the main food crops (Peters, 2000) – see Table 5.2. So there seems to be a difference between semiquantitative studies, but more importantly, there may be limited correlation with field studies.

As was discussed in the section on 'Soil Sampling, Soil Analysis and Errors' (Chapter 4, page 97), nutrient balances may suffer from serious errors as inaccuracies and biases in the individual factors accumulate in the net balance. There are virtually no nutrient-balance studies in which standard errors are given for the factors so that an impression of the variation could be gained. Notwithstanding the lack of error quantification, possible biases and discrepancies with some field studies, the Stoorvogel and Smaling study clearly showed that nutrient depletion is a problem in sub-Saharan Africa, particularly in the eastern part, and the study received much international attention. It formed the base for subsequent detailed studies, and various international research efforts to combat soil fertility decline in Africa, of which excellent summaries are given in van Reuler and Prins (1993), Buresh *et al.* (1997) and Vanlauwe *et al.* (2002).

National and district scale

Following the supra-national studies, a number of studies were conducted on the national and district level in sub-Saharan Africa. In these studies, inputs and outputs of nutrients could be more accurately quantified, and some of this more detailed work is presented in Smaling (1998). Table 5.11 gives nutrient budgets calculated for districts in Kenya and Mali.

In recent years a whole series of other studies have appeared using nutrient budgets and farmers' perception to indicate the status of soil fertility in tropical land-use systems. A series of papers entitled 'Managing Africa's soils' has been published by the International Institute for Environment and Development (ITED) including an overview book

Table 5.11. Nutrient budgets (kg/ha/year) for Kisii District (Kenya) and southern Mali. Modified from Smaling *et al.* (1993) and van der Pol and Traore (1993).

	Kisii, Kenya			Southern Mali		
	N	P	K	N	P	K
Inputs						
Inorganic fertilizers	17	12	2	7	2	2
Organic manure	24	5	25	3	1	3
Atmospheric deposition	6	1	4	10	2	6
Biological nitrogen fixation	8			3		
Outputs						
Yield	55	10	43	23[a]	3[a]	14[a]
Crop residues	6	1	13			
Leaching	41	0	9	4	0	4
Denitrification	28			12		
Water erosion	37	10	36	9	2	13
ΣInputs minus ΣOutputs	–112	–3	–70	–25	0	–20

[a] Nutrients in yield in southern Mali includes nutrients in the crop residues.

(Hilhorst and Muchena, 2000). These studies are often based on qualitative approaches to soil fertility decline and are not further considered in this book. They contribute to the awareness of soil fertility decline and nutrient mining in subsistence farming systems of sub-Saharan Africa and have resulted in several policy documents for soil fertility management (Scoones and Toulmin, 1999; Hilhorst and Toulmin, 2000).

A semiquantitative study was conducted by Lindert (2000), who compared soil analytical data from the 1930s with more recent observations in Indonesia and China. The data have been aggregated by region or land-use system and statistical analyses were conducted, which indicated a decline in the soil chemical properties in both countries. The study gives the impression that long-term changes in soil chemical properties were quantified but the data were too few and unevenly spread over the country to allow for such rigid conclusions. A more interesting example from long-term changes was prepared by Richter and Markewitz (2001), who analysed the records of the 'Calhoun' site in the southeastern USA. The study traces changes in soil properties as affected by different land-use systems but is confined to the temperate zone and is not considered further here.

Discussion and Conclusions

This chapter reviewed soil fertility decline studies based on different data types: measured changes in soil chemical properties and semiquantitative data. The soil chemical fertility declined under annual cropping in the absence of inorganic fertilizer use, but rates of change in soil chemical properties differed. A decline was not found in each of the soil chemical property in all studies. Spatial and temporal variation in soil chemical properties may exceed the changes brought about by the cropping systems and land management systems. Knowledge of the soil processes involved and models are available but what is lacking is a critical mass of data to augment both model- and process-oriented studies. The data must be regarded as indications of the order of magnitude rather than as accurate quantities. The advantage of studying long-term soil fertility trends in annual cropping systems is that the decline is not to be confused with immobilization of nutrients in the biomass which may occur in studies with perennial crops (see Chapter 6).

The second set of studies was largely based on a semiquantitative approach in which the nutrient balance stood central. The approach resulted in useful large-scale maps depicting the state of the nutrient budgets. Negative nutrient balances require some discussion. To reduce the imbalance either inputs must increase or outputs must decrease, but different measures have different effects (Janssen, 1999). For example, the effects of erosion control differ from those achieved upon reduction of leaching and gaseous losses. Less erosion would imply a less negative nutrient balance but probably no immediate yield increase. Reducing leaching losses may result in higher crop uptake and higher yields, but not necessarily in a less negative balance. Likewise, extra N fertilizer may be leached or volatilized or it may be absorbed by the crop, resulting in higher yields but not in a less negative balance. Only when a portion of added nutrients is stored in the soil will this result in a more positive nutrient balance (Janssen, 1999). So the nutrient balance provides less evidence on soil fertility decline under permanent cropping systems.

Semiquantitative studies do not show the balance between soil degradation and soil improvement, which is larger in extent than soil degradation – otherwise Malthus' prediction would have been realized (Greenland, 1994). Soil improvement has occurred in many parts of the tropics through the use of inorganic and organic fertilizers but data showing positive changes are scarcer than data showing negative changes in soil chemical properties. Table 5.12 presents some data from China where both N and P fertilizers were annually applied in combination with straw and stubble retention. Although there is some inter-annual variation and the exact amount of applied fertilizers is not known, an increase in organic C, total N and available P was found.

In the measured and the semiquantitative studies a certain degree of merry-go-round is present: permanent cropping, no fertilizers, low yields, no money, no fertilizers etc. (Fig. 5.1). There are

Table 5.12. Changes in soil chemical properties ($n = 20$) in the topsoils of Quzhou County, Hebei Province, China. Modified from Jin *et al.* (1999). Type I data.

Year of sampling	Organic C (g/kg)	Total N (g/kg)	Available P (Olsen) (mg/kg)
1980	4.6	0.51	3.8
1983	5.4	0.55	5.3
1984	5.5	0.72	6.4
1985	5.3	0.77	10.4
1986	6.3	0.66	10.3
1989	6.8	0.71	9.3
1990	7.2	0.74	9.8
1991	7.5	0.80	9.5

Fig. 5.1. Maize field near Maseno in western Kenya, heavily infested with the parasitic weed *Striga hermonthica*. Continuous cropping of maize in this densely populated area has reduced the soil chemical fertility, which increased the incidence of striga and the maize produced no yield.

relatively few hard data to prove that it occurs under all conditions but for the design of sustainable systems, information on basic rates of change is required in order to achieve a better match between inputs and outputs of nutrients. Maintenance of soil nutrient levels is an important aspect in achieving sustainable land management. Although exact quantification as well as the estimates of the spatial distribution of soil fertility decline remains to be improved, the studies have shown that soil fertility decline is a problem in many soils of the tropics. A question remaining is how soil fertility is affected in plantation agriculture, which constitutes an important economic segment in many tropical countries. That question is addressed in the next chapters.

References

Adejuwon, J.O. and Ekanade, O. (1988) A comparison of soil properties under different landuse types in a part of the Nigerian cocoa belt. *Catena* 15, 319–331.

Aina, P.O. (1979) Soil changes resulting from long-term management practises in Western Nigeria. *Soil Science Society of America Journal* 43, 173–177.

Ali, M.M., Saheed, S.M., Kubota, D., Masunaga, T. and Wakatsuki, T. (1997) Soil degradation during the period 1967–1995 in Bangladesh. 1 Carbon and nitrogen. *Soil Science and Plant Nutrition* 43, 863–878.

Arnason, T., Lambert, J.D.H., Gale, J., Cal, J. and Vernon, H. (1982) Decline of soil fertility due to intensification of land use by shifting agriculturists in Belize, Central America. *Agro-Ecosystems* 8, 27–37.

Ayanaba, A., Tuckwell, S.B. and Jenkinson, D.S. (1976) The effects of clearing and cropping on the organic reserves and biomass of tropical forests soils. *Soil Biology and Biochemistry* 8, 519–525.

Barthes, B., Kouakoua, E., Sala, G.H., Hartman, C. and Nyete, B. (1996) Short-term effects of cultivation on organic status and aggregation of a clayey Oxisol in Congo. *Canadian Journal of Soil Science* 76, 493–499.

Brams, E.A. (1971) Continuous cultivation of West African soils: organic matter diminution and effects of applied lime and phosphorus. *Plant and Soil* 35, 401–414.

Buresh, R.J., Sanchez, P.A. and Calhoun, F. (eds) (1997) *Replenishing Soil Fertility in Africa.* SSSA Special Publication Number 51. SSSA, Madison, Wisconsin.

Dalal, R.C. and Mayer, R.J. (1986) Long-term trends in fertility of soils under continuous cultivation and cereal cropping in Southern Queensland. I. Overall changes in soil properties and trends in winter cereal yields. *Australian Journal of Soil Research* 24, 265–279.

Daliparthy, J., Chatterjee, B.N. and Mondal, S.S. (1992) Changes in soil nitrogen, phosphorus, and potassium under intensive cropping in sub-humid tropics of India. *Communications in Soil Science and Plant Analysis* 23, 1871–1884.

de Moraes, J.F.L., Volkoff, B., Cerri, C.C. and Bernoux, M. (1996) Soil properties under Amazon forest and changes due to pasture installation in Rondonia, Brazil. *Geoderma* 70, 63–81.

Dubey, H.D. and Fox, R.H. (1974) Denitrification losses from humid tropical soils of Puerto Rico. *Soil Science of Society of America Proceedings* 38, 917–920.

Eswaran, H., Almaraz, R., Vandenberg, E. and Reich, P. (1997) An assessment of the soil resources of Africa in relation to productivity. *Geoderma* 77, 1–18.

Feigl, B.J., Melillo, J. and Cerri, C.C. (1995) Changes in the origin and quality of soil organic matter after pasture introduction in Rondonia (Brazil) *Plant and Soil* 175, 21–29.

Folmer, E.C.R., Geurts, P.M.H. and Francisco, J.R. (1998) Assessment of soil fertility depletion in Mozambique. *Agriculture Ecosystems and Environment* 71, 159–167.

Fujisaka, S., Bell, W., Thomas, N., Hurtado, L. and Crawford, E. (1996) Slash-and-burn agriculture, conversion to pasture, and deforestation in two

Brazilian Amazon colonies. *Agriculture Ecosystems and Environment* 59, 115–130.

Gray, L.C. (1999) Is land being degraded? A multi-scale investigation of landscape change in southwestern Burkina Faso. *Land Degradation and Development* 10, 329–343.

Greenland, D.J. (1994) Soil science and sustainable land management. In: Syers, J.K. and Rimmer, D.L. (eds) *Soil Science and Sustainable Land Management in the Tropics*. CAB International, Wallingford, pp. 1–15.

Greenland, D.J., Gregory, P.J. and Nye, P.H. (1997) Introduction and conclusions. In: Greenland, D.J., Gregory, P.J. and Nye, P.H. (eds) *Land Resources: On the Edge of the Malthusian Precipice?* The Royal Society, London, pp. 861–867.

Henao, J. and Baanante, C. (1999) *Estimating Rates of Nutrient Depletion in Soils of Agricultural Lands of Africa*. IFDC, Muscle Shoals, Alabama.

Hilhorst, T. and Muchena, F. (eds) (2000) *Nutrients on the Move – Soil Fertility Dynamics in African Farming Systems*. IIED, London.

Hilhorst, T. and Toulmin, C. (2000) *Integrated Soil Fertility Management. Policy and Best Practice Document No. 7*. Ministry of Foreign Affairs, The Hague.

Isbell, R.F. (1996) *The Australian Soil Classification*. CSIRO, Collingwood, Victoria.

Islam, K.R. and Weil, R.R. (2000) Land use effects on soil quality in a tropical forest ecosystem of Bangladesh. *Agriculture Ecosystems and Environment* 79, 9–16.

Janssen, B.H. (1999) Basics of budgets, buffers and balances of nutrients in relation to sustainability of agroecosystems. In: Smaling, E.M.A., Oenema, O. and Fresco, L.O. (eds) *Nutrient Disequilibria in Agroecosystems: Concepts and Case Studies*. CAB International, Wallingford, pp. 27–56.

Jenny, H. (1980) *The Soil Resource. Origin and Behavior*. Springer-Verlag, New York.

Jin, J.Y., Bao, L. and Zhang, W.L. (1999) Improving nutrient management for sustainable development of agriculture in China. In: Smaling, E.M.A., Oenema, O. and Fresco, L.O. (eds) *Nutrient Disequilibria in Agroecosystems: Concepts and Case Studies*. CAB International, Wallingford, pp. 157–174.

Jones, M.J. (1971) The maintenance of soil organic matter under continuous cultivation at Samaru, Nigeria. *Journal of Agricultural Science (Cambridge)* 77, 473–482.

Juo, A.S.R. and Lal, R. (1977) The effect of fallow and continuous cultivation on the chemical and physical properties of an Alfisol in Western Nigeria. *Plant and Soil* 47, 567–584.

Koutika, L.S., Andreux, F., Hassink, J., Chone, T. and Cerri, C.C. (1999) Characterization of organic matter in topsoils under rain forest and pasture in the eastern Brazilian Amazon basin. *Biology and Fertility of Soils* 29, 309–313.

Koutika, L.S., Chone, T., Andreux, F. and Cerri, C.C. (2000) Carbon decomposition of the topsoils and soil fractions under forest and pasture in the western Brazilian Amazon basin, Rondonia. *Biology and Fertility of Soils* 30, 284–287.

Lal, R. (1997a) Soil degradative effects of slope length and tillage method on Alfisols in Western Nigeria. 2. Soil chemical properties, plant nutrient loss and water quality. *Land Degradation and Development* 8, 221–244.

Lal, R. (1997b) Soil degradative effects of slope length and tillage methods on Alfisols in Western Nigeria. 1. Runoff, erosion and crop response. *Land Degradation and Development* 8, 201–219.

Lal, R. (1998) Soil quality changes under continuous cropping for seventeen seasons of an Alfisol in Western Nigeria. *Land Degradation and Development* 9, 259–274.

Lindert, P.H. (2000) *Shifting Ground. The Changing Agricultural Soils of China and Indonesia.* The MIT Press, Cambridge, Massachusetts.

Lumbanraja, J., Syam, T., Nishide, H., Mahi, A.K., Utomo, M., Sarno and Kimura, M. (1998) Deterioration of soil fertility by land use changes in South Sumatra, Indonesia – from 1970 to 1990. *Hydrological Processes* 12, 2003–2013.

McAlister, J.J., Smith, B.J. and Sanchez, B. (1998) Forest clearance: impact of landuse change on fertility status of soils from the Sao Francisco area of Niteroi, Brazil. *Land Degradation and Development* 9, 425–440.

Moberg, J.P. (1972) Some soil fertility problems in the west lake region of Tanzania, including the effects of different forms of cultivation on the fertility of some Ferralsols. *East African Agricultural and Forestry Journal* 35–46.

Nye, P.H. (1955) Some soil forming processes in the humid tropics. II. The development of the upper-slope member of the catena. *Journal of Soil Science* 6, 51–62.

Nye, P.H. and Greenland, D.J. (1960) *The Soil under Shifting Cultivation.* Commonwealth Bureau of Soils, Harpenden.

Peters, J.B. (2000) Gambian soil fertility trends, 1991–1998. *Communications in Soil Science and Plant Analysis* 31, 2201–2210.

Piccolo, M.C., Neill, C. and Cerri, C.C. (1994) Natural abundance of ^{15}N in soils along forest-to-pasture chronosequences in the western Brazilian Amazon Basin. *Oecologia* 99, 112–117.

Pieri, C. (1989) *Fertilité des Terres de Savanes.* Ministère de la Cooperation et CIRAD-IRAT, Paris.

Prasad, N., Biswas, B.C. and Tewatia, R.K. (1994) Nutrient depletion and agricultural sustainability in Punjab. *Fertiliser News* 39, 35–41.

Probert, M.E. (ed.) (1992) *A Search for Strategies for Sustainable Dryland Cropping in Semi-arid Eastern Kenya. ACIAR Proceedings No. 41.* ACIAR, Canberra.

Rhodes, E.R. (1995) Nutrient depletion by food crops in Ghana and soil organic nitrogen management. *Agricultural Systems* 48, 101–118.

Richter, D.D. and Markewitz, D. (2001) *Understanding Soil Change. Sustainability over Millennia, Centuries, and Decades.* Cambridge University Press, Cambridge.

Ruffin, E. (1832) *An Essay on Calcareous Manures.* J.W. Campbell, Petersburg, Virginia.

Ruppenthal, M., Leihner, D.E., Steinmuller, N. and Elsharkawy, M.A. (1997) Losses of organic matter and nutrients by water erosion in cassava-based cropping systems. *Experimental Agriculture* 33, 487–498.

Sanchez, P.A. (1976) *Properties and Management of Soils in the Tropics.* John Wiley & Sons, New York.

Scoones, I. and Toulmin, C. (1999) *Policies for Soil Fertility Management in Africa*. IIED, London.

Smaling, E.M.A. (ed.) (1998) Nutrient balances as indicators of productivity and sustainability in sub-Saharan African agriculture. *Agriculture Ecosystems and Environment* Vol. 71 (special issue).

Smaling, E.M.A. and Braun, A.R. (1996) Soil fertility research in sub-Saharan Africa – new dimensions, new challenges. *Communications in Soil Science and Plant Analysis* 27, 365–386.

Smaling, E.M.A., Stoorvogel, J.J. and Windmeijer, P.N. (1993) Calculating soil nutrient balances in Africa at different scales. 2. District scale. *Fertilizer Research* 35, 237–250.

Snedecor, G.W. and Cochrane, W.G. (1989) *Statistical Methods*, 8th edn. Iowa State University Press, Ames.

Sobulo, R.A. and Osiname, O.A. (1986) Soil properties and crop yields under continuous cultivation with different management systems. In: Lal, R., Sanchez, P.A. and Cummings, R.W. (eds) *Land Clearing and Development in the Tropics*. A.A. Balkema, Rotterdam, pp. 363–370.

Stoorvogel, J.J. and Smaling, E.M.A. (1990) *Assessment of Soil Nutrient Decline in Sub-Saharan Africa, 1983–2000*. Winand Staring Centre-DLO, Wageningen.

Stoorvogel, J.J., Smaling, E.M.A. and Janssen, B.H. (1993) Calculating soil nutrient balances in Africa at different scales. 1. Supra-national scale. *Fertilizer Research* 35, 227–235.

van der Pol, F. and Traore, B. (1993) Soil nutrient depletion by agricultural production in southern Mali. *Fertilizer Research* 36, 79–90.

van Reuler, H. and Prins, W.H. (eds) (1993) *The Role of Plant Nutrients for Sustainable Food Crop Production in Sub-Saharan Africa*. Dutch Association of Fertilizer Producers, Leidschendam.

Vanlauwe, B., Diels, J., Sanginga, N. and Merckx, R. (2002) (eds) *Integrated Plant Nutrient Management in Sub-Saharan Africa: From Concept to Practice*. CAB International, Wallingford, UK.

Zöbisch, M.A., Richter, C., Heiligtag, B. and Schlott, R. (1995) Nutrient losses from cropland in the central highlands of Kenya due to surface runoff and soil erosion. *Soil and Tillage Research* 33, 109–116.

Perennial Crop Plantations

6

> Every landowner should adopt for his land a system of farming that is permanent, a system under which the land becomes better rather than poorer.
>
> C.G. Hopkins (1910)

At the first General Soil Science Congress in the Dutch East Indies held in 1916 in Java, the productivity of agricultural plantations was widely discussed (Mohr, 1916). The industry had plenty of examples of misfortune including the dwindling and vanishing tobacco, cocoa, coffee and pepper cultures. Some considered that the widespread yield decline was caused by pests and diseases. The questions were raised whether crops had suffered as much where they had been healthier and stronger; and whether the incidence of pests and disease might be related to a decline in soil conditions. Rumours were also being spread that yield decline on tea and rubber plantations was common, whereas some believed that the decline in plantation crop yields was highly exaggerated. Mohr had the opinion that all these groups and rumours were a waste of time, and he noted: 'Wat geeft al dat gepraat? Waar zijn de cijfers?'.[1] He strongly advocated that data were essential for planters and their companies to investigate the relation between crop production and possible changes in soil fertility

[1] 'Why all that talking? Where are the data?'

(Mohr, 1916). More than 85 years have passed since Mohr, who was a pioneering soil scientist in the early 20th century, made these comments. As was discussed in Chapter 3, plantation agriculture has changed greatly, but the need for data on soil changes has remained – and is the main focus of this chapter.

Soil Changes

Export products such as coffee, tea, rubber and oil palm are often grown on large-scale plantations. These crops remain in the same field for many years and they require an initial high investment (Fig. 6.1). Long-term returns from such investments can only be expected if production is sustained, which requires, among other factors, that the soil remains in good condition. However, permanent cropping will affect the soil conditions and in order to sustain productivity on perennial crop plantations, it is necessary to investigate the long-term changes brought about by the crop and management practices. This chapter focuses on soil changes under perennial crops: coffee, tea, rubber and oil palm. Although there are a number of other perennial crops grown on plantations (cashew, coconuts, etc.) as well as more herbaceous crops (pineapple, tobacco, cotton, etc.), too little information was available on these crops to be included in this review. Sugarcane is also widely grown on plantations and it is discussed in Chapters 8 and 9, whereas soil changes under sisal are reviewed in Chapter 10.

Fig. 6.1. Clearing of secondary jungle for the development of an oil palm plantation in North Sumatra.

Theoretical considerations

There are a number of important differences between annual and perennial crop cultivation. Annual crops are usually sown or planted at high densities and the economic benefits are obtained within some months after they have been planted or sown. Perennial crops may take 2–4 years to come to maturity and give an economic return. They are mostly planted at relatively low densities and are long-term investments.

Ruthenberg (1972) listed 13 farm-management characteristics of perennial crop cultivation that can be regarded as advantages. A number of these are related to the soil and its management:

- Most perennial crops are characterized by a comparatively high productivity per hectare. This encourages the use of inorganic fertilizer and the conservation of soil fertility and economizing land use (Fig. 6.2).
- Some tree and shrub crops influence and conserve the soil in the same way as the forest, and permanent planting encourages the construction of terraces and other land improvements. Land damage under perennial crops is usually less than in arable farming under similar conditions.
- Various perennial crops allow the farm-management advantages of monoculture without a reduction in soil fertility, and properly husbanded pure stands of rubber, oil palm or tea may well be considered as soil-conserving types of land use.

Fig. 6.2. Terraced rubber in North Sumatra. Note dense cover crop and fertilizer patches: no erosion, no nutrient mining?

- For some perennial crops, land that is unsuitable for arable farming or would be very costly to plant can be used. Examples are steep slopes (banana, tea, rubber), rocky terrain (tea, rubber), areas with unreliable rainfall (sisal) and semi-saline soils (date palms).
- Perennial crops present considerable scope for intensification including the use of inorganic fertilizers.

Ruthenberg's arguments can be divided into those related to the nature of perennial crops and those related to the scale of operation, which is mainly an economic question. For example, cocoa in Nigeria is a smallholder crop for it needs little investment and the economics of scale have not been demonstrated in plantation production (Saylor and Eicher, 1970). A scale advantage for plantation does, however, exist when buying external inputs like inorganic fertilizers. These are often cheaper per unit nutrient for plantations than for smallholders or subsistence farmers because there are fewer middlemen between the producer and consumer. Plantations have better access to credit and can afford to buy large amounts of fertilizers or other external inputs, which generally reduces the price.

The biophysical advantages listed by Ruthenberg are not very specific and are difficult to consider at the soil-process level. Moreover, they may be hard to prove and interactions between species, management practices and agroecological conditions may overrule such universal statements about perennial crops. More detailed hypotheses have been developed for agroforestry research. Young (1997) listed 12 hypotheses related to soils and agroforestry derived from the foundation work *The Soil under Shifting Cultivation* (Nye and Greenland, 1960). Since ICRAF was established in 1978, but especially after 1991, when it joined the CGIAR, a tremendous amount of biophysical agroforestry research has been conducted. The research was structured along the lines of several hypotheses.

The agroforestry hypotheses

Agroforestry is the spatial and temporal mix of trees and crops on farms, and therefore different from permanent agricultural plantations. Nevertheless, some hypotheses are directly relevant and somewhat similar to those mentioned by Ruthenberg (1972), including the following:

- Agroforestry systems can control runoff and erosion due to a (semi) permanent soil cover.
- Agroforestry systems can maintain soil organic matter and biological activity; the decomposition of tree litter and prunings can substantially contribute to the maintenance of soil fertility.
- N-fixing trees can increase N inputs.

- Trees can increase nutrient inputs by retrieval from lower soil horizons and weathering rock; tree roots can take up nutrients that would otherwise be lost by leaching ('safety-net').
- Agroforestry systems can lead to more closed nutrient cycling.

Sanchez (1995) listed what has been proved and what remains uncertain in agroforestry research. Table 6.1 lists the main soil fertility hypotheses and their current status. There is now sufficient

Table 6.1. A summary of the main soil fertility agroforestry (AF) hypotheses and the degree to which these have been proven by experimental evidence. Modified from Sanchez (1995) and updated with information provided by C.K. Ong and M. van Noordwijk of ICRAF (personal communication, 2000).

Hypothesis	Status	
Agroforestry (AF) systems can control soil erosion	Proven	Proven in contour hedgerow systems and multistrata systems
Trees in AF systems provide deep nutrient capture from subsoil layers that are inaccessible to crop roots	Proven	Proven for deep nitrate capture in oxic subsoils. Not yet proven widely and unlikely to be relevant in other infertile subsoils
Cycling of bases accumulated by trees in AF systems and returned to the soil as litter can help reduce soil acidity	Proven	Proven for litter high in Ca and Mg in non-AF systems; such litter temporarily complexes Al in the soil solution thus decreasing soil acidity. No new evidence
Shade from tree canopy improves soil biological activity and N mineralization	Proven	Researched in the 1980s. No new evidence
N-fixing trees can substantially augment N inputs in AF systems	Proven	Limited quantification of N-fixation by legume species and subsequent biomass N accumulation and return to the soil via litter
AF can lead to more closed nutrient cycling and to more efficient use of nutrients and less leaching losses	Proven	Recently proven by ^{15}N studies in Lampung, Indonesia
AF systems maintain soil organic matter (SOM) at satisfactory levels for soil fertility	Not fully proven	There are no reliable SOM levels related to satisfactory soil fertility. SOM increases have been detected in sandy soils under alley cropping. Recent experimental evidence suggests that alley cropping slows down SOM decrease
Tree roots are as important as above-ground biomass in soil fertility maintenance	Not proven	An important research topic. Evidence is accumulating that fine root turnover is important. Overall very few data

evidence to state that agroforestry can control erosion and that trees provide deep nutrient capture and may control soil acidity. Trees may also enhance BNF and N mineralization and lead to more closed nutrient cycling and less leaching losses. The effects of trees on the maintenance of soil organic matter and the importance of tree roots remain to be quantified although developments have been made.

Since the hypotheses were formulated it has become apparent that agroforestry systems are far less homogenous than was anticipated and therefore hypotheses proved for some types of agroforestry cannot be extrapolated (M. van Noordwijk, personal communication, 2000). In other words: it is uncertain whether agroforestry systems that work in a particular agroecology will also work in a different agroecology. Likewise, it is uncertain whether a different agroforestry system will work in an agroecology where other systems have proved to function.

The question arises of whether the hypotheses tested and proved in agroforestry research apply to plantation agriculture with perennial tree crops. The main difference between an agroforestry system and a perennial crop plantation is the temporal arrangement of the crops. In agroforestry systems, annual crops are commonly planted in between woody perennials or after a fallow period with perennials. The perennial in agroforestry systems is usually not the economic crop although farmers and agroforestry researchers have combined perennial and annual crops that both yield economic produce. Since perennial crops cover the soil continuously and more completely than agroforestry systems, it is very likely that the positive effects of agroforestry systems as listed in Table 6.1 also pertain to perennial crops.

Data availability

When reviewing the available soil data under perennial plantation crops, it is necessary to understand the phases of crop development as they affect collection and interpretation of the data. With plantation tree crops, the following phases of tree crop development can be recognized – modified from Sanchez *et al.* (1985) and PORIM (1994):

1. Forest clearing and crop establishment.
2. First years after clearing up to closing of canopy and coming into production.
3. Period of maximum production.
4. Felling and harvesting the first rotation.
5. Beginning of the second rotation.

A large number of studies have focused on soil changes when the forest is cut and crops are planted (phases 1 and 2), and there are excellent summaries available (Nye and Greenland, 1960; Sanchez *et al.*, 1983; Lal, 1986). Information on soil changes between phases 1 and 4 are scarce as it requires long-term research commitment.

Another important factor to consider is that soil conditions have been improved at many agricultural plantations due to fertilizer applications, liming, irrigation or drainage. These improvements are usually not reported in scientific literature and published information on soil changes under plantation crops is not very extensive. Most research has been adaptive and has only local relevance, and research conducted at commercial research institutes is mostly reported in local journals and reports. The limited number of publications in international journals might also be an aftermath of the general negative image of agricultural plantations (see Chapter 3), which affected research interest. More probably it is because only limited research has been conducted that produces results suitable for publication in an international journal. Furthermore, research conducted on commercial plantations may not be published because plantations are private companies and they may not wish to publicize their soil management practices.

Plantation managers

Some plantation managers have a conservative attitude and consider agricultural experimentation and science as a necessary although heavy burden. Practices that have proved suitable elsewhere may be applied without critical examination or testing of the technologies involved. For example, at the sugarcane plantation in Papua New Guinea inorganic fertilizer applications were based on recommendations for sugarcane in Barbados. These recommendations have been used at the plantation since 1979 and were still in use in the late 1990s. When it was mentioned that these recommendations had no firm basis, the plantation management suggested conducting an unreplicated trial over 1 year to find out optimum fertilizer applications. They were aware of the limitations of such a trial but found a more rigorous approach too expensive, and a waste of effort. Similar experiences were noted with sisal plantation managers in Tanzania. The situation is certainly not the same for each plantation crop but differences in perception between soil scientists and users of the soil are not uncommon (Bouma, 1993). It is likely that such differences affect the data availability on soil changes and may explain in part the limited number of publications on soil changes under plantation agriculture in international journals.

The advantages of on-plantation research

Quantifying soil changes under plantation cropping using Type I data is relatively easy. Firstly, plantations have clear spatial boundaries and usually crops are planted in numbered blocks (fields) varying in size from 0.5 to 100 ha. These blocks may follow the landscape and therefore physiographic soil units, but in flat land soil-mapping units may cut across block boundaries. From the blocks, land-use history is recorded including agricultural practices such as inorganic fertilizer applications, date of planting or crop yield. A second advantage is that plantations are mostly planted with only one crop, which receives uniform management, as opposed to subsistence farmers, who have many fields of unknown sizes, with a large mix of crops and non-uniform management – in both time and space.

On many plantations, soil and plant data are collected on a routine basis at regular intervals for the evaluation of fertilizer applications or for a general assessment of the soil chemical status. Sampling and analysis is usually conducted in consultation with researchers of a nearby commercial research station. For some perennial plantation crops such as oil palm, leaf analysis has become an effective tool to evaluate the soil nutrient status and form the basis of inorganic fertilizer recommendations. As a result, less accent has been placed on soil research at oil palm plantations. The archives of agricultural plantations may contain large amounts of soil and leaf analytical records from many fields (blocks). Plantation data (soil, climate or crop data) are therefore often long term, i.e. more than 10 years (Lal, 1995). When these Type I data are coupled with information on crop husbandry practices (inorganic fertilizer applications, weeding, pesticides etc.) and production figures, it provides a useful tool to evaluate soil chemical changes under plantation crops.

Soil Erosion Under Perennial Crops

It is generally assumed that a perennial plant cover protects the soil better against erosion than an annual crop (Jacks and Whyte, 1939; Ruthenberg, 1972; Lal, 1990), although much depends on the soil, site factors (slope, rainfall etc.) and management practices. Most annual crops provide adequate cover within 30–45 days after planting and pastures provide cover within 2–6 months, but tree crops may require 2–5 years to close their canopy (Sanchez *et al.*, 1985). Soil erosion can be considerable with inappropriate land-clearing methods and because of insufficient soil cover immediately after clearance.

Lal (1990) reviewed the literature on the effects of trees on soil erosion. He concluded that surface runoff from catchments with natural forest is generally low, but is larger with increasing annual rainfall. Soil erosion and sediment transport from the catchments with natural forests is minimal (< 1 Mg/ha/year), but soil erosion increases when natural forest is changed to planted forest, or forest plantations (see Chapter 7, section on 'Soil Erosion on Forest Plantations', page 199). Erosion is greater during the initial stages of tree establishment than when the tree canopy is fully developed. A much-used solution to the problem of soil exposure during plantation establishment is to use a managed cover crop (Sanchez *et al.*, 1985). New techniques used in plantation agriculture are underplanting, advanced planting material (very large plants) and high initial density followed by thinning (W. Gerritsma, WUR, personal communication, 2000). On oil palm plantations soil erosion is checked by early cover crop establishment, strategic placement and treatment of pruned fronds and old palm trunks after felling, terracing, construction of silt pits, and mulching with empty fruit bunches (PORIM, 1994). Despite these possibilities to reduce soil losses, erosion can be a problem in oil palm and there are also reports on soil erosion in other perennial plantation crops.

Erosion under oil palm

Several erosion studies have been conducted on oil palm plantations in Malaysia. The Palm Oil Research Institute of Malaysia (PORIM, now called MPOB – Malaysian Palm Oil Board) summarized the studies as follows: minimal soil erosion before forest clearing, five to seven times greater after clearance and a subsequent decline to almost pre-clearance level when the crop is established (PORIM, 1994). Published data on soil erosion under oil palm in Malaysia are shown in Table 6.2.

Soil erosion from Oxisols ranged from 12.5 to 77.6 Mg/ha/year and depended on the slope of the site. Soil erosion on Ultisols ranged from 1.1 to 28.0 Mg/ha/year and erosion was higher in harvesting paths. Soil erosion losses in mature oil palm plantations chiefly depend on the slope of the site and soil management practices. Soil erosion under young oil palm is usually limited due to the cover crop protecting the soil, and the limited erosion is not attributable to the palms. This was also reported from rubber plantations (Morgan, 1995). As the cover crop disappears after the closure of the palm canopy, harvest paths become exposed and compacted, which enhances runoff and soil erosion. Therefore, soil erosion may not necessarily decrease when the palms get older and the canopy is closed. Chew *et al.* (1999) reviewed published soil erosion data under moist forest and tree crop systems. Erosion under mature oil palm ranged from 6.6 to 20.5 Mg/ha/year. These values are lower than those reported in Table 6.2 but the review also showed that soil erosion can be considerable at oil palm plantations in Malaysia.

Table 6.2. Soil erosion losses under oil palm in Malaysia. Adapted from PORIM (1994), based on several studies conducted in peninsular Malaysia between 1979 and 1990.

Soil order	Palm age (years)	Slope (%)	Condition	Soil erosion (Mg/ha/year)
Tropeptic Hapludox (Oxisols)	2–4	2	With legume cover crop	18.8
		5	With legume cover crop	24.0
		9	With legume cover crop	35.4
		15	With legume cover crop	50.0
	12	<5	Uncovered	12.5
Typic Hapludox (Oxisols)	2–4	2	With legume cover crop	23.5
		5	With legume cover crop	38.8
		9	With legume cover crop	57.1
		15	With legume cover crop	77.6
Orthoxic Tropudult (Ultisols)	11	5	Harvesting path	14.9
			Palm row	7.4
			Beneath row	1.1
Typic Paleudult (Ultisols)	12–16	3–5	Uncovered	28.0
			Plots with fronds cut	19.7
			Plots with extra fronds cut	16.3

The effects of erosion under oil palm are that the soil is removed from between the tertiary and quaternary feeding roots near the soil surface, in particular in the weeded circle (Fig. 6.3). Exposed roots dry up and die, so the water and nutrient uptake capacity of the root system is reduced. Although no experimental evidence is available, it is obvious that oil palms growing under these conditions undergo water deficits and nutritional deficiencies (Ferwerda, 1977). Moreover, the nutrient-use efficiency of applied fertilizers is reduced because of the lower uptake capacity of the roots.

Erosion under coffee and cocoa

Soil erosion losses can be considerable in coffee plantations that have no adequate shade or a low planting density with little natural mulch formed by litter. This is especially important for coffee grown in highlands on steep slopes and in new coffee plantations. Research in Colombia showed that annual soil N losses from unprotected areas exceeded the amount extracted by a good crop of coffee, but on well-developed coffee plantations that are adequately shaded or with a high planting density, erosion can be reduced to less than 2% of the losses that occur on unprotected plots (Bornemisza, 1982).

Fig. 6.3. Slightly eroded weeding circle around an oil palm in Malaysia (a). Note exposed dense root mat and cover crop behind the palm. Eroded soil along a road on an oil palm plantation in the East Usambaras of Tanzania (b).

In Venezuela, where since the mid-1970s the government has actively promoted the removal of shade trees from coffee plantations, low erosion losses were found (Ataroff and Monasterio, 1997). Under shaded coffee, total erosion losses were very low and ranged from 0.7 to 1.8 Mg soil/ha/year whereas under coffee without shade, erosion losses were 6.6 Mg soil/ha in the first year after the shade was removed. Erosion losses of unshaded coffee after 2 years were comparable to shaded coffee whereas in general runoff and soil loss are lower in shaded than in unshaded plantations (Beer *et al.*, 1998). The research in Venezuela showed that soil erosion correlated positively with agricultural activities (i.e. harvesting, pruning, weeding) – more people in the field: more erosion.

Under monocropping cocoa in Malaysia, soil erosion losses were 11 Mg/ha/year but losses were considerably lower when cover crops such as *Indigofera spicata* were planted (Hashim *et al.*, 1995). When the cocoa was intercropped with banana and clean weeding with herbicide was practised, soil losses of up to 70 Mg soil/ha/year were measured, which are high losses based on a general rating of tolerable soil erosion losses (Hudson, 1986).

Erosion under tea

Soil erosion can be a problem when plantations run down. This was found to occur in Sri Lanka where tea plantations have been neglected since the mid-1970s causing soil erosion of vacant patches (Botschek *et al.*, 1998). Othieno (1975) reported from Kericho, Kenya, erosion losses of up to 168 Mg soil/ha in the first year after the establishment of a tea plantation. In the second year, soil losses were up to 81 Mg/ha whereas in the third year losses were less than 7 Mg soil/ha. The severe erosion, accounting for three-quarters of the total erosion over the 3-year period of the experiment, occurred between planting and the time when the canopy had developed to about 30%. The research indicated that soil erosion in fields with young tea can be effectively controlled by either mulching or inter-row planting of oats, but these need to be cut back as oats compete with young tea for water and nutrients (Othieno, 1975).

The effects of soil erosion on soil chemical properties

Erosion can reduce the soil chemical fertility (Zöbisch *et al.*, 1995; Lal, 1997; Ruppenthal *et al.*, 1997). Moberg (1972) compared soil fertility properties of Oxisols developed from sandstone in eroded and non-eroded coffee plots and in virgin land near Lake Victoria, Tanzania. Coffee gardens where erosion occurred were more acid and had lower levels of soil fertility than non-eroded soils with coffee, the levels of which were comparable to virgin soils (Table 6.3). No information was given on the period that the soils had been under coffee.

Type I and II Data

Type I data

One of the first studies investigating long-term changes under oil palm was conducted by P.B.H. Tinker in West Africa in the late 1950s. The oil palm was planted in 1941 on acid sandy soils (Ultisols?) previously under rainforest and soil samples were taken every 5–6 years in plots of different treatments. During the first 5 years of the plantation, there was a steady and in some cases a very marked increase in soil fertility (Kowal and Tinker, 1959). Thereafter K and Mg levels decreased, but soil organic C levels remained constant. The soils were

Table 6.3. Soil chemical properties in eroded and non-eroded Oxisols under coffee and in virgin Oxisols near Lake Victoria, Tanzania. Modified from Moberg (1972). Type II data.

Land use	Sampling depth (m)	pH	Organic C (g/kg)	Available P (mg/kg)	CEC and exchangeable cations ($mmol_c$/kg)			
					CEC	Ca	Mg	K
Virgin	0–0.15	5.2	25.2	12	259	61	40	1.8
	0.15–0.30	4.2	14.5	3	249	22	8	1.0
Coffee (non-eroded)	0–0.15	5.2	25.9	33	160	52	21	3.2
	0.15–0.30	4.8	12.2	3	128	23	18	1.8
Coffee (eroded)	0–0.15	4.1	19.0	5	256	23	14	1.9
	0.15–0.30	3.9	13.1	<2	259	8	3	1.6

re-sampled in 1961 and the stored samples were re-analysed. Soil organic C levels and pH had not changed, but levels of total N and exchangeable bases, particularly K, were lower (Tinker, 1963). Organic S was only affected when the oil palm was intercropped or when the soil was tilled. This is one of the longest data-sets on soil changes under a plantation crop, but it is unfortunate that the soils were not sampled in 1940 before the forest clearance.

A Type I study was conducted by W.J. Broughton and E. Pushparajah on rubber plantations in Malaysia, and the results were reviewed by Sanchez *et al.* (1985). Soils were classified as Typic Paleudult. Soil organic C levels decreased from about 18 g/kg under forest to 10 g C/kg after 16 years of rubber cultivation. The soil reaction increased slightly after clearing due to addition of ash, slightly decreased thereafter and was at the same level as the forest after 16 years. Soil fertility was higher when a leguminous soil cover was planted with the establishment of the rubber plantation, although the soil chemical fertility remained below that of the pre-clearing levels.

Beer *et al.* (1990) measured soil organic C in cocoa ecosystems on a Typic Humitropept at CATIE, Turrialba, Costa Rica. When the cocoa was shaded with *Erythrina poeppigiana*, topsoil (0–0.15 m) organic C levels increased from 28 g/kg to 32 g/kg in 9 years. Levels in the 0.15–0.30 m soil horizon changed from 23 g C/kg to 25 g/kg over the same period. Similar changes were noted in the top- and subsoil when the cocoa was shaded with *Cordia alliodora* (Beer *et al.*, 1990). None of the changes were statistically significant, which indicates that both cocoa ecosystems were able to maintain soil organic C contents.

Before discussing Type II data studies, there is one more Type I study which contains some interesting data. Härdter *et al.* (1997) presented soil fertility properties under forest regrowth and 6-year-old oil palm in Malaysia, but soil classification was not given. In general, the soil chemical fertility was higher under oil palm than under natural regrowth of the same age owing to inorganic fertilizer applications (Table 6.4). Soil organic C levels and CEC after 6 years of oil palm cultivation were below those under the natural forest.

Type II data – oil palm and rubber

PORIM (1994) summarized changes under oil palm cultivation as follows: levels of nutrients are found to increase in the early years under oil palm after forest because of the fertilizer applications and due to the N fixation by the leguminous cover crop. Longer-term trends are less well known, but given the application of a conservative inorganic fertilizer policy, it is likely that soil nutrients decline due to palm uptake and retention exceeding fertilizer applications

Table 6.4. Soil chemical properties of topsoils under natural jungle regrowth and 6-year-old oil palm. The soils were under forest in 1980 and the forest was felled in 1981; some parts were left to regrow and other parts were planted with oil palm. Type I and II data, modified from Härdter *et al.* (1997).

	Forest regrowth			Oil palm	
	1980	1981	1986	1981	1986
Organic C (g/kg)	24.9	22.0	20.2	21.1	21.8
Available P (mg/kg)	2.2	3.4	3.0	3.5	35.8
CEC ($mmol_c$/kg)	103	53	75	39	88
Exchangeable Mg ($mmol_c$/kg)	3.1	2.2	2.6	5.2	6.9
Exchangeable K ($mmol_c$/kg)	2.4	0.7	1.8	0.8	4.1

(PORIM, 1994). Pushparajah (1998) reported that soil organic C levels increased after 20 years on oil palm plantations but no reference data (forest) are given.

In the coastal plain of southwestern Nigeria, Aweto (1987) sampled sandy, deeply weathered soils (Oxisols) under a rubber plantation. Soil organic C levels were 14 g/kg under primary forest and 12 g C/kg in soils under 18-year-old rubber stands. The organic C contents of soils under rubber increased over time as the quality of the litter produced by rubber trees increased because of more standing biomass. Exchangeable cations, particularly K, were considerably lower under rubber, but the pH was similar in soils under forest and rubber. Correlations between age of the rubber plantation and soil properties were reasonably high for pH, exchangeable Ca, Mg and K, but no correlations were found between the age of the rubber stand and organic C or total N in the soil (Aweto, 1987).

Duah-Yentumi *et al.* (1998) found that Ultisols at Kade (Ghana) under 40-year-old rubber had significantly lower C contents when compared with soils under virgin forest or 20-year-old cocoa. Differences in soil organic C contents between the forest and cocoa were less than 3 g C/kg. Both rubber and cocoa were cultivated as pure stands and received no inorganic fertilizer or manure. The pH of the soils under virgin forest and rubber was about the same (pH 5.0), whereas the pH of Ultisols under cocoa was about 0.5 unit higher.

Type II data – coffee and cocoa

Lumbanraja *et al.* (1998) sampled Kanhapludults under coffee plantations and primary forest in South Sumatra. The plantation extent had increased from 0% to 60% of the area between 1970 and 1990, largely at the expense of the forest. Transmigration and the increased human population were the major driving forces in land-use change. The coffee plantations were less than 20 years old and topsoil organic C levels were 29 g/kg as compared to 60 g/kg in the soils under forest. All exchangeable cations were lower under coffee than under forest. Cultivated land under crops and vegetables had soil organic C levels of 15 g/kg. Similar differences were found in other soil orders in West Lampung with soil organic C levels under forest twice as high as under coffee. Soil organic C levels under arable crops were nearly half of the levels of soils under coffee; so arable cropping had even more negative effects than coffee cultivation on the soil organic C levels.

Kunu and Hartemink (1997) sampled Haplustands under primary forest and coffee gardens in the southern highlands of Papua New Guinea. The coffee gardens were established from primary forest years before the sampling and some gardens had been intercropped with food crops during the first years. None of the gardens had received inorganic fertilizers and the coffee was growing poorly. There was little difference in the soil chemical properties between forest and coffee, although exchangeable Ca and the base saturation were significantly higher under coffee. It was a residual effect of the ash addition after the forest was slashed and burned. All soils had extremely low levels of available P, but high total P levels, due to the high P retention capacity, which is common in the volcanic soils of the southern highlands (Kunu and Hartemink, 1997).

Several studies have been conducted in Nigeria investigating the effects of growing cocoa on soil chemical properties. Near Ibadan, Ekanade *et al.* (1991) found soil organic C contents under forest of 29 g/kg, whereas this was 19 g C/kg under cocoa. Available P was much higher under cocoa than under forest whereas exchangeable K was lower. In Oyo State (Nigeria), Adejuwon and Ekanade (1987) found topsoil organic C levels of 26 g/kg under forest and 19 g C/kg in the topsoils under cocoa. All major nutrients and the pH were lower under cocoa compared to soils under forest.

In another study in the Oyo State of Nigeria, Adejuwon and Ekanade (1988) sampled a large number of soils under forest, and soils that had been under 10–15 years of cocoa. Soils were classified as Alfisols and mean annual rainfall at the sites was about 1300 mm. The soil pH under forest was 6.8 whereas the pH under cocoa had decreased to 5.5. Soil organic C levels were 27 g/kg under forest but only 13 g C/kg under cocoa. Also total N and levels of exchangeable cations were much reduced when the soils under forest were compared with those under cocoa (Adejuwon and Ekanade, 1988).

In southern Nigeria it was found that soil organic C under secondary forest was about 35 g/kg; under 10-year-old cocoa levels were 25 g C/kg (Ogunkunle and Eghaghara, 1992). Total N and most other soil properties were about the same under cocoa and secondary forest.

Ekanade (1988) sampled 60 cocoa plots of ages from 1 to 55 years and collected 30 soil samples under forest on Alfisols in southwestern Nigeria. Soil organic C was on average 26 g/kg under forest and 19 g/kg under cocoa; soil reaction was 6.8 under forest and 5.9 under cocoa. All soil chemical properties were significantly lower under cocoa.

These five studies from West Africa consistently show that soil C levels under cocoa cultivation remained below those in the soils under natural forest. Furthermore the long-term data show that soil organic C equilibrium data under cocoa settle below those of the soils under natural forest.

Other studies containing Type II data

In some studies the age of the plantation is not given so rates of change cannot be calculated. Table 6.5 presents mean values of soil chemical properties under forest and cocoa on Alfisols in Nigeria. Soils under cocoa were more acid and the soil chemical fertility was significantly lower under cocoa compared with soils under forest (Ekanade, 1988). Differences are similar to those found in other studies with cocoa in West Africa and soil organic C levels under cocoa remain below that of the natural forest.

In the cocoa areas of Oyo State, Nigeria, Adejuwon and Ekanade (1987) took 300 soil samples under cocoa and 260 in natural forest. Mean annual rainfall in the area is about 1500 mm and the soils are predominantly Alfisols (Egbeda association). The results for two soil depths are given in Table 6.6. Soil pH was lower in the topsoils under cocoa but slightly higher in the subsoils when compared with natural forest. Organic C was lower in the topsoils. The CEC and exchangeable cations were lower under cocoa. The range of values was large

Table 6.5. Mean values of soil chemical properties under forest and cocoa on Alfisols in Nigeria. Type II data, modified from Ekanade (1988).

	Forest (n = 30)	Cocoa (n = 60)	Difference
pH ($CaCl_2$)	6.8	5.9	P<0.01
Organic C (g/kg)	25.2	18.5	P<0.01
Available P (mg/kg)	16	12	P<0.01
CEC ($mmol_c$/kg)	196	125	P<0.01
Exchangeable Ca ($mmol_c$/kg)	122	83	P<0.01
Exchangeable Mg ($mmol_c$/kg)	57	28	P<0.01
Exchangeable K ($mmol_c$/kg)	9	4	P<0.05

Table 6.6. Mean values (range in parentheses) of soil chemical properties under forest and cocoa on Alfisols in Nigeria. Type II data, modified from Adejuwon and Ekanade (1987).

Sampling depth:	0–0.15 m				0.15–0.45			
Land use:	Forest		Cocoa		Forest		Cocoa	
pH	6.8	(5.4–7.6)	5.9	(4.6–7.4)	4.7	(3.1–6.0)	5.0	(4.0–6.0)
Organic C (g/kg)	26	(13–37)	19	(12–30)	13	(6–23)	12	(6–20)
Available P (mg/kg)	14	(8–23)	12	(6–20)	13	(4–28)	9	(5–16)
CEC ($mmol_c$/kg)	144	(90–191)	95	(52–174)	78	(31–200)	62	(27–129)
Exchangeable Ca ($mmol_c$/kg)	90	(54–112)	53	(20–118)	47	(14–113)	41	(15–73)
Exchangeable Mg ($mmol_c$/kg)	38	(21–53)	20	(14–38)	29	(9–63)	20	(6–47)
Exchangeable K ($mmol_c$/kg)	7	(4–15)	4	(2–14)	5	(2–8)	3	(2–5)

and this might reflect the fact that cocoa plantings of different ages and on different landforms were sampled. The data provide some insight into differences and changes that may be expected when the natural forest is converted to perennial cropping.

Summary of Type I and Type II data

Table 6.7 summarizes the absolute change in soil properties under perennial plantation crops based on Type I and Type II data. The table shows pH increases in Ultisols under perennial cropping but a decrease in soil organic C, N and exchangeable cations. Soil organic C decreased in most soils except in the Inceptisols under cocoa.

Rates of Change in Soil Chemical Properties

In the previous sections, studies were reviewed in which soil fertility data were given for individual crops and soil orders. Here rates of change in soil properties are calculated for different soil orders. These rates of change are given as a value and percentage per year (Table 6.8).

Except for the Inceptisols under cocoa, most perennial crops seem to reduce soil organic C, and in particular total N levels. Exchangeable cations and available P data were too few for any trend to be observed among soil orders. From the absolute changes in soil properties, changes in percentage per year were calculated (Table 6.9). The relative decrease in soil organic C was high in the Alfisols under cocoa and the Ultisols under coffee. This is further discussed in Chapter 11 (page 315).

Table 6.7. Changes in soil chemical properties on perennial crop plantations in the tropics. Type I and II data.[a]

Soil order	Sampling depth (m)	Tree sp.[b]	Period[c] (years)	pH	C (g/kg)	N (g/kg)	P (mg/kg)	CEC and exchangeable cations ($mmol_c$/kg) CEC	Ca	Mg	K	Reference
TYPE I												
Inceptisols	0–0.15	Coc	9	nd	+4.3	nd	nd	nd	nd	nd	nd	(1)
Inceptisols	0.15–0.30	Coc	9	nd	+2.0	nd	nd	nd	nd	nd	nd	(1)
Inceptisols	0–0.15	Coc	9	nd	+1.8	nd	nd	nd	nd	nd	nd	(1)
Inceptisols	0.15–0.30	Coc	9	nd	+1.4	nd	nd	nd	nd	nd	nd	(1)
Ultisols	0–0.15	Oil	10	+0.4	+0.06	–0.04	nd	nd	+0.4	+0.8	–0.35	(2)
Ultisols	0.15–0.45	Oil	10	+0.2	–0.04	–0.14	nd	nd	–4.6	–2.6	–0.57	(2)
Ultisols	0–0.15	Rub	16	+0.1	–8.00	–0.60	nd	nd	–3.5	–0.8	–0.40	(3)
TYPE II												
Alfisols	0–0.10	Coc	7	0.6	–10.5	0	–0.5	nd	–6.0	+1.0	–1.0	(4)
Alfisols	0–0.10	Coc	13	–1.3	–14.0	–1.5	nd	–65.0	–48.9	–11.2	–2.4	(5)
Alfisols	0–0.15	Coc	50	0.0	–2.3	nd	+0.7	–21.0	–16.0	–1.0	–1.0	(6)
Oxisols	0–0.10	Rub	18	–0.9	+1.2	+0.8	nd	nd	–6.0	–5.0	–1.1	(7)
Oxisols	0.10–0.30	Rub	18	–0.4	–1.2	+0.1	nd	nd	–2.2	–2.4	–0.1	
Ultisols	0–0.15	Coc	20	+0.5	–2.6	–0.4	nd	nd	nd	nd	nd	(8)
Ultisols	0–0.20	Cof	20	+0.5	–31.9	–3.2	–2.5	–318	–83.7	–1.7	–1.2	(9)
Ultisols	0.20–0.40	Cof	20	+0.1	–14.9	–1.1	–0.5	–57.0	–22.5	–5.0	+0.7	
Ultisols	0–0.15	Rub	40	<–0.1	–11.7	–1.4	nd	nd	nd	nd	nd	(8)

[a] Calculated from the following references: (1) Beer *et al.* (1990); (2) Tinker (1963); (3) Sanchez *et al.* (1985); (4) Ogunkunle and Eghaghara (1992); (5) Adejuwon and Ekanade (1988); (6) Ekanade (1988); (7) Aweto (1987); (8) Duah-Yentumi *et al.* (1998); (9) Lumbanraja *et al.* (1998).

[b] Coc, cocoa; Cof, coffee; Rub, rubber; Oil, oil palm.

[c] Period between two soil samplings.

nd, No data.

Table 6.8. Absolute change (unit/year) in soil chemical properties on perennial crop plantations. Type I and II data.[a]

Soil order	Sampling depth (m)	Data type	Tree sp.[b]	Period[c] (years)	pH	C (g/kg)	N (g/kg)	P (mg/kg)	CEC and exchangeable cations ($mmol_c$/kg) CEC	Ca	Mg	K
Alfisols	0–0.10	II	Coc	7	+0.1	–1.5	0	–0.1	nd	–0.9	0.1	–0.1
Alfisols	0–0.10	II	Coc	13	–0.1	–1.1	<–0.1	nd	–5.2	–3.9	–0.9	–0.2
Alfisols	0–0.15	II	Coc	50	0	<–0.1	nd	0	–0.4	–0.3	<–0.1	<–0.1
Inceptisols	0–0.15	I	Coc	9	nd	+0.5	nd	nd	nd	nd	nd	nd
Inceptisols	0.15–0.30	I	Coc	9	nd	+0.2	nd	nd	nd	nd	nd	nd
Inceptisols	0–0.15	I	Coc	9	nd	+0.2	nd	nd	nd	nd	nd	nd
Inceptisols	0.15–0.30	I	Coc	9	nd	+0.2	nd	nd	nd	nd	nd	nd
Ultisols	0–0.15	I	Oil	10	<+0.1	<+0.1	<–0.1	nd	nd	+0.1	+0.1	<–0.1
Ultisols	0.15–0.45	I	Oil	10	<+0.1	<–0.1	<–0.1	nd	nd	–0.5	–0.3	–0.1
Ultisols	0–0.15	I	Rub	16	<–0.1	–0.50	<–0.1	nd	nd	–0.2	–0.1	<–0.1
Ultisols	0–0.15	II	Coc	20	0	–0.1	<–0.1	nd	nd	nd	nd	nd
Ultisols	0–0.20	II	Cof	20	0	–1.6	–0.2	–0.1	–15.9	–4.2	–0.1	–0.1
Ultisols	0.20–0.40	II	Cof	20	0	–0.7	–0.1	<–0.1	–2.9	–1.1	–0.3	0
Ultisols	0–0.15	II	Rub	40	<–0.1	–0.3	<–0.1	nd	nd	nd	nd	nd
Oxisols	0–0.10	II	Rub	18	–0.1	+0.1	0	nd	nd	–0.3	–0.3	–0.1
Oxisols	0.10–0.30	II	Rub	18	<–0.1	–0.1	0	nd	nd	–0.1	–0.1	<–0.1

[a] Calculated from the references listed in Table 6.7.
[b] Coc, cocoa; Cof, coffee; Oil, oil palm; Rub, rubber.
[c] Period of cultivation.
nd, No data.

Table 6.9. Relative change (% per year) in soil chemical properties on perennial crop plantations. Type I and II data.[a]

Soil order	Sampling depth (m)	Data type	Tree sp.[b]	Period[c] (years)	pH	C (g/kg)	N (g/kg)	P (mg/kg)	CEC and exchangeable cations ($mmol_c$/kg) CEC	Ca	Mg	K
Alfisols	0–0.10	II	Coc	7	+1	–4	0	–1	nd	–1	+1	–5
Alfisols	0–0.10	II	Coc	13	–2	–4	–4	nd	–4	–5	–4	–3
Alfisols	0–0.15	II	Coc	50	0	<–0.5	nd	0	–0	–1	<–0.5	–1
Inceptisols	0–0.15	I	Coc	9	nd	+2	nd	nd	nd	nd	nd	nd
Inceptisols	0.15–0.30	I	Coc	9	nd	+1	nd	nd	nd	nd	nd	nd
Inceptisols	0–0.15	I	Coc	9	nd	+1	nd	nd	nd	nd	nd	nd
Inceptisols	0.15–0.30	I	Coc	9	nd	+1	nd	nd	nd	nd	nd	nd
Ultisols	0–0.15	I	Oil	10	+1	+1	–1	nd	nd	0	1	–5
Ultisols	0.15–0.45	I	Oil	10	<+0.5	<–0.5	–3	nd	nd	–2	–6	–7
Ultisols	0–0.15	I	Rub	16	<–0.5	–3	–2	nd	nd	–5	–4	–3
Ultisols	0–0.15	II	Coc	20	+1	<–0.5	–1	nd	nd	nd	nd	nd
Ultisols	0–0.20	II	Cof	20	+1	–3	–3	–3	–4	–4	–1	–1
Ultisols	0.20–0.40	II	Cof	20	0	–3	–2	–2	–2	–3	–2	+3
Ultisols	0–0.15	II	Rub	40	<–0.5	–1	–1	nd	nd	nd	nd	nd
Oxisols	0–0.10	II	Rub	18	–1	+1	+4	nd	nd	–3	–5	–3
Oxisols	0.10–0.30	II	Rub	18	<–0.5	–1	+1	nd	nd	–3	–4	–1

[a] Calculated from the references listed in Table 6.7.
[b] Coc, cocoa; Cof, coffee; Oil, oil palm; Rub, rubber.
[c] Period of cultivation.
nd, No data.

Semiquantitative Studies

Perennial plantation crops have been grown for many decades and considerable research efforts have been made to increase productivity of these crops. Nutrient balances have not received much attention. For example, it was not until the early 1980s that a N balance was available for coffee and cocoa because available data for N cycling in coffee and cocoa plantations were scarce (Robertson, 1982). In the early 1980s, a preliminary N balance was developed by a 'Work Group on Coffee and Cocoa Plantations' (Table 6.10).

Nitrogen fixed by shade trees was estimated for coffee plantations using data from Mexican sites. Wet deposition was estimated to range from 5 to 15 kg/ha/year; dry deposition was considered insignificant since precipitation is high in the coffee and cocoa regions of South America. Data from inorganic fertilizer inputs were probably the most reliable and precise. The only N output known is that removed by the harvest and all other outputs are essentially unknown. It was expected that denitrification under coffee and cocoa is probably small since the crops are mostly grown on well-drained soils. Leaching, however, was expected to be a major output and losses through volatilization could also be significant, taking into account the amount of inorganic fertilizers applied on some plantations. Erosion losses are thought to be low except in coffee grown on steep slopes without shade and when the coffee is young (Roskoski *et al.*, 1982).

A more complete nutrient balance study was conducted for cocoa ecosystems at CATIE (Costa Rica). Soils at the site are Typic Dystropepts and annual rainfall is high. The input and output of N, P and K of cocoa with different shade trees is presented in Table 6.11. Inputs of N and P exceeded the outputs, but input and output of K was about the same. The K removal with the beans is approximately equivalent to the amount applied with inorganic K fertilizers. Not all the outputs are quantified and it is therefore not possible to arrive at firm conclusions.

Table 6.10. Nitrogen fluxes in coffee and cocoa ecosystems in Latin America (kg/ha/year). Modified from Roskoski *et al.* (1982).

		Coffee	Cocoa
Input	N fixation by *Rhizobia*	1–50	9–15
	N fixation by free-living bacteria	> 0.5	nd
	Wet deposition	5–15	5–15
	Inorganic fertilizers	50–300	30–60
	Total	55–400	44–90
Output	Harvest	30–60	20–40
	Volatilization, leaching, denitrification, erosion	nd	nd

nd, No data.

Table 6.11. Nutrient input and output under 6- to 10-year-old cocoa with *Cordia alliodora* or *Erythrina poeppigiana* as shade tree. Calculated from Fassbender *et al.* (1991) based on various sources (Fassbender *et al.*, 1988; Heuveldop *et al.*, 1988; Imbach *et al.*, 1989).

		Cocoa with *Cordia alliodora*			Cocoa with *Erythrina poeppigiana*		
		N	P	K	N	P	K
Input	Wet deposition	5	<0.5	3	5	<0.5	3
	Inorganic fertilizers	88	34	32	88	34	32
	Total	93	34	35	93	34	35
Output	Yield	23	5	34	28	5	30
	Leaching	6	1	2	6	1	2
	Total	29	6	36	34	6	32
Difference		+64	+28	−1	+59	+28	3

For example, fertilizer-use efficiency is not taken into account. If for K fertilizer an efficiency of 50% is assumed (the other 50% being leached) then the partial balance is negative. Similarly, if N fertilizer-use efficiency is assumed to be 30%, the output of N largely exceeds N input.

The N difference in the balance is far less a problem in the cocoa with the *Erythrina poeppigiana* system because *Erythrina* is an N-fixer and turnover rates of the prunings are fast (Fassbender *et al.*, 1991). Nitrogen fixation by *Erythrina* is estimated to be about 60 kg N/ha/year but in these cocoa ecosystems it was found that litter productivity is a more important shade tree characteristic than the capacity to fix N (Beer, 1988).

A partial nutrient balance was developed for oil palm on Oxisols in Côte d'Ivoire (Roth *et al.*, 1986). Nutrient input and output were measured for both smallholder and large-scale oil palm plantations (Table 6.12). Inorganic fertilizer use (mainly K) is twice as high on plantations than on smallholders' oil palm fields. If a 100% recovery is assumed, the K balance is positive for both oil palm grown by smallholders and plantations, but a lower, more realistic, recovery will result in a negative K balance for oil palm grown by smallholders. The table also shows that nutrient accumulation by the biomass was considerable compared with the amount of nutrients removed with the harvest. Although the higher applications of K fertilizers yielded a positive K balance, it was concluded that the high applications of KCl fertilizer were responsible for accelerated soil acidification in these Oxisols. The soil pH on the plantations was 0.1 to 0.2 units lower than in soils under oil palm from smallholders (Roth *et al.*, 1986). Given the fact that these were chemically poor soils (pH ranged from 4.0 to 4.4) such an acidification trend is unfavourable.

Table 6.12. Nutrient balances for smallholder and large-scale oil palm plantations on Oxisols in Côte d'Ivoire. The oil palm was about 14 years old. Modified from Roth *et al.* (1986).

	Smallholders					Plantation				
	N	P	K	Ca	Mg	N	P	K	Ca	Mg
Input										
Wet deposition	23	2	5	7	30	23	2	5	7	30
Inorganic fertilizers	1	0	54	0	0	10	0	108	1	0
Output										
Harvest	29	4	29	6	8	31	4	33	5	9
Difference	–5	–2	+30	+1	+22	+2	–2	+80	+3	+21
Biomass accumulation	44	7	32	15	8	44	7	32	15	8

Soil-process-oriented Studies

Most studies focusing on soil chemical changes under plantation crops were based on soil samples taken at different periods (Type I) or under different land-use systems (Type II). These studies are essentially descriptive and although differences between land-use systems were quantified and possible causes discussed, they do not distinguish and quantify the underlying processes. Few studies have looked at the soil processes involved, such as, for example, leaching or denitrification. Information on such losses in the humid tropics is scanty despite the importance for the management of nutrients and the overall sustainability of the cropping systems. Part of the explanation is that perennial crops are complex systems to study and require long-term observations to unravel trends and patterns in soil behaviour.

Some leaching studies have been conducted with coffee and cocoa in Costa Rica, cocoa in Venezuela and Brazil, and oil palm in Nigeria and Malaysia. In Costa Rica, management systems under Arabica coffee are changing from coffee grown beneath a tree overstorey to systems where shade is removed and inorganic N fertilizers are applied (Babbar and Zak, 1995). Leaching losses were investigated in Udands under shaded and unshaded coffee plantations that were both fertilized with 300 kg N/ha/year. Rainfall exceeded 2000 m/year. Annual leaching losses were almost three times greater in unshaded plantations (24 kg NO_3/ha/year) than in shaded plantations (9 kg NO_3/ha/year), whereas denitrification rates were higher in soils of shaded plantations. It was concluded that for both shaded and unshaded coffee plantations, the potential for N loss via leaching was small compared with that for denitrification (Babbar and Zak, 1995). This is contrary to the report by Roskoski *et al.* (1982) and Bornemisza (1982), who emphasized that NO_3 leaching is one of the main channels of N loss from coffee plantations in Guatemala.

Differences in the amount of N leached are probably related to differences in soil type, rainfall and soil management practices.

In the cocoa-growing regions of south Bahia, Brazil, a lysimeter study was conducted on a 30–40-year-old cocoa plantation with Tropudalfs as dominant soil orders (Santana and Cabala-Rosand, 1982). Leaching losses in unfertilized and fertilized plots (40 kg N, 40 kg P and 50 kg K/ha) were compared. The NH_4-N and NO_3-N losses were proportional to the amount of rainfall. Less N was leached from the fertilized plots than from the unfertilized plots. This was possibly because the application of P and K increased the development of cocoa roots, thus increasing nutrient-absorbing surfaces and decreasing the amounts of N available for leaching. No data were given in kg/ha or as a percentage of applied nutrients, but it was concluded that N leaching losses were small and do not seriously affect N availability to the cocoa (Santana and Cabala-Rosand, 1982). Another study with cocoa on Alfisols in Bahia showed mean annual losses of 22 kg N, 0.9 kg P, 17 kg K, 86 kg Ca and 18 kg Mg/ha (de Oliveira Leite, 1985). These are fairly high losses particularly when it is considered that the cocoa was unfertilized.

A long-term and soil-process-oriented study in shaded cocoa ecosystems was conducted at CATIE. Inorganic fertilizer input was 120 kg N, 29 kg P and 33 kg K/ha/year. Leaching losses were calculated from nutrient concentrations in lysimetric capsules at 1 m depth and the volume of percolated water (Table 6.13). Losses of N, P and K were very small and seem to have no management significance when compared with inorganic fertilizer inputs, or with the total soil reserves on the experimental site. Leaching losses of Ca and Mg were larger under cocoa with *Erythrina poeppigiana* as shade tree, but both were only a fraction of the total Ca and Mg reserves in the soil. Leaching losses in the cocoa ecosystems were about equivalent to natural or artificial forest ecosystems (Imbach *et al.*, 1989).

Table 6.13. Annual leaching losses (kg/ha ±1 SD), inorganic fertilizer inputs (kg/ha) and soil nutrient reserves (kg/ha for 0–0.45 m) under cocoa with different shade trees at Turrialba, Costa Rica. Modified from Imbach *et al.* (1989) and Alpízar *et al.* (1986).

	Cocoa with *Erythrina poeppigiana* as shade tree	Cocoa with *Cordia alliodora* as shade tree	Inorganic fertilizer inputs	Soil nutrient reserves[a]
N	5.3 ± 0.2	5.2 ± 0.3	120	8800
P	0.5 ± 0.1	0.5 ± 0.1	29	3400
K	1.5 ± 0.1	1.2 ± 0.1	33	650
Ca	26.9 ± 3.0	6.5 ± 0.5	0	2800
Mg	19.7 ± 1.2	5.7 ± 0.4	0	600

[a] Averaged and rounded data from soils under cocoa with *Erythrina* or *Cordia* as shade trees.

The losses calculated for the cocoa at CATIE were much lower than those reported by de Oliveira Leite (1985) in Brazil. On the other hand, reports from cocoa on Psamments in the lowlands of Venezuela showed that leaching losses under traditional shaded cocoa were unimportant, although in such light-textured soils leaching may be large when inorganic fertilizers are being applied (Aranguren *et al.*, 1982). Again, if and how much leaching will occur seems to depend considerably on the site conditions and crop management practices.

In Nigeria, a leaching study with inorganic fertilizers was conducted on an oil palm plantation with Rhodic Paleudult as the dominant soil order (Omoti *et al.*, 1983). Annual leaching losses of N, K and S applied as fertilizers were found to be moderate (10–15% of applied nutrients) whereas losses of Ca and Mg were excessive (60–170% of applied nutrients). The losses represented less than 15% of the initial amounts present in the soil up to 0.60 m depth, and there was no difference between 4- and 22-year-old oil palm plantations. The study showed that rates of inorganic fertilizer applications were inadequate for good vegetative growth of oil palm and optimum bunch yield (Omoti *et al.*, 1983).

Comparable results were found with oil palm in Malaysia. A lysimeter study in oil palm was conducted by S.F. Foong and the results were summarized by Chew *et al.* (1999). Leaching losses of N applied to young palms (1–4 years) was 17%, but losses decreased to less than 3% in mature oil palms. Leaching losses of P were below 2% of the applied P, and less than 3% of the applied K was leached under mature oil palms. The leaching losses were found to be low, particularly when the inorganic fertilizers were broadcast (Chew *et al.*, 1999). Low leaching losses under oil palm were also noted by PORIM (1994) and minimal contamination of groundwater by applied nutrients is likely to occur under oil palm.

Despite the obvious differences in crop phenology and soil management, very few studies have compared leaching losses under perennial and annual crops in the tropics. A leaching experiment was conducted at CATIE with two contrasting cropping systems: (i) a mixed perennial cropping system composed of *Cordia alliodora* (a timber species), cocoa and plantain; and (ii) an annual monocropping system with maize (Seyfried and Rao, 1991). Losses of Ca, Mg and K were significantly greater by 2 to 15 times in the maize system and NO_3 losses from the maize plots was 56 kg/ha compared with 1 kg NO_3/ha in the mixed perennial plots. The difference was explained by the much larger nutrient retention and uptake capabilities of the perennial crops (Seyfried and Rao, 1991). Research conducted in West Africa also concluded that leaching losses under annual crops are probably higher than under oil palm (Omoti *et al.*, 1983), whereas Imbach *et al.* (1989) concluded that leaching losses in cocoa ecosystems are much lower than under annual crops. These studies provide evidence for the 'safety-net' theory of tree crops whereby nutrients leached to a deeper soil horizon can be taken up by tree roots at great depths (van Noordwijk, 1989; Sanchez, 1995). However, leaching under tree crops may occur and a leaching study on Oxisols near Manaus, Brazil showed that leaching losses cannot be completely avoided in cropping systems with trees (Schroth *et al.*, 1999).

Environmental Impact

Many plantation crops require the use of insecticides and fungicides to control pests and diseases, and herbicides are also widely used. Repeated use of agrochemicals could be a point of concern, and in particular the heavy-metal input into the soil. There is public concern on the potential impact of heavy metals and pesticides on crop quality and the contamination of surface and subsurface soil. Although these concerns may be identified correctly, the number of environmental studies in tropical soils, and on agricultural plantations in particular, is very limited.

Wilcke *et al.* (1998) sampled soils under coffee and forest near San José, Costa Rica, and analysed for heavy metals (Al, Cd, Cu, Fe, Mn, Pb and Zn). The soils under coffee are subject to heavy-metal inputs as a result of the regular use of fungicides and inorganic fertilizers, and possibly because of atmospheric deposition originating from the densely populated and industrialized Valle Central surrounding San José. Cadmium, Pb and Zn concentrations in all soils were low or comparable to background concentrations in temperate soils. Total Cu concentrations were high in soils under coffee but also high in soils under forest. It was concluded that the influence of the parent material on the metal concentrations in the soil was more pronounced than that of agrochemical inputs (Wilcke *et al.*, 1998).

Although the Costa Rica study showed no marked increase in heavy metals of soils under coffee plantations, it is likely that the continuous use of agrochemicals may result in a build-up of heavy metals comparable to some of the problems encountered in soils under high-input temperate agriculture (Tinker, 1997). Some recent research on banana plantations in Costa Rica has indeed shown that that fungicides, nematicides and insecticides were found in surface waters and sediments near the plantation (Castillo *et al.*, 2000). Also research on tea plantations near Kyushu, Japan, showed that the heavy use of inorganic fertilizers at the plantations markedly increased nitrate and sulphate in neighbouring river waters (Ii *et al.*, 1997). The rates of pesticide accumulation and nitrate leaching and their environmental impact are different in the humid tropics because of the longer growing season, higher temperatures, etc., which affect many soil processes. It is likely that more attention will be given to these aspects due to the increasing environmental awareness, although at present little direct evidence exists that inorganic fertilizers and pesticides are a source of significant contamination.

Discussion and Conclusions

This section reviewed the available literature on soil erosion and soil fertility changes under perennial plantation crops in the tropics. Hard data on soil erosion losses collected over many seasons and in different regions are scarce. Most papers reiterate that soil erosion under perennial crops is relatively low, provided the crops are well managed. Measured data show that erosion could be high in oil palm (PORIM, 1994), cocoa (Hashim *et al.*, 1995), and tea (Othieno, 1975) – particularly during establishment. Agroforestry research has accumulated more evidence confirming lower erosion in land-use systems with tree crops than sole perennial plantation crops research (Sanchez, 1995; Young, 1997). The available data confirm that soil erosion is lower in land-use systems with perennial crops than under annual cropping. Common sense also provides the impression that under the same agroecological conditions, land-use systems with perennial crops form better protection against soil erosion than annual crops, simply because they cover the soil the whole year through. Erosion is lower under mature crops than in young plantings because of the complete ground cover. However, harvesting paths and open patches in mature plantations may result in high soil erosion losses. The risks of soil erosion under plantation crops differ among crop species depending upon their canopy characteristics and the soil management systems used (Lal, 1979, 1990). As soil erosion has pronounced effects on the soil chemical fertility, the fertility of the soil should be taken into account when interpreting erosion losses. In general, tolerable losses from acid and poor-fertility soils should be set at a much lower level than losses from inherently fertile soils.

Changes in soil chemical properties under perennial crops were found to be different. This is related to the fact that soils, crops and climates were different (Fig. 6.4). Nevertheless a slight decline was found in most soil chemical properties and in most soils. Various studies indicated that the original C and N levels under natural forest are not attained again in perennial cropping systems, although levels of P and exchangeable cations, in particular K, may be much higher in soils under perennial crops due to the use of inorganic fertilizers. The change in soil chemical properties may reflect the decrease in nutrient stocks of the soil, but it also reflects immobilization of nutrients in the biomass. Therefore it is more difficult to assess soil fertility decline and its causes in perennial crops than in annual cropping systems.

Fig. 6.4. Complex systems to study: cocoa under kapok shade trees on Oxisols in northeast Tanzania (left), and arabica coffee partly shaded with *Casuarina* trees on Andisols near Kainantu in the highlands of Papua New Guinea.

The few studies discussed here have shown that leaching losses under perennial plantation crops can be considerable, although the losses are consistently smaller than under annual crops. This is related to the fact that tree crops grow the whole year through in the humid tropics whereas in annual cropping there may be periods when there is no crop or the crop is either too young or too old to take up nutrients from the soil solution. The absence of a crop, in combination with the onset of the seasonal rainfall inducing the N priming effect, may result in leaching losses: no crop, no uptake. Tree crops generally provide a 'safety-net' but an adequate supply of all nutrients for a dense root mat is essential to reduce nutrient leaching and enable deep uptake of leached nutrients. Overall these processes can be better influenced in land-use systems with perennial crops than with annual crops.

Different crops in different agroecological zones give different changes in soil chemical properties. Because of the many factors involved and the lack of abundant data, some of the observations reported here are somewhat speculative. In order to be definitive about the findings, rigorous comparisons are needed on well-characterized soils. Type I experiments, in which tree crop systems can be compared with annual crops are needed to support these observations. Only then can it be demonstrated conclusively to what degree plantation agriculture with perennial crops can improve or maintain soil properties in the humid tropics (Sanchez *et al.*, 1985).

References

Adejuwon, J.O. and Ekanade, O. (1987) Edaphic component of the environmental degradation resulting from the replacement of tropical rain forests by field and tree crops in SW Nigeria. *The International Tree Crops Journal* 4, 269–282.

Adejuwon, J.O. and Ekanade, O. (1988) A comparison of soil properties under different landuse types in a part of the Nigerian cocoa belt. *Catena* 15, 319–331.

Alpízar, L., Fassbender, H.W., Heuveldop, J., Fölster, H. and Enríquez, G. (1986) Modelling agroforestry systems of cacao (*Theobroma cacao*) with laurel (*Cordia alliodora*) and poro (*Erythrina poeppigiana*) in Costa Rica. I. Inventory of organic matter and nutrients. *Agroforestry Systems* 4, 175–189.

Aranguren, J., Escalante, G. and Herrera, R. (1982) Nitrogen cycle of tropical perennial crops under shade trees. II Cacao. *Plant and Soil* 67, 259–269.

Ataroff, M. and Monasterio, M. (1997) Soil erosion under different management of coffee plantations in the Venezuelan Andes. *Soil Technology* 11, 95–108.

Aweto, A.O. (1987) Physical and nutrient status of soils under rubber (*Hevea brasiliensis*) of different ages in South-Western Nigeria. *Agricultural Systems* 23, 63–72.

Babbar, L.I. and Zak, D.R. (1995) Nitrogen loss from coffee agroecosystems in Costa Rica – leaching and denitrification in the presence and absence of shade trees. *Journal of Environmental Quality* 24, 227–233.

Beer, J. (1988) Litter production and nutrient cycling in coffee (*Coffea arabica*) or cocoa (*Theobroma cacao*) plantations with shade trees. *Agroforestry Systems* 7, 103–114.

Beer, J., Bonneman, A., Chavez, W., Fassbender, H.W., Imbach, A.C. and Martel, I. (1990) Modelling agroforestry systems of cacao (*Theobroma cacao*) with laurel (*Cordia alliodora*) or poro (*Erythrina poeppigiana*) in Costa Rica. V. Productivity indices, organic material models and sustainability over ten years. *Agroforestry Systems* 12, 229–249.

Beer, J., Muschler, R., Kass, D. and Somarriba, E. (1998) Shade management in coffee and cacao plantations. *Agroforestry Systems* 38, 139–164.

Bornemisza, E. (1982) Nitrogen cycling in coffee plantations. *Plant and Soil* 67, 214–246.

Botschek, J., Neu, A., Skowronek, A. and Jayakody, A.N. (1998) Agricultural suitability of degraded Acrisols and Lixisols of former tea lands in Sri Lanka. *Zeitschrift fur Pflanzenernährung und Bodenkunde* 161, 627–632.

Bouma, J. (1993) Soil behavior under field conditions – differences in perception and their effects on research. *Geoderma* 60, 1–14.

Castillo, L.E., Ruepert, C. and Solis, E. (2000) Pesticide residues in the aquatic environment of banana plantation areas in the North Atlantic Zone of Costa Rica. *Environmental Toxicology and Chemistry* 19, 1942–1950.

Chew, P.S., Kee, K.K. and Goh, K.J. (1999) Cultural practices and their impact. In: Singh, G., Huan, L.K., Leng, T. and Kow, D.L. (eds) *Oil Palm and the Environment – a Malaysian Perspective.* Malaysian Oil Palm Growers' Council, Kuala Lumpur, pp. 55–81.

de Oliveira Leite, J. (1985) Interflow, overland flow and leaching of natural nutrients on an Alfisol slope of Southern Bahia, Brazil. *Journal of Hydrology* 80, 77–92.

Duah-Yentumi, S., Ronn, R. and Christensen, S. (1998) Nutrients limiting microbial growth in a tropical forest soil of Ghana under different management. *Applied Soil Ecology* 8, 19–24.

Ekanade, O. (1988) The nutrient status of soils under peasant cocoa farms of varying ages in southwestern Nigeria. *Biological Agriculture and Horticulture* 5, 155–167.

Ekanade, O., Adesina, F.A. and Egbe, N.E. (1991) Sustaining tree crop production under intensive land use: an investigation into soil quality differentiation under varying cropping patterns in Western Nigeria. *Journal of Environmental Management* 32, 105–113.

Fassbender, H.W., Alpízar, L., Heuveldop, J., Fölster, H. and Enríquez, G. (1988) Modelling agroforestry systems of cacao (*Theobroma cacao*) with laurel (*Cordia alliodora*) and poro (*Erythrina poeppigiana*) in Costa Rica. III. Cycles of organic matter and nutrients. *Agroforestry Systems* 6, 49–62.

Fassbender, H.W., Beer, J., Heuveldop, J., Imbach, A., Enríquez, G. and Bonneman, A. (1991) The year balances of organic matter and nutrients in agroforestry systems at CATIE, Costa Rica. *Forest Ecology and Management* 45, 173–183.

Ferwerda, J.D. (1977) Oil palm. In: Alvim, P.deT. and Kozlowski, T.T. (eds) *Ecophysiology of Tropical Crops*. Academic Publishers, New York, pp. 351–382.

Härdter, R., Chow, W.Y. and Hock, O.S. (1997) Intensive plantation cropping, a source of sustainable food and energy production in the tropical rain forest areas in Southeast Asia. *Forest Ecology and Management* 91, 93–102.

Hashim, G.M., Ciesiolka, C.A.A., Yusoff, W.A., Nafis, A.W., Mispan, M.R., Rose, C.W. and Coughlan, K.J. (1995) Soil erosion processes in sloping land in the east coast of Peninsular Malaysia. *Soil Technology* 8, 215–233.

Heuveldop, J., Fassbender, H.W., Alpízar, L., Enríquez, G. and Fölster, H. (1988) Modelling agroforestry systems of cacao (*Theobroma cacao*) with laurel (*Cordia alliodora*) and poro (*Erythrina poeppigiana*) in Costa Rica. II. Cacao and wood production, litter production and decomposition. *Agroforestry Systems* 6, 37–48.

Hopkins, C.G. (1910) *Soil Fertility and Permanent Agriculture*. Ginn and Company, Boston.

Hudson, N. (1986) *Soil Conservation*. B.T. Batsford, London.

Ii, H., Hirata, T., Matsuo, H., Nishikawa, M. and Tase, N. (1997) Surface water chemistry, particularly concentrations of NO_3^- and DO and delta ^{15}N values, near a tea plantation in Kyushu, Japan. *Journal of Hydrology* 202, 341–352.

Imbach, A.C., Fassbender, H.W., Borel, R., Beer, J. and Bonneman, A. (1989) Modelling agroforestry systems of cacao (*Theobroma cacao*) with laurel (*Cordia alliodora*) and cacao with poro (*Erythrina poeppigiana*) in Costa Rica. IV. Water balances, nutrient inputs and leaching. *Agroforestry Systems* 8, 267–287.

Jacks, G.V. and Whyte, R.O. (1939) *The Rape of the Earth – a World Survey of Soil Erosion*. Faber and Faber, London.

Kowal, J.M.L. and Tinker, P.B.H. (1959) Soil changes under a plantation established from high secondary forest. *Journal of the West African Institute for Oil Palm Research* II, 376–389.

Kunu, T. and Hartemink, A.E. (1997) Soil chemical properties under primary forest and coffee in the Kutubu area of Papua New Guinea. *Papua New Guinea Journal of Agriculture, Forestry and Fisheries* 40, 1–5.

Lal, R. (1979) A brief review of erosion research in the humid tropics of south east Asia. In: Greenland, D.J. and Lal, R. (eds) *Soil Conservation and Management in the Humid Tropics.* John Wiley & Sons, Chichester, pp. 203–212.

Lal, R. (1986) Conversion of tropical rainforest: agronomic potential and ecological consequences. *Advances in Agronomy* 39, 173–264.

Lal, R. (1990) *Soil Erosion in the Tropics – Principles and Management.* McGraw-Hill, New York.

Lal, R. (1995) Development in East Africa. *Geoderma* 67, 159–164.

Lal, R. (1997) Soil degradative effects of slope length and tillage method on Alfisols in Western Nigeria. 2. Soil chemical properties, plant nutrient loss and water quality. *Land Degradation and Development* 8, 221–244.

Lumbanraja, J., Syam, T., Nishide, H., Mahi, A.K., Utomo, M., Sarno and Kimura, M. (1998) Deterioration of soil fertility by land use changes in South Sumatra, Indonesia – from 1970 to 1990. *Hydrological Processes* 12, 2003–2013.

Moberg, J.P. (1972) Some soil fertility problems in the west lake region of Tanzania, including the effects of different forms of cultivation on the fertility of some Ferralsols. *East African Agricultural and Forestry Journal* July, 35–46.

Mohr, E.C.J. (1916) Opkomst en ondergang van cultures, een humuskwestie? In: den Berger, L.G., van der Elst, P. and Mohr, E.C.J. (eds) *Algemeen Nederlandsch–Indisch Bodemcongres met Demonstratieve Tentoonstelling. Djocja, 25 t/m 28 October.* E. Fuhri & Co, Soerabaia.

Morgan, R.P.C. (1995) *Soil Erosion and Conservation.* Longman, Harlow, UK.

Nye, P.H. and Greenland, D.J. (1960) *The Soil under Shifting Cultivation.* Commonwealth Bureau of Soils, Harpenden, UK.

Ogunkunle, A.O. and Eghaghara, O.O. (1992) Influence of land use on soil properties in a forest region of Southern Nigeria. *Soil Use and Management* 8, 121–125.

Omoti, U., Ataga, D.O. and Isenmila, A.E. (1983) Leaching loss of nutrients in oil palm plantations determined by tension lysimeters. *Plant and Soil* 73, 365–376.

Othieno, C.O. (1975) Surface run-off and soil erosion on fields of young tea. *Tropical Agriculture* 52, 299–308.

PORIM (1994) *Environmental Impacts of Oil Palm Plantations in Malaysia.* Palm Oil Research Institute of Malaysia, Malaysia.

Pushparajah, E. (1998). The oil palm – a very environmental friendly crop. *The Planter (Kuala Lumpur)* 74, 63–72.

Robertson, G.P. (1982) Regional nitrogen budgets: approaches and problems. *Plant and Soil* 67, 73–79.

Roskoski, J.P., Bornemisza, E., Aranguren, J., Escalante, G. and Santana, M.B.M. (1982) Report of the work group on coffee and cocoa plantations. *Plant and Soil* 67, 403–407.

Roth, C.H., Wildhagen, H. and Meyer, B. (1986) Comparaison du bilan d'éléments nutritifs de plantations paysannes et industrielles de palmiers à huile sur les sols ferrallitiques du bassin sedimentaire du sud de la Côte d'Ivoire. *Plant and Soil* 91, 257–267.

Ruppenthal, M., Leihner, D.E., Steinmuller, N. and Elsharkawy, M.A. (1997) Losses of organic matter and nutrients by water erosion in cassava-based cropping systems. *Experimental Agriculture* 33, 487–498.

Ruthenberg, H. (1972) *Farming Systems in the Tropics.* Clarendon Press, Oxford.

Sanchez, P.A. (1995) Science in agroforestry. *Agroforestry Systems* 30, 5–55.

Sanchez, P.A., Villachica, J.H. and Bandy, D.E. (1983) Soil fertility dynamics after clearing a tropical rainforest in Peru. *Soil Science Society of America Journal* 47, 1171–1178.

Sanchez, P.A., Palm, C.A., Davey, C.B., Szott, L.T. and Russel, C.E. (1985) Tree crops as soil improvers in the humid tropics? In: Cannell, M.G.R. and Jackson, J.E. (eds) *Attributes of Trees as Crop Plants.* Institute of Terrestrial Ecology, Huntingdon, UK, pp. 79–124.

Santana, M.B.M. and Cabala-Rosand, P. (1982) Dynamics of nitrogen in a shaded cacao plantation. *Plant and Soil* 67, 271–281.

Saylor, R.G. and Eicher, C.K. (1970) Plantations in Nigeria: lessons for West African development. In: Bunting, A.H. (ed.) *Change in Agriculture.* Gerald Duckworth & Co, London, pp. 498–503.

Schroth, G., da Silva, L.F., Seixas, R., Teixeira, W.G., Macedo, J.L.V. and Zech, W. (1999) Subsoil accumulation of mineral nitrogen under polyculture and monoculture plantations, fallow and primary forest in a ferralitic Amazonian upland soil. *Agriculture Ecosystems and Environment* 75, 109–120.

Seyfried, M.S. and Rao, P.S.C. (1991) Nutrient leaching loss from two contrasting cropping systems in the humid tropics. *Tropical Agriculture* 68, 9–18.

Tinker, P.B. (1997) The environmental implications of intensified land use in developing countries. In: Greenland, D.J., Gregory, P.J. and Nye, P.H. (eds) *Land Resources: on the Edge of the Malthusian Precipice?* The Royal Society, London, pp. 1023–1033.

Tinker, P.B.H. (1963) Changes occurring in the sedimentary soils of southern Nigeria after oil palm plantation establishment. *Journal of the West African Institute for Oil Palm Research* IV, 66–81.

van Noordwijk, M. (1989) Rooting depth in cropping systems in the humid tropics in relation to nutrient use efficiency. In: Heide, J. van der (ed.) *Nutrient Management for Food Crop Production in Tropical Farming Systems.* Institute of Soil Fertility, Haren, pp. 129–144.

Wilcke, W., Kretzschmar, S., Bundt, M., Saborio, G. and Zech, W. (1998) Aluminum and heavy metal partitioning in a horizons of soils in Costa Rican coffee plantations. *Soil Science* 163, 463–471.

Young, A. (1997) *Agroforestry for Soil Management*, 2nd edn. CAB International, Wallingford, UK.

Zöbisch, M.A., Richter, C., Heiligtag, B. and Schlott, R. (1995) Nutrient losses from cropland in the central highlands of Kenya due to surface runoff and soil erosion. *Soil and Tillage Research* 33, 109–116.

Forest Plantations

7

> In the tropics, more than anywhere else, there are silvicultural, economic, social and environmental benefits which make tree-planting and plantation forestry attractive.
>
> Julian Evans (1992)

Research on soil changes under forest plantations (man-made forests) is of relevance to agricultural plantations with perennial tree crops. Both systems consist of tree crops cultivated for prolonged periods. The main difference is the time of harvest of the produce, i.e. at the end of the cycle in forest plantations and during the tree's life in agricultural plantations. This has implications for the pattern of nutrient cycling, and the main drain of nutrients occurs in agricultural plantations during the crop life when the produce is harvested, whereas in forest plantations most nutrients are removed when the trees are cut and harvested. Several studies have focused on soil changes under forest plantations and these are summarized in this chapter. At the end of this chapter some of the differences and similarities in perennial crop and forest plantations are discussed.

History, Extent and Present Status

The first forest plantations in the tropics were started in 1680 in Sri Lanka when teak was successfully introduced (Evans, 1992). Throughout the 18th and 19th centuries, colonization and settlements in the tropics led to many introductions being made. In 1829 teak was planted in Java, in 1840 in India, in 1871 in Bangladesh and in Trinidad in the 1880s. *Eucalyptus* and wattles were planted in 1843 in India, and *Eucalyptus* was introduced in South America in 1823. Before 1900 there was no need to plant trees extensively as an industrial source. Extensive planting of industrial tree crops was made in the period 1900–1945 in South Africa, Australia (Queensland), India, Brazil and Kenya. Many countries increased the rate of planting after the Second World War for various purposes, including industrial, protection, and firewood (Evans, 1992).

Unlike agricultural plantations there are reasonably accurate data on the extent of forest plantations. Between 1945 and 1965 the area of forest plantations in the tropics increased fivefold, but the increase was small compared with the absolute increase in forest plantations between 1965 and 1990. In 1965, there were about 6.7 million ha of forest plantations, in 1980 20.9 million ha, and the total area for 1990 was estimated to be 42.6 million ha (Evans, 1992). Brown *et al.* (1997) estimated the total plantation area in the tropics to be 44 million in 1990.

In the 1980s, a shift took place towards the planting for social and environmental objectives (Evans, 1986; Wiersum, 1997). Annual planting rates in the 1980s were about 1 million ha and a major improvement was the development of clonal plantings (Webster and Watson, 1988). Other estimates on planting rates in the tropics range from 1.2 to 3 million ha/year (Sawyer, 1993). Evans (1986) estimated that the area of new plantings in the 1980s was probably more than twice the figure of the early 1970s, but only 10% of the rate of tropical deforestation. Tropical deforestation is about 17 million ha/year, and between 1960 and 1990 developing countries lost one-fifth of their natural tropical forest cover (Sedjo and Botkin, 1997).

Monocropping is common on forest plantations and about 85% of all plantation forestry is dominated by two genera (*Eucalyptus* and pines) and one species (teak) (Sawyer, 1993). Globally, about 30% of the industrial wood is harvested from old growth (Amazon, Russia, Canada, Indonesia, Malaysia), 36% from second growth (North America, Russia, Europe) and 34% from industrial plantations, of which 24% are in the temperate regions and 10% are in the tropics (Sedjo and Botkin, 1997). Wood from forest plantations in the tropics has many uses including timber, woodchips, building material, or furniture. Plantation forests are also being established for wood to provide energy in the form of fuel wood or indirectly as wood gas or charcoal. Other reasons are for poles and other domestic products, watershed protection and management and dune stabilization, (Sawyer, 1993). An increasing area in the tropics is planted for afforestation and rehabilitation purposes, and for the sequestration of C (see the section on 'Environmental Impact' on pages 215, 218 and 219).

Soil Erosion on Forest Plantations

Soil erosion losses in tropical rainforests are generally low because of the constant vegetation cover, the different vegetation strata intercepting rainwater, and the litter layer protecting the soil. Soil erosion may take place when treefall gaps arise following strong winds or landslides. Accelerated erosion may occur when fires destroy plant cover and the forest floor. The soil is exposed to increased kinetic energy of raindrops, which enhances soil movement. Fires may also decrease water infiltration in the soil due to the plugging of surface pores and increased fire-induced water repellency (Fisher and Binkley, 2000).

In most situations it is found that soil erosion increases when the natural forest is cut or when selective logging is practised (Whitmore, 1985; Bruenig, 1996). Evans (1992) listed various studies on soil erosion in the tropics. Erosion rates under natural forest in Indonesia were absent, but reached 48 Mg soil/ha when the trees were cut and the litter removed. The largest soil erosion losses can be expected when the natural forest is cleared and a plantation established.

In Indonesia, soil erosion on teak as taungya system was 5.2 Mg/ha when the trees were 1 year old but it decreased to 1 Mg/ha when the trees were 3 or 4 years old (Evans, 1992). In some forest plantations, local people remove litter for fuel wood, which may result in high rates of soil erosion. In the USA, where soil erosion has received much attention since the dustbowls of the 1930s, there is an increasing interest in planting short-rotation woody crops like sweetgum (*Liquidambar styraciflua*) (Malik *et al.*, 2000). However, soil erosion occurs in such short-rotation plantations. In an experiment in Alabama, losses of up to 11 Mg/ha/year were recorded but soil erosion losses were reduced when a cover-crop strip was introduced between the rows of the trees (Malik *et al.*, 2000).

Some plantation forest species, such as *Eucalyptus* and teak, produce litter that decomposes very slowly (O'Connell and Sankaran, 1997). As a result, there is a thick litter layer but hardly any undergrowth in such plantation (Fig. 7.1). The litter may release toxic substances depressing or killing young seedlings of grasses and other tree species. There is not much danger of soil erosion in such plantations when the canopy is closed but when the trees are young or the leaf litter is burned, soil erosion may occur.

Type I and II Data

There is fair a body of recent literature on soil changes under forest plantations in the temperate regions (e.g. Knoepp and Swank, 1994; Wang *et al.*, 1996; Falkiner and Polglase, 1997; Cromack *et al.*, 1999;

Fig. 7.1. Sheet erosion under young *Eucalyptus* trees on old volcanic soils in Gran Canaria (a), and gully erosion in Ultisols under *Eucalyptus* near Bukavu in Congo (b).

Noble *et al.*, 1999; Piatek and Allen, 1999). Several studies have been conducted in tropical regions, although the number of studies has not kept pace with the area increase of forest plantations. Most studies contain Type II data and the majority of the studies have focused on five species (*Pinus, Gmelina, Eucalyptus, Acacia, Tectona*) and some studies were conducted with less common species (e.g. *Casuarina, Larix*).

Type I data

On Sabah, Malaysia, Nykvist (1997) reported a decrease in soil fertility in a 4-year-old *Acacia mangium* plantation on Ultisols. Soil organic C and levels of major nutrients had decreased, and amounts were much lower in the plantation than before planting, owing to plant uptake and leaching. An important drain of nutrients was through log removal, but this could be reduced if the trunks were debarked in the forest, for the bark is more rich in nutrients than the wood itself (Nykvist *et al.*, 1994).

Parrotta (1992) measured soil properties under *Albizzia lebek* and 4.5-year-old natural regrowth on degraded calcareous Tropopsamments in Puerto Rico. Soil organic C and total N were significantly higher in plantation plots than under natural regrowth, whereas available P and exchangeable cations were similar. From the same site, Parrotta (1999) reported changes in soil properties under the fast-growing trees *Casuarina equisetifolia*, *Eucalyptus robusta* and *Leucaena leucocephala* as compared to natural regrowth. After 7.5 years, soil organic C levels were similar in the planted tree fallows and in natural regrowth, but total N content was significantly higher in forests with N-fixers (*Casuarina, Leucaena*). Cations and available P were similar under plantation forests and natural regrowth (Parrotta, 1999).

On abandoned pasture land in the Atlantic lowlands of Costa Rica, Fisher (1995) sampled soils before the establishment of forests and 4 years later. The soils were classified as Typic Tropohumults and had not been influenced by volcanic ash. In total 11 different species were planted, of which five species were N-fixing. The species had different effects on soil chemical properties. *Pinus* significantly acidified the soil whereas the pH under *Gmelina* increased from 4.5 to 4.8 between 1987 and 1992. Soil organic C increased significantly in the plots with *Acacia*, but no changes in the C levels were found in the soils under *Pinus*, *Gmelina* or *Inga*. Total N increased significantly under *Inga* and *Acacia* but decreased under *Pinus*. In the N-fixing tree plots soil N levels increased because more organic N was returned to the soil beneath these species. It was concluded that significant amelioration of degraded soils occurs in a short time beneath the majority of species (Fisher, 1995).

An extensive data-set has been published on soil changes in forest plantations in the Hunan Province of southern China (Dai *et al.*, 1998). Soils were sampled in the early 1960s and re-sampled in 1994. Dominant soil orders were Dystrochrepts, Hapludults and Rhodudults. At the majority of the sites, sampling depths in 1994 differed from the depths sampled in the 1960s. At three sites (all Ultisols) topsoil chemical properties could be compared as they were from the same depth, but subsoil values could not be compared. The pH had not changed in Hapludults over a period of 34 years but in the Rhodudults the pH declined from 5.5 in 1959, to 4.5 in 1994. In two of the three soils, organic C was increased but in Rhodudults soil organic C had decreased from 43.9 g/kg in 1959 to 31.2 g C/kg in 1994 (Dai *et al.*, 1998).

In New Zealand, soils under pastures were planted with *Pinus radiata* in 1975, and the soils were sampled at the time of planting, and again in 1998 (Yeates *et al.*, 2000). Soils were volcanic and classified as Typic Udivitrands. Annual rainfall was about 1490 mm. Between 1975 and 1998 topsoil pH had declined from 4.9 to 4.7 and the pH in the subsoil also declined. Soil organic C and total N under *Pinus* declined up to 0.60 m depth. Levels of plant-available P increased between 1975 and 1998 but exchangeable cations, particularly K, decreased. It was concluded that the changes in soil chemical properties reflect the change from pasture to forest (Yeates *et al.*, 2000).

A summary of the studies on soil changes on forest plantations using Type I data is presented in Table 7.1.

Type II data – West Africa

Various studies on the effects of forest plantations have been conducted in Nigeria. In Zaria in the northern part of the country, Alfisols

Table 7.1. Changes in the soil chemical fertility on forest plantations in the tropics. Type I data.

Soil order	Sampling depth (m)	Period[a] (years)	Tree sp.[b]	pH	C (g/kg)	N (g/kg)	P (mg/kg)	CEC and exchangeable cations ($mmol_c$/kg)				Reference[c]
								CEC	Ca	Mg	K	
Andisol	0–0.18	23	Pin	−0.2	−2.5	−0.5	+385	+119	−40.4	−2.0	−2.7	(1)
	0.18–0.30	23		−0.5	−1.5	−0.2	+121	+4	−3.7	+0.1	−1.1	
	0.30–0.40	23		−0.1	−18.4	−1.3	+114	−44	−11.4	−0.2	−3.7	
Ultisol	0–0.15	4	Pin	−0.2	−0.1	−0.3	+1	nd	−0.2	−0.2	−0.1	(2)
Ultisol	0–0.15	4	Gme	+0.3	−0.1	−0.2	<−0.5	nd	+0.2	−0.2	−0.1	(2)
Ultisol	0–0.15	4	Ing	−0.1	−73.3	+0.4	+2	nd	−0.3	−0.1	−0.1	(2)
Ultisol	0–0.15	4	Aca	0	+0.1	+0.3	+1	nd	+0.3	−0.1	0.1	(2)
Ultisol	0–0.10	32	Pin	+0.3	+31.3	nd	nd	129	nd	nd	nd	(3)
Ultisol	0–0.15	34	Pin	0	+1.3	nd	nd	−19	nd	nd	nd	(3)
Ultisol	0–0.10	35	Pin	−1.0	−4.3	nd	nd	−104	nd	nd	nd	(3)
Psamments	0–0.20	5	Alb	+0.1	+2.6	+0.2	+9	nd	−8.0	−0.6	+0.1	(4)

[a] Period between two soil samplings.
[b] Pin, *Pinus*; Gme, *Gmelina arborea*; Ing, *Inga edulis*; Alb, *Albizzia lebek*.
[c] Calculated from the following references: (1) Yeates *et al.* (2000); (2) Fisher (1995); (3) Dai *et al.* (1998); (4) Parrotta (1992).
nd, No data.

under a 25-year-old *Eucalyptus* plantation and secondary savanna woodland (>80 years old) were sampled (Jaiyeoba, 1995). Mean annual rainfall at the site is about 1100 mm, and the soils are derived from Basement complex and Quaternary deposits. No differences were found in the topsoil organic C contents, but the subsoils in the savannah woodland contained higher levels of soil organic C. Total N was lower in both topsoils and subsoils under *Eucalyptus* whereas exchangeable K was lower in the topsoils only. Also the pH was lower in the soils of the *Eucalyptus* plantations, and overall the soils under *Eucalyptus* had a lower soil chemical fertility when compared with soils under savanna woodland (Jaiyeoba, 1995).

Jaiyeoba (1998) also measured soil properties under *Pinus oocarpa* and *Eucalyptus camaldulensis* plantations of different ages (2–21 years). Compared with secondary savannah woodland, organic C declined during the first years of plantation establishment, but gradually increased and reached an equilibrium level after about 10 years. Most plant nutrients followed the same pattern and the difference between the two youngest plantations and the oldest plantations, and the natural vegetation, were significant. Soil pH decreased over time, and overall there was little difference in the soil chemical properties of the *Pinus* and eucalyptus plantations (Jaiyeoba, 1998).

In southern Nigeria, Typic Kandiudalfs under 7-year-old *Gmelina* had soil organic C levels of 18.3 g/kg vs. 24.0 g C/kg under secondary forest (Ogunkunle and Eghaghara, 1992). Also the available P and exchangeable Ca and Mg levels were lower in soils under *Gmelina*, although little differences were found in the soil pH under *Gmelina* and secondary forest. Okoro *et al.* (2000) sampled different monoculture plantation species on Paleudults in the lowland rainforest belt of southwestern Nigeria. Under 28-year-old *Tectona grandis* (teak), the topsoil pH had increased from 6.0 to 6.2 whereas the subsoil pH had decreased by 0.7 units. Soil organic C had decreased from 12.5 to 10.6 g/kg and also levels of exchangeable cations had declined. Soils of 28-year-old *Gmelina arborea* plantations had acidified and exchangeable cations had decreased. Levels of soil organic C were slightly increased when these soils were compared with natural (secondary) forest (Okoro *et al.*, 2000).

In northern Senegal, 18-year-old plantations with *Acacia senegal* on red sandy soils (Psamments?) had significantly higher soil organic C, N and K contents than 3-year-old plantations (Deans *et al.*, 1999). In this study it was found that harvesting the wood at the end would be less threatening to the soil nutrient budgets than regular large-scale fodder harvesting. The threat was less serious for N than for P and K (Deans *et al.*, 1999). Soil organic C and N levels were measured in *Pinus* and *Eucalyptus* plantations and natural savannah near Pointe Noire in Congo (Trouve *et al.*, 1996). Soils at the site were classified as

Psamments and clay contents were less than 75 g/kg. Under natural savannah, soil organic C was 6.5 g/kg compared to 8.8 g C/kg under 19-year-old eucalyptus plantations and 6.7 g/kg in the soils under 28-year-old *Pinus* plantations. Total N levels were lowest under *Pinus* and similar in soils of the savannah and under eucalyptus. The silt fraction had higher total C contents than the clay fraction (Trouve *et al.*, 1996); whereas it is mostly found that the C fraction decreases with increasing soil particle size (Hassink, 1995; Schnitzer, 2000).

Type II data – Latin America

Several studies have been conducted in Central and South America, where there are extensive areas of forest plantations. Smith *et al.* (1998) measured soil chemical properties in natural forest and on forest plantations in Curuá-Una forest reserve in eastern Pará, Brazil. Average yearly rainfall at the sampling sites was 1900 mm and the soils were classified as Typic Haplustox. Topsoil organic C levels were not significantly different between soils under natural forest and *Pinus caribaea*, *Euxylophora paraenis* or *Carapa guianensis* plantations. The pH was significantly lower in the soils under *Euxylophora* compared with the other two plantation species and natural forest. Total N was lower in the soils under *Pinus*, but extractable P was the same in the soils of the forest plantations and natural forest (Smith *et al.*, 1998).

In the western Amazon basin of Brazil, McGrath *et al.* (2001) compared soil properties under 6-year-old plantation agroforests and adjacent native forest. The soils at the site are classified as Oxisols and Ultisols with low effective CECs and high Al saturation. Exchangeable Ca and Mg were significantly greater in the soils under agroforests, as was the pH, which resulted in a higher ECEC and lower Al saturation in the soils under agroforests. Soil organic C and total N were similar in soils under agroforests and native forests (McGrath *et al.*, 2001).

Lilienfein *et al.* (2000) compared soils under 20-year-old *Pinus radiata* and under native cerrado vegetation south of Brasilia, Brazil. Soils were classified as Acrustox and the vegetation was characterized by open grassland with a 15–40% cover of 3–5-m-high trees. Soil organic C was slightly lower under *Pinus*, particularly in the subsoil. Compared with cerrado vegetation, soils under *Pinus* had been acidified by 0.4–0.6 pH units. Differences were larger in the subsoils. As a result of the acidification, there were increased leaching losses of plant nutrients from the topsoil (Lilienfein *et al.*, 2000).

Soil changes after abandoned pastures have been planted with forest plantation species were measured in the Atlantic humid lowlands of Costa Rica (Montagnini and Porras, 1998). Annual rainfall at the site is about 4000 mm and the soils were Fluventic Dystropepts derived from volcanic alluvium. The area had been cleared in the mid-1950s and was grazed until 1981. The soils were too poor for cultivation of bananas or other commercial crops commonly grown in the region. Three plantations were established with four species grown either as sole crop or in a mixture, and one plot with natural regrowth. Sampling interval was 4 years. In most soils, organic C was slightly lower under the tree crops than under natural regrowth. Also exchangeable K was lower in most soils in the forest plantations. This reduction was mostly caused by the rapid uptake and accumulation of nutrients in the tree biomass. In some of the mixed plots, soil conditions were less affected by the tree crops, which suggests that in mixed conditions it may take longer to deplete soil nutrients than in monospecific stands of fast-growing species (Montagnini and Porras, 1998).

Type II data – Asia and Oceania

In the Gazipur District, Dhaka Division, of Bangladesh, Islam and Weil (2000) measured soil chemical properties in a Typic Paleudult under natural forest and in soils cropped for 7 years with *Acacia auriculiformis* and *A. minijiri*. No difference was found in the pH and in both soils the pH was around 5.0. Organic C levels under natural forest were around 8.4 g/kg compared with 12.8 g C/kg under *Acacia* sp. Also N levels were higher under *Acacia* than under natural forest. The authors explained this by the N-fixing capacity of *Acacia* (Islam and Weil, 2000).

In Hawaii, measurements were made in afforested plots on Andisols that had been under sugarcane since the beginning of the 20th century, and were planted with *Eucalyptus* in the 1980s (Bashkin and Binkley, 1998). Total organic C in soils (0–0.55 m) under natural forest was 130 Mg C/ha. Total C under eucalypts was 114 Mg C/ha whereas under permanent sugarcane organic C was 113 Mg C/ha. Organic C contents differed significantly by depth, but the total C content in the soils under *Eucalyptus* and sugarcane was the same (Bashkin and Binkley, 1998).

A review of soil changes under *Pinus radiata* plantations in Australia showed that soil organic C was lower under *Pinus* than under adjacent original forest (Turner and Lambert, 2000). There was a rapid decline in soil organic C until about 12 years after plantation establishment. Soil organic C then stabilized with some indication of increases after about 20 years of establishment. It was concluded that the results have significant implications for fast-growing, short-rotation plantations (<15 years) for pulpwood or biofuel, and soil organic C decline can be expected to continue over subsequent rotations (Turner and Lambert, 2000). In New Zealand, soils under *Pinus radiata* plantations were compared with soils under indigenous forest (Ross *et al.*, 1999). The soils had been under pinus for 22 years and were classified as Typic Udivitrand. Annual rainfall at the site was 1500 mm. The soil pH was higher under *Pinus* in both the 0–0.10 m and the 0.10–0.20 m soil horizons. Contrary to the results found in Australia (Turner and Lambert, 2000), soil organic C and total N were not increased after 20 years under *Pinus* (Ross *et al.*, 1999). Much seems to depend on the initial organic C level whether a decrease or increase is found.

Table 7.2 summarizes changes in soil chemical properties on forest plantations in the tropics as determined by Type II data.

Rates of Change in Soil Chemical Properties

The absolute change in soil units per year was calculated for the different soils and tree species (Table 7.3). The pH changes in most soils were less than 0.1 unit but in more than two-thirds of the soils there was a slight pH increase. Changes in organic C were variable, and the largest decrease in soil organic C was found in an *Inga edulis* plantation on Ultisols. Under *Albizzia lebek* a considerable increase in soil organic C was found but the data were from a degraded site which was reforested. Annual changes in total N were equal to or less than 0.1 g/kg in all soils. Phosphorus changes were variable although a decrease was found in most soils. Forest plantations on Oxisols increased exchangeable cations whereas cations were decreased or changed little in Ultisols.

From the absolute changes in soil properties, changes in percentage per year were calculated (Table 7.4). Relatively large increases occurred in the exchangeable cations of Oxisols and the highest increase was 49% per year. Also the pH and organic C and N increased considerably in Psamments under forest plantations. The relative changes in soil properties seem to suggest that they are larger in soils that are chemically poor (Psamments, Oxisols) and this is further investigated in Chapter 11.

Semiquantitative and Soil-process-oriented Studies

It is generally perceived that without inorganic fertilizer applications, intensively managed tree plantations have a negative balance (Mackensen and Folster, 2000). The number of nutrient-balance studies on forest plantations is limited. Recently a study was conducted in which the costs of the inorganic fertilizers to compensate for management-dependent nutrient losses were calculated for a plantation in East Kalimantan, Indonesia. Different scenarios were taken into

Table 7.2. Changes in soil chemical properties on forest plantations in the tropics. Type II data.

Soil order	Sampling depth (m)	Period[a] (years)	Tree sp.[b]	pH	C (g/kg)	N (g/kg)	P (mg/kg)	CEC and exchangeable cations ($mmol_c/kg$)				Reference[c]
								CEC	Ca	Mg	K	
Alfisols	0–0.10	7	Gme	−0.1	−12.8	+1.0	−1.1	nd	−4.0	−1.0	+4.0	(1)
Alfisols	0–0.10	13	Gme	+1.5	−5.7	−0.6	−1.1	nd	+25.1	+5.7	−0.4	(2)
Alfisols	0.10–0.20	13	Gme	+0.7	−2.7	−0.4	−0.2	nd	+13.7	−0.3	−0.4	
Alfisols	0.20–0.40	13	Gme	+0.6	+0.7	+0.3	+0.3	nd	+13.1	+0.8	+0.8	
Andisols	0.10–0.20	22	Pin	+0.5	−34.0	−1.5	nd	nd	nd	nd	nd	(3)
Andisols	0.10–0.20	22	Pin	+0.7	−16.0	−0.2	nd	nd	nd	nd	nd	
Oxisols	0–0.20	6	Pin	+0.6	+0.9	+0.1	nd	nd	+14.6	+1.8	+0.1	(4)
Oxisols	0.20–0.40	6	Pin	+0.4	nd	nd	nd	nd	nd	nd	nd	
Oxisols	0–0.30	10	Cor	nd	+5.3	nd	nd	nd	+10.0	+4.5	+0.7	(2)
Oxisols	0–0.30	10	Cae	nd	−2.5	nd	nd	nd	+10.5	+5.0	+0.8	
Oxisols	0–0.30	10	Dal	nd	+2.7	nd	nd	nd	+2.5	+1.0	+0.2	
Oxisols	0–0.15	20	Pin	−0.2	−0.1	nd	nd	nd	nd	nd	nd	(5)
Oxisols	0.15–0.30	20	Pin	−0.4	−1.9	nd	nd	nd	nd	nd	nd	
Oxisols	0.30–0.45	20	Pin	−0.5	+0.1	nd	nd	nd	nd	nd	nd	
Oxisols	0–0.20	23	Eux	+0.5	+5.9	+0.5	0	nd	nd	nd	nd	(6)
Oxisols	0–0.20	36	Pin	+0.1	−6.9	−0.9	+0.1	nd	nd	nd	nd	
Oxisols	0–0.20	36	Car	+0.4	−7.4	−0.4	−0.1	nd	nd	nd	nd	
Psamments	0–0.05	19	Euc	+2.3	+0.1	nd	nd	nd	nd	nd	nd	(7)
Psamments	0–0.05	28	Pin	+0.2	−0.1	nd	nd	nd	nd	nd	nd	

Continued

Table 7.2. *Continued.*

Soil order	Sampling depth (m)	Period[a] (years)	Tree sp.[b]	pH	C (g/kg)	N (g/kg)	P (mg/kg)	CEC and exchangeable cations ($mmol_c$/kg) CEC	Ca	Mg	K	Reference[c]
Ultisols	0–0.20	11	Gme	+0.2	+0.7	nd	−1.4	nd	0	−0.2	−0.1	(2)
Ultisols	0.20–0.50	11	Gme	−0.1	+1.2	nd	+1.4	nd	+0.1	−0.1	+0.1	
Ultisols	0.50–1.00	11	Gme	−0.1	0	nd	−0.1	nd	−0.1	−0.5	0	
Ultisols	0–0.02	28	Tec	+0.5	−5.8	−3.1	+53.6	nd	−27.6	−3.2	−0.6	(8)
Ultisols	0.02–0.15	28	Tec	+0.2	−1.9	+0.2	−25.8	nd	−16.8	−1.6	−0.4	
Ultisols	0.15–0.30	28	Tec	−0.7	+3.5	−0.2	+1.2	nd	−4.0	0	−0.1	
Ultisols	0–0.02	28	Gme	+0.3	+8.9	−2.4	+47.8	nd	−24.4	−2.4	−0.1	(2)
Ultisols	0.02–0.15	28	Gme	−0.8	+0.9	+0.7	−29.9	nd	−15.2	−1.6	−0.4	
Ultisols	0.15–0.30	28	Gme	−0.7	+3.2	0	−5.2	nd	−4.0	0	−0.1	

[a] Period between two soil samplings.
[b] Gme, *Gmelina arborea*; Pin, *pinus*; Cor, *Cordia trichotoma*; Cae, *Caesalpinia echinata*; Dal, *Dalbergia nigra*; Eux, *Euxylophora paraensis*; Car, *Carapa guianensis*; Euc, *Eucalyptus*; Tec, *Tectona grandis*.
[c] Data from the following references: (1) Ogunkunle and Eghaghara (1992); (2) Sanchez *et al.* (1985); (3) Ross *et al.* (1999); (4) McGrath *et al.* (2001); (5) Lilienfein *et al.* (2000); (6) Smith *et al.* (1998); (7) Trouve *et al.* (1996); (8) Okoro *et al.* (2000).
nd, No data.

Table 7.3. Absolute change (unit/year) in soil fertility (topsoils only) on forest plantations. Type I and II data.[a]

Soil order	Data type	Tree sp.[b]	Period[c] (years)	pH	C (g/kg)	N (g/kg)	P (mg/kg)	CEC and exchangeable cations ($mmol_c$/kg)			
								CEC	Ca	Mg	K
Alfisols	II	Gme	7	<−0.1	−1.8	+0.1	−0.2	nd	−0.6	−0.1	+0.6
Alfisols	II	Gme	13	<+0.1	−0.4	<−0.1	−0.1	nd	+1.9	+0.4	0
Andisols	I	Pin	23	<−0.1	−0.1	<−0.1	+16.7	+5.2	−1.8	−0.1	−0.1
Oxisols	II	Pin	6	+0.1	+0.2	<+0.1	nd	nd	+2.4	+0.3	<+0.1
Oxisols	II	Cor	10	nd	+0.5	nd	nd	nd	+1.0	+0.5	+0.1
Oxisols	II	Cae	10	nd	−0.3	nd	nd	nd	+1.1	+0.5	+0.1
Oxisols	II	Dal	10	nd	+0.3	nd	nd	nd	+0.3	+0.1	<+0.1
Oxisols	II	Pin	20	<−0.1	<−0.1	nd	nd	nd	nd	nd	nd
Oxisols	II	Eux	23	<+0.1	+0.3	<+0.1	<−0.1	nd	nd	nd	nd
Oxisols	II	Pin	36	<+0.1	−0.2	<−0.1	<+0.1	nd	nd	nd	nd
Oxisols	II	Car	36	<+0.1	−0.2	<−0.1	<−0.1	nd	nd	nd	nd
Psamments	I	Alb	5	<+0.1	+0.6	<+0.1	+2.0	nd	−1.8	−0.1	<+0.1
Psamments	II	Euc	19	<+0.1	<+0.1	nd	nd	nd	nd	nd	nd
Psamments	II	Pin	28	<+0.1	<−0.1	nd	nd	nd	nd	nd	nd
Ultisols	I	Pin	4	<−0.1	<−0.1	−0.1	+0.2	nd	0	−0.1	<−0.1
Ultisols	I	Gme	4	<+0.1	<−0.1	<−0.1	−0.1	nd	0	0	<−0.1
Ultisols	I	Ing	4	<+0.1	−18.3	+0.1	+0.5	nd	−0.1	0	<−0.1
Ultisols	I	Aca	4	0	<+0.1	+0.1	+0.2	nd	+0.1	0	<+0.1
Ultisols	II	Gme	11	<+0.1	<+0.1	nd	−0.1	nd	0	<−0.1	<−0.1
Ultisols	II	Tec	28	<+0.1	<−0.1	<+0.1	−0.9	nd	−0.6	−0.1	<−0.1
Ultisols	II	Tec	28	<−0.1	<+0.1	<+0.1	−1.1	nd	−0.5	−0.1	<−0.1

Continued

Table 7.3. *Continued.*

Soil order	Data type	Tree sp.[b]	Period[c] (years)	pH	C (g/kg)	N (g/kg)	P (mg/kg)	CEC and exchangeable cations (mmol$_c$/kg)			
								CEC	Ca	Mg	K
Ultisols	I	Pin	32	<+0.1	+1.0	nd	nd	+4.0	nd	nd	nd
Ultisols	I	Pin	34	0	<+0.1	nd	nd	−0.6	nd	nd	nd
Ultisols	I	Pin	35	<−0.1	−0.1	nd	nd	−3.0	nd	nd	nd

[a] Calculated from the references listed in Tables 7.1 and 7.2.
[b] Gme, *Gmelina arborea*; Pin, *Pinus* sp.; Cor, *Cordia trichotoma*; Cae, *Caesalpinia echinata*; Dal, *Dalbergia nigra*; Eux, *Euxylophora paraensis*; Car, *Carapa guianensis*; Alb, *Albizzia lebek*; Euc, *Eucalyptus*; Ing, *Inga edulis*; Aca, *Acacia mangium*; Tec, *Tectona grandis*.
[c] Period of cultivation.
nd, No data.

Table 7.4. Relative change (%/year) in soil chemical properties (topsoils only) on forest plantations. Type I and II data.[a]

Soil order	Data type	Tree sp.[b]	Period[c] (years)	pH	C (g/kg)	N (g/kg)	P (mg/kg)	CEC and exchangeable cations ($mmol_c$/kg)			
								CEC	Ca	Mg	K
Alfisols	II	Gme	7	<–0.5	–5	+7	–3	nd	–1	–1	+19
Alfisols	II	Gme	13	+3	–2	–1	–2	nd	+13	+12	–2
Andisols	I	Pin	23	<–0.5	<–0.5	–1	+16	+32	–3	–2	–2
Oxisols	II	Pin	6	+2	+1	+1	nd	nd	+49	+9	+1
Oxisols	II	Cor	10	nd	+3	nd	nd	nd	+14	+9	+14
Oxisols	II	Cae	10	nd	–2	nd	nd	nd	+15	+10	+16
Oxisols	II	Dal	10	nd	+2	nd	nd	nd	+4	+2	+4
Oxisols	II	Pin	20	<–0.5	<–0.5	nd	nd	nd	nd	nd	nd
Oxisols	II	Eux	23	<+0.5	<–0.5	+1	<–0.5	nd	nd	nd	nd
Oxisols	II	Pin	36	<+0.5	<–0.5	–1	<+0.5	nd	nd	nd	nd
Oxisols	II	Car	36	<+0.5	<–0.5	<–0.5	<–0.5	nd	nd	nd	nd
Psamments	I	Alb	5	<+0.5	+4	+6	+4	nd	–1	–1	+3
Psamments	II	Euc	19	+2	+1	nd	nd	nd	nd	nd	nd
Psamments	II	Pin	28	<+0.5	–1	nd	nd	nd	nd	nd	nd
Ultisols	I	Pin	4	–1	–1	–2	+3	nd	–4	–4	–2
Ultisols	I	Gme	4	+2	–1	–2	–2	nd	+4	–5	–1
Ultisols	I	Ing	4	–1	–25	+3	+10	nd	–6	–3	–4
Ultisols	I	Aca	4	0	+1	+3	+3	nd	+5	–3	+1
Ultisols	II	Gme	11	<+0.5	+1	nd	–2	nd	0	–3	–3
Ultisols	II	Tec	28	<+0.5	–1	+2	–2	nd	–3	–2	–2
Ultisols	II	Tec	28	<–0.5	<+0.5	+6	–3	nd	–3	–2	–2

Continued

Table 7.4. *Continued.*

Soil order	Data type	Tree sp.[b]	Period[c] (years)	pH	C (g/kg)	N (g/kg)	P (mg/kg)	CEC and exchangeable cations ($mmol_c$/kg)			
								CEC	Ca	Mg	K
Ultisols	I	Pin	32	<0.5	+2	nd	nd	+2	nd	nd	nd
Ultisols	I	Pin	34	0	<0.5	nd	nd	<−0.5	nd	nd	nd
Ultisols	I	Pin	35	−1	−1	nd	nd	−1	nd	nd	nd

[a] Calculated from the references listed in Tables 7.1 and 7.2.

[b] Gme, *Gmelina arborea*; Pin, *Pinus* sp.; Cor, *Cordia trichotoma*; Cae, *Caesalpinia echinata*; Dal, *Dalbergia nigra*; Eux, *Euxylophora paraensis*; Car, *Carapa guianensis*; Alb, *Albizzia lebek*; Euc, *Eucalyptus*; Ing, *Inga edulis*; Aca, *Acacia mangium*; Tec, *Tectona grandis*.

[c] Period of cultivation.

nd, No data.

account and inorganic fertilizer compensation, for nutrient losses incurred due to tree harvest, was found to increase establishment costs by 18–33% and total investment costs by 9–15%. It was concluded that strategies should be developed that reduce management-dependent nutrient losses and that these should include: low-impact management alternatives to slash-and-burn; soil-conserving harvesting techniques; and appropriate site selection. Furthermore it was advised that nutrient costs should be considered in investment calculations for industrial forest plantations (Mackensen and Folster, 2000). Sensible as this may be, it would have been of greater value if an attempt was made to quantify nutrient flows and pools in order to develop management strategies that reduce nutrient losses.

A partial nutrient balance was constructed for pure larch (*Larix gmelinii*) plantations in northeastern parts of China (Liu *et al.*, 1998). The research was conducted on 'dark brown forest soils with abundant organic matter and high soil fertility'. The soils had been degraded and the main cause was the slow decomposition of the litter, of which more than 23 Mg litter/ha had accumulated in the 20-year-old plantations. The accumulation of litter reduced nutrient cycling in pure larch plantations and caused a reduction in soil chemical fertility (Liu *et al.*, 1998). Input of nutrients with the rainfall and output with runoff were measured (Table 7.5). Although this is an incomplete nutrient budget, inputs exceed outputs of all major nutrients except Ca. Much more is known about input–output relationships of nutrients of hardwood plantations in the temperate regions, and a useful overview of studies is given by Ranger and Turpault (1999).

Not many soil-process-oriented studies have been conducted in forest plantations in the tropics. A leaching losses study was conducted on sandy Ultisols at Jarí, Pará, Brazil by C.E. Russell and the work was summarized by Sanchez *et al.* (1985). Annual leaching losses in virgin rainforest are low but not absent (Table 7.6). When the forest is logged and the debris burnt, leaching losses increase dramatically. Leaching losses of cations under young *Pinus* plantations were considerable, but leaching in 10.5-year-old *Pinus* plantations was similar to the virgin forest.

Table 7.5. Partial nutrient budgets (kg/ha/year) of larch plantations in China. Modified from Liu *et al.* (1998).

Nutrient	Input by rainfall	Output with runoff	Difference
N	12.8	4.0	+8.8
P	1.0	0.3	+0.7
K	5.6	1.5	+4.1
Ca	8.5	11.0	−2.5
Mg	3.0	2.5	+0.5

Table 7.6. Annual leaching losses in forests and *Pinus* plantations at Jarí, Pará, Brazil. Modified from Sanchez *et al.* (1985) based on data from C.E. Russell.

	Leaching losses (kg/ha/year)			
Site	P	Ca	Mg	K
Rainforest – virgin	0.04	16.7	8.1	12.7
Rainforest – logged and burned	0.21	103.7	146.6	37.1
Pinus plantation – 0.5 year old	0.16	89.4	74.7	89.9
Pinus plantation – 10.5 years old	0.05	12.1	6.1	9.6

Environmental Impact

When evaluating the environmental impact of forest plantations, much depends on whether the plantation is compared with natural forest or with degraded lands, which obviously would give very different conclusions. In many parts of the tropics, forest plantations are being planted for the rehabilitation of deforested watersheds (Grainger, 1988). In such deforested areas there may be massive erosion having adverse on-site and off-site effects. Trees can stabilize soils through their extensive root systems, increase soil organic matter through enhanced litter production and through fine root turnover (Parrotta, 1992). Reforestation is beneficial for the widespread *Imperata* grasslands in South-East Asia, and in the 1980s it was estimated that there was more than 2000 million ha of tropical lands in various stages of degradation, which in theory could be reforested (Grainger, 1988).

Despite the large increase in forest plantations there have been few reports on the environmental aspects of plantation forestry. An IUCN report by Sawyer (1993) provides an overview of environmental factors but contains no hard data. There are some detailed case studies and Lamb (1990) reviewed the history and effects of large-scale pulpwood logging in Papua New Guinea. It was concluded that much damage has been done to the biophysical and socio-economic environment but no information is given on the effects of the pulpwood logging on the soil (Fig. 7.2).

Cannell (1999) reviewed the environmental impacts of forest monocultures. It was concluded that evaporation from planted forest monocultures is greater than from short vegetation because of greater interception loss. Water loss from conifer forests is usually greater than from deciduous hardwoods, but evapotranspiration from *Eucalyptus* in the dry tropics is often no greater than from native

Fig. 7.2. Felled *Acacia mangium* plantation in the Gogol Valley of the Madang Province (Papua New Guinea). Note massive removal of topsoil and tracks made by the heavy machinery used for cutting and transporting the trees.

hardwoods. Compared with short vegetation, forests can significantly increase the transfer of acidifying pollutants from the air to the soil and surface waters, and conifers are more likely to enhance acidification than are hardwoods. Furthermore, it was concluded that there are normally sufficient plantation management options available to make most plantation landscapes the home of a rich diversity of flora and fauna (Cannell, 1999). A recent study in Tasmania confirmed that many ground-dwelling invertebrates were found at least as commonly in plantations as in native forests (Bonham *et al.*, 2002). An increasing number of studies challenge the contention that plantations necessarily have vastly lower biodiversity than surrounding native forests (Bowyer, 2001).

Carbon sequestration

At present there is an increased interest in forest plantations because they may save further cutting of natural forests and produce wood for both local use and export (McCullough, 1999; Wright *et al.*, 2000). Moreover, forest plantations are regarded as an option for sequestration of atmospheric CO_2 (Cairns and Meganck, 1994; Montagnini and Porras, 1998). In the tropics, sequestration by trees is much faster than in the temperate zone (Dabas and Bhatia, 1996). It is assumed that plant growth will be increased due to the 'CO_2 fertilization' effect and current estimates of annual terrestrial plant uptake of C owing to this effect range from 8% to 33% of the annual fossil-fuel emissions (Davidson and Hirsch, 2001).

The quantities of C sequestered by forest plantations depend on the management system. Mixtures of species generally give higher biomass per hectare and thus fix more C (Montagnini and Porras, 1998). The net effect on soil C from afforestation of cultivated lands depends on the new C gained, but also on C lost from the previous management, as was shown in reforested sugarcane fields in Hawaii (Bashkin and Binkley, 1998). Recent research in *Pinus* plantations in the USA has shown that estimates of increases in C sequestration of forests, which is expected to partially compensate for increasing CO_2 in the atmosphere, are unduly optimistic (Oren *et al.*, 2001). The experiment was conducted on acid soils (Ultic Alfisols) and shortage of water and nutrients limited the long-term response of trees to increased levels of CO_2. There was also no significant increase in soil organic C attributable to higher concentrations of CO_2. A fast turnover of leaf litter was found, which appeared to constrain the potential size of this C sink (Schlesinger, 1999; Schlesinger and Lichter, 2001). So the initial C level as well as availability of water and nutrients affects the C sequestration of forest plantations.

The argument that plantations sequester C is also being advocated by plantation agriculturists (Aweto, 1995; Pushparajah, 1998) because many plantation crops are woody species, such as rubber, cocoa and oil palm. In Malaysia it has been estimated that the net fixation of C by oil palm can be equivalent to that of a lowland rainforest. The total above-ground C stock of mature oil palm is three to four times less than above-ground biomass of the rainforest (Henson, 1999). At the end of the crop cycle most of the CO_2 in the oil palm vegetation will be released when the trunks are burned or decomposing. Rubber is a slightly different story as an increasing amount of rubber wood is used for industrial and other purposes so the C fixed in the wood is not immediately released at the end of the crop cycle. However, if *Imperata* grasslands, which are quite extensive in South-East Asia (Menz *et al.*, 1998), are planted with rubber or oil palm there may be net C sequestration.

It seems that C sequestration by agricultural and forest plantations is an area with little hard data but lots of trumpeting. A reduction in deforestation yields much greater benefits for a reduction in the CO_2 emission than expanding plantation silviculture, but deforestation reduction is much more uncertain (Fearnside, 2000).

Discussion and Conclusions

This chapter focused on soil erosion and soil nutrient changes in forest plantations in the tropics. Soil erosion losses in forest plantations are likely to occur when the trees are immature, but in mature forest plantations soil erosion may occur in gaps or after fires have burned the undergrowth and litter layer. Given the fact that undergrowth is often absent in forest plantations (Fig. 7.1) and the heavy machinery used for the removal of the trees at harvest, there is considerable potential for soil erosion, which is determined by management practices and site conditions.

This review has shown that under certain conditions forest plantations have favourable effects on the soil: trees improve soil fertility by reducing losses and optimizing nutrient gains. This is a confirmation of the general agroforestry hypothesis (Sanchez, 1995; Young, 1997). It occurs mostly when degraded soils are planted with trees. Although that is an obvious advantage over agricultural plantations, the risk for soil fertility decline is of a similar magnitude, as many nutrients are removed with the harvest of the wood. Nutrient removal is considerably lower when only the stems are harvested and the bark, leaves, twigs and small branches are returned to the soil.

The effects of trees on soils are species specific. For example, in Brazil it was found that soil acidification under *Gmelina* was lower than under *Pinus*. It seems that *Gmelina* acts as a Ca accumulator under acid soil conditions (Sanchez *et al.*, 1985). This review has shown that soil acidification under *Pinus* was common, and it is generally found that conifers (e.g. *Pinus*) are more likely to enhance soil acidification than are hardwoods (Cannell, 1999; Dommergues and Subba Rao, 2000). Different species have different effects on soil processes, resulting in different rates of change. Comprehensive studies such as those of Bruijnzeel (1983) in Central Java and Waterloo (1994) in Fiji are useful to unravel the effects of different species on different soils in relation to the productivity of forest plantations.

Some forest plantations are established on soils that were degraded, so it not surprising that the soil chemical fertility improved compared with the original degraded level. Forest plantations may be established on soils that are less suitable for agriculture and with an intrinsic low fertility. Amelioration effects of plantation forests on soils generally occur during the period immediately following canopy closure (the fallow-enrichment phase), while during the maximum-production phase a deterioration of site quality may occur, as nutrients are taken up by trees and litter accumulates on the floor due to unfavourable conditions for decomposition (Sanchez *et al.*, 1985; Montagnini and Porras, 1998). As discussed in the section on 'Interpretation of the Results' (Chapter 4, page 123), the time of observation has much influence on whether a change in soil chemical properties is measured. Long-term observations are essential for detecting soil changes under forest plantations, so as to distinguish the ameliorating effect from the soil fertility decline effect.

Soil organic C and C sequestration

An important issue in research on forest plantations is the effect on soil organic C and the sequestration of C. There is little evidence of a loss of soil C in forest plantations, according to Cannell (1999). This review has shown that in quite a number of studies soil organic C decreased under forest plantations. On *Pinus* plantations it took at least 20 years before the declining trend in soil organic C was stopped, whereas other studies showed that there was no increase in soil organic C after 22 years. Rates of change in soil organic C are affected by the initial C level, species (litter fall etc.) and sites. The net decline in soil organic C is not offset by plantation growth under traditional plantation management, and plantations established specifically for C accumulation will need to use silvicultural systems which are modified to maximize C accumulation (Turner and Lambert, 2000). Research in the USA has shown that C sequestration may be limited in *Pinus* plantations because of water and nutrient becoming limiting factors (Oren *et al.*, 2001).

That forest plantations are a good sink for C is disputable (Cannell, 1999; Chapela *et al.*, 2001). Averaged over time (exceeding a rotation time), an area covered with plantations that are managed for maximum yield of wood will normally contain substantially less C in the trees than the same area of unmanaged forest species with similar growth characteristics. In most instances, the difference in C storage in the trees is so large that it is unlikely that it can be compensated for by the C stored in the harvested wood products or by intensive management of plantations (Cannell, 1999).

An important part of the net C sequestration takes place in the soil, and increases in soil organic C are a way to counterbalance the increased atmospheric CO_2 levels (Schlesinger, 1999). Soil organic C increases have been reported in some studies. Part of the explanation of higher soil organic C levels is in the nature of the litter and roots of the trees. Sanger *et al.* (1997) compared the phenolic and carbohydrate signatures of soil profiles under grass, spruce and ash stands. An undecomposed phenolic fraction from lignin and a hydrolysable carbohydrate fraction from plants had accumulated in the soils under ash and spruce. The lignin and carbohydrate fractions in the grassland soil were more decomposed than in the ash or spruce profiles, suggesting that soils under forest had rapidly accumulated what would be considered chemically recalcitrant but unstable C compounds. These C fractions appear to have been stabilized against further microbial decomposition in the mineral soil (Sanger *et al.*, 1997). Although this research was in temperate soils, the same explanation may apply for tropical soils with forest plantations: organic matter fractions differ largely between soils under forest and grassland. Also in tropical regions there is increasing evidence that changes in land use and the planting of trees affects the composition of whole soil and the texture (Trouve *et al.*, 1996; Barrios *et al.*, 1997). Decomposition of C is affected by clay content, mineralogy, pH, aerobic conditions and many other soil and agroecological factors (DeLaune *et al.*, 1981; Hassink, 1992; Torn *et al.*, 1997), and overall forest-derived C is more stable and harder to decompose (van Noordwijk *et al.*, 1997).

The capacity of forest plantations to sequester C is also influenced by the species and the type of management and in particular the harvesting method. A recent review on the effects of forest management showed that forest harvesting, on average, had little or no effect on soil C and N (Johnson and Curtis, 2001). Significant effects of harvest type and species were noted, with sawlog harvesting causing increases in soil C and N, and whole-tree harvesting causing decreases (Fig. 7.3). The positive effects of sawlog harvesting appeared to be restricted to coniferous species. Evidence is accumulating that there is greater soil C sequestration under N_2-fixing species compared with non-fixers (Resh *et al.*, 2002). This is mainly due to the retention of older soil C under N_2-fixing trees.

Concluding remarks

Plantation forests will become increasingly significant in the world's future timber supply (Sedjo, 2001). It is important that they be managed well from production, environmental, social and economic perspectives (Brown *et al.*, 1997). Few data are available on the long-term productivity of forest plantations but this review has shown that the soil chemical fertility under forest plantations is affected, which influ-

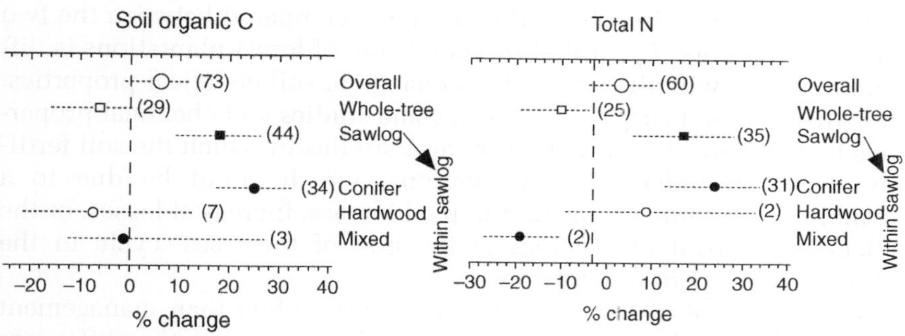

Fig. 7.3. Effects of different harvesting techniques on soil organic C and total N contents in A horizons under different tree species. Bars indicate 99% confidence intervals and the number of studies is shown in parentheses. Modified from Johnson and Curtis (2001).

ences long-term productivity. In many parts of the world there are forest plantations reaching second or even third rotation, and there is evidence of a build-up in numbers of potential diseases and pest species (Evans, 1986). However, there is no evidence to suggest that yield decline, sometimes called second-rotation decline, is widespread. Where growth has been depressed in second or third rotations it is as much due to bad harvesting practices as to any inherent nutrient removals causing exhaustion of the site (Evans, 1986, 2001).

Differences Between Perennial Crops and Forest Plantations

How relevant are studies on soil changes in forest plantations for agricultural plantations – the main subject of this book. Both perennial and forest plantations consist of tree crops that remain in the field for many years. Erosion may be problematic in both types of land use when the trees are immature but no information is available on the long-term effects of soil erosion on productivity. Terracing, which is not uncommon in coffee, rubber and oil palm plantations, is rarely done in forest plantations. An advantage of forest plantations over agricultural plantations is that there are usually no people in such forests, so soil erosion of pathways is absent. Fires may be more common in forest plantations, which is an important indirect cause of soil erosion.

Some forest plantations may be planted on inferior soils compared with soils used for agricultural plantations. This is likely given the pressure on the land in the humid tropics by which the best soils will be used for commercial and food crops, and the poorer soils for growing trees. Moreover, in some areas forest plantations are established on degraded soils for watershed protection or for other purposes in which protection is more important than production. This has implications if changes in soil fertility are to be compared between the two land-use systems. The initial soil condition of forest plantations is different, which will affect the rate of change in soil chemical properties. It is therefore not surprising that in some studies soil chemical properties had improved. But there were also studies in which the soil fertility declined under forest plantations, which could be due to a combination of immobilization in the biomass, increased losses, or the effect of removal of the logs at the end of the trees' cycle in the absence of nutrient replenishment.

Management of forest plantations is less subtle than management of agricultural plantations, and one could argue that the difference between forestry and agriculture is as large as the differences between agriculture and horticulture. Intensive harvesting and site preparation including maximum harvest removal of aerial biomass, can result in a serious depletion of soil organic matter and nutrient reserves, from which recovery on a short-rotational basis is incomplete (Fisher, 1981).

Harvesting techniques for the logs affect the soil C and N stocks in forest plantations (Johnson and Curtis, 2001; Parfitt *et al.*, 2002). The time or frequency of the harvest of the produce is an important difference from agricultural plantations. In agricultural plantations, there is a seasonal drain of nutrients after the crop is mature. This drain can be very high but is partly compensated for by regular inorganic fertilizer applications. In forest plantations, inorganic fertilizer applications may be given at the time of planting but it is uncommon that the mature trees receive inorganic fertilizers. Forest plantations mimic the natural forest, in which nutrient cycling is fairly closed (Sanchez *et al.*, 1985). Unless thinning of trees has occurred during the first years after planting, the major drain of nutrients is at harvest. If only the stems were removed from the field, nutrient losses could be restricted, but soil disturbance and complete removal of the above-ground biomass is likely to induce considerable nutrient losses (Nykvist, 1997). This also occurs in perennial crop plantations when, for example, oil palm or rubber have come to the end of the economic cycle and are slashed. For the rubber this may imply an additional loss of nutrients when the wood is removed, but oil palm above-ground biomass is usually not removed.

In summary, the dynamics of soil nutrient contents in agricultural plantations are faster than in forest plantations as there is more frequent disturbance through removal, fertilizer applications, pruning etc. These differences in dynamics affect the assessment of declining soil chemical properties in agricultural and forest plantations. The data from forest plantations are variable and the effects of forest plantations are species specific, and much seems to depend on management factors. They do provide, however, a fair insight into soil changes under tree crops and the effects will be further discussed in Chapter 11.

References

Aweto, A.O. (1995) Organic carbon diminution and estimates of carbon dioxide release from plantation soil. *The Environmentalist* 15, 10–15.

Barrios, E., Kwesiga, F., Buresh, R.J. and Sprent, J.I. (1997) Light fraction soil organic matter and available nitrogen following trees and maize. *Soil Science Society of America Journal* 61, 826–831.

Bashkin, M.A. and Binkley, D. (1998) Changes in soil carbon following afforestation in Hawaii. *Ecology* 79, 828–833.

Bonham, K.J., Mesibov, R. and Bashford, R. (2002) Diversity and abundance of some ground-dwelling invertebrates in plantation vs. native forests in Tasmania, Australia. *Forest Ecology and Management* 158, 237–247.

Bowyer, J.L. (2001) Environmental implications of wood production in intensively managed plantations. *Wood and Fibre Science* 33, 318–333.

Brown, A.G., Nambiar, E.K.S. and Cossalter, C. (1997) Plantations for the tropics – their role, extent and nature. In: Nambiar, E.K.S. and Brown, A.G. (eds) *Management of Soil, Nutrients and Water in Tropical Plantation Forests.* ACIAR, Canberra, pp. 1–23.

Bruenig, E.F. (1996) *Conservation and Management of Tropical Rainforests – an Integrated Approach to Sustainability.* CAB International, Wallingford, UK.

Bruijnzeel, L.A. (1983) Hydrological and biogeochemical aspects of man-made forests in south-central Java, Indonesia. PhD dissertation, Free University, Amsterdam, The Netherlands.

Cairns, M.A. and Meganck, R.A. (1994) Carbon sequestration, biological diversity, and sustainable development: integrated forest management. *Environmental Management* 18, 13–22.

Cannell, M.G.R. (1999) Environmental impacts of forest monocultures: water use, acidification, wildlife conservation, and carbon storage. *New Forests* 17, 239–262.

Chapela, I.H., Osher, L.J., Horton, T.R. and Henn, M.R. (2001) Ectomycorrhizal fungi introduced with exotic pine plantations induce soil carbon depletion. *Soil Biology and Biochemistry* 33, 1733–1740.

Cromack, K., Miller, R.E., Helgerson, O.T., Smith, R.B. and Anderson, H.W. (1999) Soil carbon and nutrients in a coastal Oregon Douglas-fir plantation with red alder. *Soil Science Society of America Journal* 63, 232–239.

Dabas, M. and Bhatia, S. (1996) Carbon sequestration through afforestation: role of tropical industrial plantations. *Ambio* 25, 327–330.

Dai, Z.H., Liu, Y.X., Wang, X.J. and Zhao, D.W. (1998) Changes in pH, CEC and exchangeable acidity of some forest soils in southern China during the last 32–35 years. *Water, Air, and Soil Pollution* 108, 377–390.

Davidson, E.A. and Hirsch, A.I. (2001) Fertile forest experiments. *Nature* 411, 431–433.

Deans, J.D., Diagne, O., Lindley, D.K., Dione, M. and Parkinson, J.A. (1999) Nutrient and organic-matter accumulation in *Acacia senegal* fallows over 18 years. *Forest Ecology and Management* 124, 153–167.

DeLaune, R.D., Reddy, C. and Patrick, W.H. (1981) Organic matter decomposition in soil as influenced by pH and redox conditions. *Soil Biology and Biochemistry* 13, 533–534.

Dommergues, Y.R. and Subba Rao, N.S. (2000) Introduction of N_2-fixing trees in non-N_2-fixing tropical plantations. In: Subba Rao, N.S. and Dommergues, Y.R. (eds) *Microbial Interactions in Agriculture and Forestry.* Science Publishers, Enfield, pp. 131–154.

Evans, J. (1986) Plantation forestry in the tropics – trends and prospects. *The International Tree Crops Journal* 4, 3–15.

Evans, J. (1992) *Plantation Forestry in the Tropics,* 2nd edn. Clarendon Press, Oxford.

Evans, J. (2001) Tropical hardwood plantations. In: Last, F.T. (ed.) *Tropical Crop Ecosystems.* Vol. 19 *Ecosystems of the World.* Elsevier, Amsterdam, pp. 217–227.

Falkiner, R.A. and Polglase, P.J. (1997) Transport of phosphorus through soil in an effluent-irrigated tree plantation. *Australian Journal of Soil Research* 35, 385–397.

Fearnside, P.M. (2000) Uncertainty in land-use change and forestry sector mitigation options for global warming: plantation silviculture versus avoided deforestation. *Biomass and Bioenergy* 18, 457–486.

Fisher, R.F. (1981) Impact of intensive silviculture on soil and water quality in a coastal lowland. In: Lal, R. and Russell, E.W. (eds) *Tropical Agricultural Hydrology.* John Wiley & Sons, Chichester, pp. 299–309.

Fisher, R.F. (1995) Amelioration of degraded rain forest soils by plantations of native trees. *Soil Science Society of America Journal* 59, 544–549.

Fisher, R.F. and Binkley, D. (2000) *Ecology and Management of Forest Soils,* 3rd edn. John Wiley & Sons, Chichester.

Grainger, A. (1988) Estimating areas of degraded tropical lands requiring replenishment of forest cover. *The International Tree Crops Journal* 5, 31–61.

Hassink, J. (1992) Effects of soil texture and structure on carbon and nitrogen mineralization in grassland soils. *Biology and Fertility of Soils* 14, 126–134.

Hassink, J. (1995) Decomposition rate constants of size and density fractions of soil organic matter. *Soil Science Society of America Journal* 59, 1631–1635.

Henson, I.E. (1999) Comparative ecophysiology of oil palm and tropical rain forest. In: Singh, G., Huan, L.K., Leng, T. and Kow, D.L. (eds) *Oil Palm and the Environment – a Malaysian Perspective.* Malaysian Oil Palm Growers' Council, Kuala Lumpur, pp. 9–39.

Islam, K.R. and Weil, R.R. (2000) Land use effects on soil quality in a tropical forest ecosystem of Bangladesh. *Agriculture Ecosystems and Environment* 79, 9–16.

Jaiyeoba, I.A. (1995) Changes in soil properties related to different land uses in part of the Nigerian semi-arid savannah. *Soil Use and Management* 11, 84–89.

Jaiyeoba, I.A. (1998) Changes in soil properties related to conversion of savannah woodland into pine and *Eucalyptus* plantations, northern Nigeria. *Land Degradation and Development* 9, 207–215.

Johnson, D.W. and Curtis, P.S. (2001) Effects of forest management on soil C and N storage: meta analysis. *Forest Ecology and Management* 140, 227–238.

Knoepp, J.D. and Swank, W.T. (1994) Long-term soil chemistry changes in aggrading forest ecosystems. *Soil Science Society of America Journal* 58, 325–331.

Lamb, D. (1990) *Exploiting the Tropical Rain Forest. An Account of Pulpwood Logging in Papua New Guinea.* UNESCO, Paris.

Lilienfein, J., Wilcke, W., Ayarza, M.A., Vilela, L., Lima, S.D. and Zech, W. (2000) Soil acidification in *Pinus caribaea* forests on Brazilian savanna Oxisols. *Forest Ecology and Management* 128, 145–157.

Liu, S.R., Li, X.M. and Niu, L.M. (1998) The degradation of soil fertility in pure larch plantations in the northeastern part of china. *Ecological Engineering* 10, 75–86.

Mackensen, J. and Folster, H. (2000) Cost-analysis for a sustainable nutrient management of fast growing-tree plantations in East-Kalimantan, Indonesia. *Forest Ecology and Management* 131, 239–253.

Malik, R.K., Green, T.H., Brown, G.F. and Mays, D. (2000) Use of cover crops in short rotation hardwood plantations to control erosion. *Biomass and Bioenergy* 18, 479–487.

McCullough, R.B. (1999) Four common myths about plantation forestry. *New Forests* 17, 111–118.

McGrath, D.A., Duryea, M.L. and Cropper, W.P. (2001) Soil phosphorus availability and fine root proliferation in Amazonian agroforests 6 years following forest conversion. *Agriculture Ecosystems and Environment* 83, 271–284.

Menz, K., Magcale-Macandog, D. and Wayan Rusastra, I. (eds) (1998) *Improving Smallholder Farming Systems in* Imperata *Areas of Southeast Asia: Alternatives to Shifting Cultivation.* Monograph No. 52. ACIAR, Canberra.

Montagnini, F. and Porras, C. (1998) Evaluating the role of plantations as carbon sinks: an example of an integrative approach from the humid tropics. *Environmental Management* 22, 459–470.

Noble, A.D., Little, I.P. and Randall, P.J. (1999) The influence of *Pinus radiata, Quercus suber,* and improved pasture on soil chemical properties. *Australian Journal of Soil Research* 37, 509–526.

Nykvist, N. (1997) Uptake of nutrients in a plantation of *Acacia mangium* in relation to decrease in soil amounts. *Journal of Sustainable Forestry* 4, 131–139.

Nykvist, N., Grip, H., Sim, B.L., Malmer, A. and Wong, F.K. (1994) Nutrient losses in forest plantations in Sabah, Malaysia. *Ambio* 23, 210–215.

O'Connell, A.M. and Sankaran, K.V. (1997) Organic matter accretion, decomposition and mineralisation. In: Nambiar, E.K.S. and Brown, A.G. (eds) *Management of Soil, Nutrients and Water in Tropical Plantation Forests.* ACIAR, Canberra, pp. 443–480.

Ogunkunle, A.O. and Eghaghara, O.O. (1992) Influence of land use on soil properties in a forest region of Southern Nigeria. *Soil Use and Management* 8, 121–125.

Okoro, S.P.A., Aighewi, I.T. and Osagie, C.O. (2000) Effect of selected monoculture plantation species on the humid tropical soils of southern Nigeria. *Indian Journal of Agricultural Sciences* 70, 105–109.

Oren, R., Ellsworth, D., Johnsen, K., Phillips, N., Ewers, B., Maier, C., Schäfer, K., McCarthy, H., Hendrey, G., McNulty, S. and Katul, G. (2001) Soil fertility limits carbon sequestration by forest ecosystems in a CO_2-enriched atmosphere. *Nature* 411, 469–472.

Parfitt, R.L., Salt, G.J. and Hill, L.F. (2002) Clear-cutting reduces nitrate leaching in a pine plantation of high natural N status. *Forest Ecology and Management* 170, 43–53.

Parrotta, J.A. (1992) The role of plantation forests in rehabilitating degraded tropical ecosystems. *Agriculture, Ecosystems and Environment* 41, 115–133.

Parrotta, J.A. (1999) Productivity, nutrient cycling, and succession in single- and mixed-species plantations of *Casuarina equisetifolia, Eucalyptus robusta,* and *Leucaena leucocephala* in Puerto Rico. *Forest Ecology and Management* 124, 45–77.

Piatek, K.B. and Allen, H.L. (1999) Nitrogen mineralization in a pine plantation fifteen years after harvesting and site preparation. *Soil Science Society of America Journal* 63, 990–998.

Pushparajah, E. (1998) The oil palm – a very environmental friendly crop. *The Planter (Kuala Lumpur)* 74, 63–72.

Ranger, J. and Turpault, M.P. (1999) Input–output nutrient budgets as a diagnostic tool for sustainable forest management. *Forest Ecology and Management* 122, 139–154.

Resh, S.C., Binkley, D. and Parrotta, J.A. (2002) Greater soil carbon sequestration under nitrogen-fixing trees compared with *Eucalyptus* species. *Ecosystems* 5, 217–231.

Ross, D.J., Tate, K.R., Scott, N.A. and Feltham, C.W. (1999) Land-use change, effects on soil carbon, nitrogen and phosphorus pools and fluxes in three adjacent ecosystems. *Soil Biology and Biochemistry* 31, 803–813.

Sanchez, P.A. (1995) Science in agroforestry. *Agroforestry Systems* 30, 5–55.

Sanchez, P.A., Palm, C.A., Davey, C.B., Szott, L.T. and Russel, C.E. (1985) Tree crops as soil improvers in the humid tropics? In: Cannell, M.G.R. and Jackson, J.E. (eds) *Attributes of Trees as Crop Plants*. Institute of Terrestrial Ecology, Huntingdon, pp. 79–124.

Sanger, L.J., Anderson, J.M., Little, D. and Bolger, T. (1997) Phenolic and carbohydrate signatures of organic matter in soils developed under grass and forest plantations following changes in land use. *European Journal of Soil Science* 48, 311–317.

Sawyer, J. (1993) *Plantations in the Tropics. Environmental concerns.* IUCN/UNEP/WWF, Gland, Switzerland.

Schlesinger, W.H. (1999) Carbon and agriculture – carbon sequestration in soils. *Science* 284, 2095.

Schlesinger, W.H. and Lichter, J. (2001) Limited carbon storage in soil and litter of experimental plots under increased atmospheric CO_2. *Nature* 411, 466–469.

Schnitzer, M. (2000) A lifetime perspective on the chemistry of soil organic matter. *Advances in Agronomy* 68, 1–58.

Sedjo, R.A. (2001) The role of forest plantations in the world's future timber supply. *Forestry Chronicle* 77, 221–225.

Sedjo, R.A. and Botkin, D. (1997) Using forest plantations to spare natural forests. *Environment* 39, 14–20.

Smith, C.K., Gholz, H.L. and Oliveira, F.D. (1998) Soil nitrogen dynamics and plant-induced soil changes under plantations and primary forest in lowland Amazonia, Brazil. *Plant and Soil* 200, 193–204.

Torn, M.S., Trumbore, S.E., Chadwick, O.A., Vitousek, P.M. and Hendricks, D.M. (1997) Mineral control of soil organic carbon storage and turnover. *Nature* 389, 170–173.

Trouve, C., Disnar, J.R., Mariotti, A. and Guillet, B. (1996) Changes in the amount and distribution of neutral monosaccharides of savanna soils after plantation of pinus and eucalyptus in the Congo. *European Journal of Soil Science* 47, 51–59.

Turner, J. and Lambert, M. (2000) Change in organic carbon in forest plantation soils in eastern Australia. *Forest Ecology and Management* 133, 231–247.

van Noordwijk, M., Cerri, C., Woomer, P.L., Nugroho, K. and Bernoux, M. (1997) Soil carbon dynamics in the humid tropical forest zone. *Geoderma* 79, 187–225.

Wang, X.J., Smethurst, P.J. and Holz, G.K. (1996) Nitrogen mineralisation indices in ferrosols under eucalypt plantations of north-western Tasmania – association with previous land use. *Australian Journal of Soil Research* 34, 925–935.

Waterloo, M.J. (1994) Water and nutrient dynamics of *Pinus caribaea* plantation forests on former grassland soils in Southwestern Viti Levu, Fiji. PhD dissertation, Free University, Amsterdam, The Netherlands.

Webster, C.C. and Watson, G.A. (1988) Tree crops of the humid tropics. *TAA Newsletter* 8, 11–19.

Whitmore, T.C. (1985) *Tropical Rain Forests of the Far East.* Oxford University Press, Oxford.

Wiersum, K.F. (1997) Indigenous exploitation and management of tropical forest resources: an evolutionary continuum in forest–people interactions. *Agriculture Ecosystems and Environment* 63, 1–16.

Wright, J.A., DiNicola, A. and Gaitan, E. (2000) Latin American forest plantations: opportunities for carbon sequestration, economic development, and financial returns. *Journal of Forestry* 98, 20–23.

Yeates, G.W., Hawke, M.F. and Rijkse, W.C. (2000) Changes in soil fauna and soil conditions under *Pinus radiata* agroforestry regimes during a 25-year tree rotation. *Biology and Fertility of Soils* 31, 391–406.

Young, A. (1997) *Agroforestry for Soil Management,* 2nd edn. CAB International, Wallingford, UK.

Sugarcane Plantations 8

> There is a growing feeling within the sugar industry that the scientific achievements of the past fifty years are not reflected in a production increase commensurate with the improvements made.
>
> N.J. King, R.W. Mungomery and C.G. Hughes (1953)

Sugar is an important commodity and more than 60% of the world production comes from sugarcane (interspecific hybrids of *Saccharum officinarum*). Throughout the tropics, smallholders and subsistence farmers cultivate sugarcane but most of the sugar traded at the world market comes from large-scale plantations. Sugarcane is grown as a ratoon crop, which means the whole above-ground biomass is harvested each year. Harvests may continue for a number of years (ratoons) but yields decline with ratooning, and after some years the land is ploughed and new sugarcane is planted. Sugarcane is highly productive but demanding crop and this chapter focuses on soil chemical changes that occur under sugarcane cropping in the tropics. A detailed case study of a sugarcane plantation in Papua New Guinea is presented in chapter 9.

Soil Erosion Under Sugarcane

Commercial sugarcane is grown on a wide range of landforms and soils. Many of the larger and long-established industries have used deep soils derived from fertile volcanic materials; examples include large sections of Hawaii, Cuba, Brazil (São Paulo) and Mauritius. Other long-established industries have traditionally selected fertile alluvial soils as in the Caribbean and in East Java (Blackburn, 1984). Irrigation is often required for high-yielding sugarcane.

Sugarcane cultivation is prone to accelerated soil erosion losses. Some soils under sugarcane are heavy textured, for example Vertisols, which are erodible owing to their extremely low water infiltration rates after wetting (Ahmad, 1996). When planted, after harvesting, or with excessive furrow irrigation, soil erosion may occur even if the land is nearly flat. In other soils, compaction may have occurred because of the use of heavy machinery when the soils were wet (Trouse and Humbert, 1961; Georges *et al.*, 1985; Braunack *et al.*, 1993; McGarry *et al.*, 1996). Compaction may be accompanied by surface sealing, reducing rainwater infiltration and increasing the likelihood of runoff and erosion.

Soil physical changes and in particular compaction under sugarcane have been fairly well documented, but there is limited data on soil erosion under sugarcane. On most sugarcane plantations erosion measures are taken (drains, bunds, ridges, strip cropping, etc.) and on heavy clay soils strip tillage has proven successful to control erosion (de Boer, 1997). There are some examples from the literature in which soil erosion under sugarcane was high.

A soil erosion study was conducted in the sugarcane areas of Australia, where the industry is largely confined to the narrow, high rainfall coastal zone from near Grafton in northern New South Wales, to Mossman in north Queensland (Johnson *et al.*, 1997). In north Queensland, soil erosion was monitored at seven sites with slopes ranging from 5% to 18% (Prove *et al.*, 1995). Soils were classified as Oxisols and annual rainfall was about 3331 mm. The study showed that soil erosion losses from conventionally cultivated ratoons was in the range of 47 to 505 Mg/ha/year with an average soil loss of 148 Mg/ha/year. Although the variation was considerable and largely explained by the variation in the rainfall, the average erosion rate is very high. No-tillage practices reduced the rates of erosion to less than 15 Mg/ha/year, and the effect of no-tillage on erosion reduction was larger than the effect of a groundcover from trash harvesting. Analyses of *in situ* and eroded soil indicated that sediment from the no-tillage practice may be transported further from the erosion site and carry a more mobile fraction of nutrients (Prove *et al.*, 1995). Although in many areas in north Queensland soil erosion is a problem, the precise rate and impact of sediment delivery to estuarine and marine environment remains poorly understood (Johnson *et al.*, 1997).

In Louisiana (USA) soil erosion losses under sugarcane were much lower and on average about 17 Mg/ha (Bengtson *et al.*, 1998). Rainfall at the site was lower than in the Australian study and ranged from 1300 to 1600 mm/year. Annual N losses by erosion were 18 kg/ha, P losses were 14 kg/ha and about 104 kg K/ha was lost by erosion. The amounts of nutrients lost depended on the application rates of the inorganic fertilizers.

Type I and II Data

A considerable portion of the world sugarcane is grown with a high degree of mechanization and heavy tractors and machinery are used for tillage and harvesting. Also large amounts of biomass are annually removed with the harvest. As a consequence, soil changes are likely to occur under sugarcane cultivation and studies have been conducted in various parts of the world to investigate these changes (Table 8.1). The table shows that a considerable number of studies have focused on soil chemical and physical changes and there were only few studies that looked into soil biological changes. Here only studies on soil chemical changes are reviewed.

Type I data

A Type I data study was conducted in Fiji, where soil chemical properties were monitored over a 6-year period on a Haplic Acrustox (Masilaca *et al.*, 1985). The soils were under native vegetation before the planting of the sugarcane. In one-third of the Oxisols the pH had declined from 5.5 to 4.6 but in other soils the pH decline was much lower (Table 8.2). Soil acidification was a result of the applications of N fertilizer (sulphate of ammonia), which were on average 150 kg N/ha/year. A decrease in exchangeable K from 2.6 to 1.8 $mmol_c$/kg was found, but P levels were increased in two of the three topsoils. There were also increases in the bulk density of the soil, which was negatively correlated with the soil organic C contents (Masilaca *et al.*, 1985). The increase in bulk density affects the nutrient availability to the sugarcane crop and this was discussed in Chapter 4 ('Effects of Bulk Density' on page 120).

Schroeder *et al.* (1994) measured soil pH over a 5-year period on South African sugarcane farms, but soil classification was not given. On soils derived from sedimentary rocks a pH decline of 0.4 unit was found, and these soils had received about 140 kg N/ha/year. Soil acidification based on Type II data was also reported from sugarcane areas in Hawaii (Humbert, 1959), Puerto Rico (Abruña-Rodriguez and Vicente-Chandler, 1967) and Florida (Coale, 1993) but these studies contained no information on years of cultivation nor what soils were studied.

Table 8.1. Summary of studies focusing on changes in soil chemical, physical and biological properties under sugarcane cultivation in the tropics.

		Soil property investigated				
Soil order	Country	Chemical	Physical	Biological	Data[a]	Reference[b]
Alfisols	Australia	√			Type II	(1) (2)
	Brazil	√	√		Type II	(3)
	India	√			TE	(4)
	Swaziland	√	√	√	Type II	(10)
Fluvents	Australia		√		TE	(5)
	Australia	√			Type II	(1) (2)
	Hawaii		√		TE	(6) (7)
	India		√		TE	(9)
	Papua N. Guinea	√	√		Type I and II	Chapter 9
Inceptisols	Australia	√			Type II	(1) (2)
Oxisols	Brazil	√	√		Type II	(3)
	Fiji	√	√		Type I	(9)
	Hawaii		√		TE	(6) (7)
	Swaziland	√	√	√	Type II	(10)
Spodosols	Australia		√	√	Type II	(11) (12)
Ultisols	Indonesia	√			Type I	(13)
Vertisols	Papua N. Guinea	√	√		Type I and II	Chapter 9
Not specified/ various	Australia	√	√	√	Type II	(14) to (19)
	India		√		Type II	(20)
	India	√			Type I	(21)
	Philippines	√			Type I	(22)
	South Africa		√		TE	(23)
	South Africa	√			Type I	(24)
	Trinidad		√		TE	(25)

[a] This column describes what type of data were collected: Type I are data where soil dynamics are followed with time on the same site; Type II are data where different land uses were sampled simultaneously; TE, treatment–effect study (see pages 86–89).
[b] References: (1) Bramley *et al.* (1996); (2) Skjemstad *et al.* (1999); (3) Caron *et al.* (1996); (4) Sundara and Subramanian (1990); (5) Braunack *et al.* (1993); (6) Trouse and Humbert (1961); (7) Juang and Uehara (1971); (8) Rao and Narasimham (1988); (9) Masilaca *et al.* (1985); (10) Henry and Ellis (1995); (11) McGarry *et al.* (1996); (12) McGarry *et al.* (1997); (13) Sitompul *et al.* (2000); (14) King *et al.* (1953); (15) Moody and Aitken (1995); (16) Moody and Aitken (1997); (17) Maclean (1975); (18) Wood (1985); (19) Garside *et al.* (1997); (20) Srivastava (1984); (21) Yadav and Singh (1986); (22) Alaban *et al.* (1990); (23) Swinford and Boevey (1984); (24) Schroeder *et al.* (1994); (25) Georges *et al.* (1985).

Soil acidification was also reported for the VMC milling district in the Philippines (Alaban *et al.*, 1990). Soil pH had declined from 5.0 to 4.7 over a 19-year period under sugarcane. The decline in pH was accompanied by a significant decrease in organic C from 13.3 to 9.9

Table 8.2. Changes in soil chemical properties at sugarcane plantations in Fiji. Soils were Oxisols and had been under sugarcane for 6 years. Type I data, modified from Masilaca *et al.* (1985).

Site	Sampling depth (m)	pH	C (g/kg)	N (g/kg)	P (mg/kg)	CEC and exchangeable cations ($mmol_c$/kg) CEC	Ca	Mg	K
A	0–0.12	−0.7	−26.9	−2.1	+62.0	−96.0	−19.9	−1.3	−1.8
	0.3–0.4	−0.8	+3.8	−0.2	+3.0	+0.3	+2.4	−0.1	−0.1
	0.7–0.8	−0.6	−0.6	0	−1.0	−18.0	+0.2	−0.2	−0.2
B	0–0.12	−0.3	−14.2	−2.2	−2.0	−38.0	−26.9	−10.6	−1.1
	0.3–0.4	+0.1	+1.3	+0.1	+1.0	+5.0	−3.3	−1.4	−0.2
	0.7–0.8	−0.1	+0.5	−0.2	+2.0	+7.0	−2.9	−2.2	0
C	0–0.12	+0.2	−17.0	−0.3	+64.0	−3.0	−29.6	+1.8	+0.5
	0.3–0.4	+0.1	+7.8	+0.3	+6.0	+35.0	−1.4	−0.4	−0.9
	0.7–0.8	+0.1	−0.2	0	−4.0	+28.0	0	−0.3	−0.2

g/kg. Also available P and levels of exchangeable cations decreased (Table 8.3). The decrease in these soil fertility parameters was accompanied by a decrease in sugarcane yield (Alaban *et al.*, 1990).

Trash harvesting is currently being practised by sugarcane farmers on the island of Negros, the Philippines, to increase soil organic C under sugarcane cultivation (Mulkins, 2000). The harvesting is done manually, and 3 months before harvest dead leaves are usually removed from the cane stalk (detrashed) and left to decompose on the soil. Although not much evidence has been collected, it seems that yields are increasing by this harvesting method (Mulkins, 2000).

Type II data – Australia

In north Queensland there is widespread evidence that sugarcane yield has been declining and higher yields are usually obtained on soils that have not been cultivated before (Wood, 1985; Lawes *et al.*, 2000). In the

Table 8.3. Changes in soil chemical properties on sugarcane plantations in the Philippines. Type I data, modified from Alaban *et al.* (1990).

Sampling period	pH	Organic C (g/kg)	Available P (mg/kg)	Exchangeable cations ($mmol_c$/kg) Ca	Mg	K
1969–1970	5.0	13.3	27.3	85.7	11.6	3.7
1988–1989	4.7	9.9	17.3	47.4	11.1	3.4

past decades, a number of Type II data studies have been conducted to investigate the effects of changing soil properties on sugarcane yields. In the Australian literature, Type II data are termed samples from 'paired sites' or 'paired sampling' (Wood, 1985), sampling 'old and new soils' (Bramley *et al.*, 1996), comparing 'cropped and undeveloped' land (Moody and Aitken, 1995), or comparing 'virgin and cultivated' soils (King *et al.*, 1953).

Bramley *et al.* (1996) sampled different sites in the main sugar-producing regions in north Queensland. The soils had been cultivated with sugarcane for approximately 20 years or more. Soil orders were Inceptisols (Dystropepts, Ustropepts, Tropaquepts), Alfisols (Natrustalfs, Haplustalfs) and an Entisol (Fluvents). There was no consistent effect of time under sugar monoculture on soil chemical properties across all sites, when either the distributions of properties through the soil profile or property values at specific depths were considered (Bramley *et al.*, 1996). In other words: soil fertility decline differed between soil orders and depths. Organic C declined only in the Fluvents and no significant changes were found in the other soil orders. A significant decline in soil reaction was only found in the Ustropepts. No clear relationship between soil changes and soil orders was found.

Skjemstad *et al.* (1999) looked at the same sites as Bramley *et al.* (1996) and found that very little change occurred in total soil organic C and in the light fraction (<1.6 Mg/m^3). Old, well-established sugarcane sites (20–70 years) had soils with lower soil organic C levels in the subsoils relative to virgin soils. No difference was found between Ustropepts, Natrustalfs and Fluvents and it appeared that sugarcane production did not lead to an overall decline in total organic C in the soil profile (Skjemstad *et al.*, 1999) confirming the observations of Bramley *et al.* (1996).

Type II data – Brazil and Swaziland

In the sugarcane areas of São Paulo, Brazil, Caron *et al.* (1996) sampled a Typic Haplorthox and Typic Paleudalf under primary forest and 20-year-old sugarcane. Topsoil organic C levels were 34 g/kg in the Alfisol under forest and 16 g C/kg under sugarcane. In the Oxisols under forest, there was 45 g C/kg compared with 30 g C/kg in the soils under sugarcane. The difference in organic C levels between forest and sugarcane extended to 1.2 m in the Oxisol and up to 0.9 m in the Alfisol. The decrease in soil organic C was accompanied by a significant decrease in soil pH in both soil orders, but the drop in pH was larger in Alfisols (Caron *et al.*, 1996).

Henry and Ellis (1995) investigated soil changes under sugarcane in northeast Swaziland. Two contrasting soil orders were compared: Oxisols derived from basalt and which had been under sugar for 18 years, and 'duplex' soils (Natraqualfs) derived from sandstone and these soils were poorly drained and had sodic properties. The Alfisols had been under paddy rice for 25 years and had been under sugarcane for 15 years when the soil samples were taken. In Oxisols, the difference in organic C between sugarcane and uncultivated soils was only 2 g C/kg. The difference in exchangeable K between cropped and uncultivated Oxisols was 5.3 $mmol_c$/kg but levels of available P were much higher in the soils under sugarcane. A similar picture emerged from the Alfisols, with a strong decrease in soil organic C and total N, and an increase in available P. Exchangeable K in soils under sugarcane was about half the values found in uncultivated soils in both Oxisols and Alfisols. Changes in soil chemical properties were accompanied by a degradation of soil physical and biological properties (Henry and Ellis, 1995).

A summary of Type II data studies is given in Table 8.4. Not all studies could be included as in some publications the data were presented in graphs and not in tables (e.g. Bramley *et al.*, 1996).

Type II data – soil organic matter dynamics

Cerri and Andreux (1990) measured different C fractions under forest and at a sugarcane plantation 180 km east of São Paulo, Brazil. Soils were classified as Typic Haplorthox. The natural abundance of the isotope ^{13}C was used to identify organic C sources in the soil and to determine the changes in soil organic matter when forest is cleared and sugarcane is planted. The approach is based on the difference in the natural ^{13}C abundance that exists between plants having different photosynthetic pathways, mainly C3 (Calvin cycle: forest) and C4 (Hatch–Slake cycle: sugarcane). The $^{13}C/^{12}C$ ratio of C3 plants is lower than in C4 plants (Cerri and Andreux, 1990).

Table 8.5 presents the variations in C content in soils under forest and sugarcane. Total C levels after 50 years of sugarcane cultivation were 46% of the levels under forest. After 12 years of sugarcane cultivation more than 80% of the soil organic C still originated from the forest but after 50 years the forest C formed 55% of the total C contents in the topsoil. The rate of increase in C originating from sugarcane was slower than the decrease in C that had originated from the forest.

The data of Table 8.5 were used in a regression model for soil organic matter dynamics (Fig. 8.1). The decline in forest-derived organic matter continued during the 50 years spanned by the investigation, and the apparent equilibrium value of total soil organic C is

Table 8.4. Changes in soil chemical properties on sugarcane plantations in the tropics.[a] Type II data.

Soil order	Period[b] (years)	Sampling depth (m)	pH	C (g/kg)	N (g/kg)	P (mg/kg)	CEC and exchangeable cations ($mmol_c$/kg)			
							CEC	Ca	Mg	K
Alfisols	14	0–0.15	–0.9	–0.8	–0.4	+19	–11.0	nd	nd	–3.4
Alfisols	14	0.15–0.30	–3.3	–0.4	–0.4	+10	–20.0	nd	nd	–4.0
Alfisols	14	0.45–0.60	–5.7	+0.3	–0.5	+2	–19.0	nd	nd	–2.6
Alfisols	20	0–0.05	nd	–3.7	nd	nd	nd	nd	nd	nd
Alfisols	20	0.50–0.60	nd	+0.6	nd	nd	nd	nd	nd	nd
Alfisols	20	0–0.15	–1.5	–18.0	nd	nd	nd	nd	nd	nd
Alfisols	20	0.15–0.50	–0.5	–7.0	nd	nd	nd	nd	nd	nd
Fluvents	90	0–0.05	nd	+1.1	nd	nd	nd	nd	nd	nd
Fluvents	90	0.50–0.60	nd	+1.5	nd	nd	nd	nd	nd	nd
Inceptisols	35	0–0.05	nd	–6.6	nd	nd	nd	nd	nd	nd
Inceptisols	35	0.50–0.60	nd	+9.1	nd	nd	nd	nd	nd	nd
Oxisols	17	0–0.15	–0.5	–2.0	–0.1	+4	–3.0	nd	nd	–5.3
Oxisols	17	0.15–0.30	–0.2	+0.4	–0.1	+6	+1.0	nd	nd	–5.7
Oxisols	17	0.45–0.60	+0.4	–0.4	–0.1	+1	–0.4	nd	nd	–5.7
Oxisols	20	0–0.20	–0.3	–15.0	nd	nd	nd	nd	nd	nd
Oxisols	20	0.20–0.40	0.8	–11.0	nd	nd	nd	nd	nd	nd

[a] Calculated from the references listed in Table 8.1.
[b] Period during which sugarcane was cultivated as compared to the uncultivated site.
nd, No data.

Table 8.5. Carbon content of soils under forest and after 12 and 50 years of sugarcane cultivation (Mg/ha, 0–0.20 m depth). Modified from Cerri and Andreux (1990). Type II data.

		Sugarcane	
	Forest	Soils under 12 years of sugarcane	Soils under 50 years of sugarcane
Total C	71.9	44.6	38.5
Stable C originating from the forest	71.9	36.0	21.0
C originating from the sugarcane		8.6	17.3

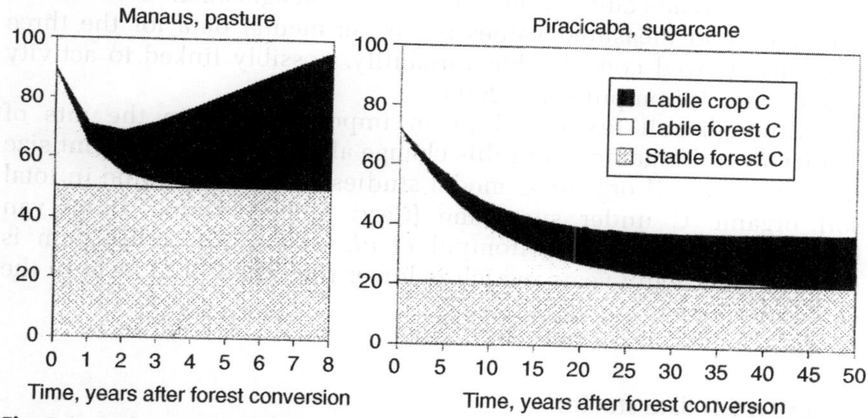

Fig. 8.1. Soil organic matter dynamics in Oxisols after forest conversion to pastures and sugarcane. Graphs based on fitted regression models. From van Noordwijk *et al.* (1997) based on work by C.C. Cerri in Brazil.

based on a balance between gradual build-up of sugarcane-derived organic matter and decay of forest-based organic matter. For comparison, soil data from pastures showed a larger stable C pool, a more rapid decline of labile forest C but also a much faster accumulation of labile crop C, which returned the total soil organic C levels to that of the forest before deforestation after about 7 years (van Noordwijk *et al.*, 1997).

Some of the differences between the pasture and sugarcane patterns can be explained by the lower annual input of C under sugarcane (0.96 Mg C/ha) compared with the pasture (7.5 Mg C/ha) and differences in soil mineralogy and climate (Cerri and Andreux, 1990). Soil texture plays a role and 12 years after conversion from forest to sugarcane, the majority of the C derived from sugarcane is in the coarse sand fraction. About 90% of the C in the clay fraction still has the forest signature after 12 years, whereas after 50 years

70% of the forest-derived C persisted in the clay fraction (Vitorello *et al.*, 1989). These data illustrate the importance of clay–organic matter linkages as a C protection mechanism (Hassink, 1992; van Noordwijk *et al.*, 1997).

On a Grossarenic Kandiudult in Sumatra, Indonesia, Sitompul *et al.* (2000) modelled soil organic matter dynamics under sugarcane. A submodel of the CENTURY model was used (Parton *et al.*, 1989), and soil organic matter was divided into particle size fractions. Rates of change differed between the fractions but the sum of light, intermediate and heavy fractions of macro organic matter (150 μm–2 mm size) showed a decline of about 250 g C/m^2 to about 100 g C/m^2 after 10 years of sugarcane cultivation. There was good agreement between the simulated and measured values but experimental data for the three fractions showed considerable variability, possibly linked to activity of soil fauna (Sitompul *et al.*, 2000).

Although soil texture plays an important role in the rate of change in soil organic C and this change also differs for different size fractions, the soil organic C model studies showed a decline in total soil organic C under sugarcane (Cerri and Andreux, 1990; van Noordwijk *et al.*, 1997; Sitompul *et al.*, 2000). An equilibrium is reached after many years, which is lower than the initial level in the soil under forest.

Other Type II data studies

There have been various studies in Australia in which no information was given on the type of soils that were sampled, or samples from different soil types were mixed in order to assess changes in soil properties of a whole sugarcane district (e.g. Maclean, 1975; Wood, 1985).

King *et al.* (1953) was one of the first to compare soil chemical properties of uncultivated soils with those that had been under sugarcane for 22 years in the Bundaberg area. The cultivated soils contained 22 g C/kg whereas organic C levels in the virgin soil were 48 g C/kg. Total N contents of the soils under sugarcane were also less than half of the N contents in virgin soils. Maclean (1975) measured soil chemical properties in the Goondi Mill area in north Queensland. In total 18 samples were aggregated from various soil types. A small (0.2 pH units) but significant difference was found in topsoil pH between sugarcane and uncultivated land (Table 8.6). Also differences in topsoil P, Ca and Mg levels were statistically significant. In the subsoil (0.15–0.30 m) available P and exchangeable Mg were significantly lower, but in the deeper subsoil (>0.30 m) there was no significant difference between soils under sugarcane and uncultivated soils (Maclean, 1975).

Table 8.6. Soil chemical properties under sugarcane and in adjacent uncultivated soils. Mean data of various soil types. Type II data, modified from Maclean (1975).

Land use	Sampling depth (m)	pH	P (mg/kg)	Exchangeable cations ($mmol_c$/kg)		
				Ca	Mg	K
Sugarcane	0–0.15	4.6	30.5	5.9	2.2	1.3
	0.15–0.30	4.7	13.8	6.0	2.0	0.7
	0.30–0.45	5.0	7.9	7.2	2.1	0.6
	0.45–0.60	5.1	8.7	7.7	2.4	0.5
Uncultivated	0–0.15	4.8	14.2	11.9	5.6	1.4
	0.15–0.30	4.9	6.8	7.7	3.2	0.8
	0.30–0.45	5.1	7.1	7.0	2.6	0.6
	0.45–0.60	5.1	7.1	6.3	2.3	0.5

The study by Maclean (1975) inspired Wood (1985) to investigate the causes of yield decline in the Herbert Valley near Ingham in north Queensland. Wood (1985) sampled cultivated and adjacent uncultivated land at 19 sites, and the sites covered a range of different soil types. The cultivated sites had been cropped with sugarcane for at least 30 years whereas the uncultivated sites were road reserves, cleared land or forest areas.

A slightly lower pH was found under sugarcane and only the difference in soil reaction in the 0.20–0.30 m soil horizon was significant (Table 8.7). Organic C levels in soils under sugarcane were less than half of the levels in uncultivated soils. Exchangeable cations and the CEC were also significantly lower in soils under sugarcane. Soils under sugarcane had significantly higher levels of available P owing to

Table 8.7. Changes in soil chemical properties on sugarcane plantations in north Queensland, Australia. Average data of various soil types. Sugarcane was cultivated for at least 30 years. Type II data, modified from Wood (1985).

Land use	Sampling depth (m)	pH	C (g/kg)	P (mg/kg)	CEC and exchangeable cations ($mmol_c$/kg)			
					CEC	Ca	Mg	K
Sugarcane	0–0.10	5.0	7.0	35	37.0	15.2	7.3	2.0
	0.10–0.20	4.9	6.5	26	37.0	15.5	5.1	1.4
	0.20–0.30	4.9	5.6	15	39.0	17.1	5.6	1.1
	0.30–0.40	5.0	4.0	9	41.3	18.7	8.1	1.0
Uncultivated	0–0.10	5.2	15.0	14	56.3	32.8	14.1	2.9
	0.10–0.20	5.2	8.1	8	47.5	26.1	12.3	1.6
	0.20–0.30	5.1	5.9	7	46.8	23.1	12.4	1.3
	0.30–0.40	5.1	4.9	3	51.7	25.0	15.3	1.3

high application rates of P fertilizers. It was concluded that many soil management practices such as excessive cultivation, insufficient fallowing, the burning of crop residues and the application of large amounts of N fertilizers are believed to be partially responsible for the decline in sugarcane yield (Wood, 1985).

The aforementioned studies in Australia had considerable impact on the way sugarcane was cultivated and several practices evolved to improve soil conditions and to avoid the negative impact of sugarcane cultivation. Less tillage was being practised and preharvest burning (Fig. 8.2) was replaced by trash harvesting in most sugarcane districts (Dick and Hurney, 1986). Preharvest burning was only possible when there was dry weather (Wood, 1991) and it may cause about 30% of the annual N removal in a sugar crop (Valdivia, 1982).

There were also several measures taken to improve the resource management following major concerns about the downstream effects of sugarcane cultivation (Bunn *et al.*, 1997; Keating and Wilson, 1997). In addition, the question of whether sugarcane is a source or sink of greenhouse gases has received attention in the Australian sugarcane industry (Weier, 1998).

Type II data – yield effects

Quite a number of studies on sugarcane plantations have been initiated to investigate the effects of soil changes on sugarcane yield, but few studies include a detailed analysis of this relationship. Most studies present

Fig. 8.2. Preharvesting burning of sugarcane in north Queensland, Australia. This was common practice up to the 1980s, when trash-harvesting was introduced.

soil chemical data and then subsequently speculate about the relation between the observed changes and the yield pattern. Clearly, the relation is complex and there are many factors involved that may explain trends and patterns in yield as well as those in soil chemical properties. A simple but useful way of starting to investigate such relationships is to present yield data and soil data from the same sugarcane block, but that combination is rarely presented. The only example that comes somewhat near is that from Muchovej *et al.* (2000), who measured soil chemical properties in 'good' and 'poor' spots in a sugarcane field in Florida, USA. In the study area, sugarcane fields often exhibited areas of localized reduced growth and poor development, which can comprise up to 25% of a field. Dominant soil orders in the area were Spodosols and the results of the investigation are presented in Table 8.8.

Soil pH, organic C and exchangeable cations were significantly higher for the areas of good sugarcane growth. No significant differences were detected for the micronutrients analysed (data not shown here). Nutrients, organic C and microbial populations were reduced with increasing depth, but pH was unaffected. Although moisture appeared to be an important factor in the areas of reduced growth, a lower or higher water-table was not consistently associated with low-yielding areas in the field. Not surprisingly, the authors concluded that additional data would be needed to obtain a more accurate diagnosis for the areas of localized reduced sugarcane growth (Muchovej *et al.*, 2000), but the data in Table 8.8 suggest that differences in soil chemical properties may be an important explanation for the differences in sugarcane growth.

Table 8.8. Soil chemical properties from Spodosols under 'good' and 'poor' sugarcane growth at two sites in Florida, USA. Modified from Muchovej *et al.* (2000).

Sampling depth (m)	Soil chemical property	Site I		Site II	
		Good	Poor	Good	Poor
0–0.15	pH	6.4	5.8	6.8	5.9
	Organic C (g/kg)	7.3	4.6	5.8	4.5
	Available P (mg/kg)	200	166	38	23
	Exchangeable Ca ($mmol_c$/kg)	162.7	90.7	61.9	23.5
	Exchangeable Mg ($mmol_c$/kg)	10.2	10.2	1.8	1.6
	Exchangeable K ($mmol_c$/kg)	3.1	2.0	2.0	1.4
0.15–0.30	pH	6.7	6.3	6.6	6.3
	Organic C (g/kg)	5.3	2.9	3.5	2.7
	Available P (mg/kg)	55	54	40.0	27.9
	Exchangeable Ca ($mmol_c$/kg)	35.9	21.9	30.8	11.9
	Exchangeable Mg ($mmol_c$/kg	3.2	3.4	1.5	0.5
	Exchangeable K ($mmol_c$/kg)	1.5	1.0	2.3	0.8

Rates of Change in Soil Chemical Properties

From the information presented in the previous section the rates of change in units per year, and in percentage per year were calculated (Tables 8.9 and 8.10). A decrease was found in the pH of most soils and for most depths. Soil organic C declined in all topsoils but in many studies a slight C increase in the subsoils was found. Total N decreased in most soils but available P levels increased, possibly because of inorganic fertilizer applications. Exchangeable cations, particularly K, decreased in most topsoils and subsoils.

A relatively large decrease was found in the pH of Alfisols and in some of the Oxisols (Table 8.10). Soil organic C decrease was relatively large in the topsoils of the Oxisols (5–7% per year) but organic C increases in the subsoil were also large (up to 13% per year). In the topsoils of Oxisols, annual N losses of up to 10% were recorded, whereas P increases ranged from 5% to 15% per year. The decrease in exchangeable K ranged from 3% to 10% per year.

Semiquantitative Studies

In the early 1980s, a 'Work Group on Sugarcane' summarized the problems with nutrient balances in sugarcane as follows: a general N balance for this crop is difficult to construct because of widely differing agronomic practices and growing conditions and also a lack of knowledge concerning certain processes (Ruschel and Vose, 1982). Agronomic complications include a growing period that ranges from 10 to 22 months; yields that range from 50 to 120 Mg/ha; burning at harvest that may disturb soil organic N accumulation by destroying potential litter, but that is not always practised; and different recycling practices for postharvest products such as filter cake. Nevertheless, it was concluded that sugarcane is a crop that has been widely studied and it was found possible to make some generalizations about the N balance (Ruschel *et al.*, 1982). The group made a partial N balance for the sugarcane in some Latin American and Caribbean countries (Table 8.11). Denitrification and losses by erosion were not considered in these balances because the data were not available (Ruschel *et al.*, 1982).

A wide range of values were found for the sugarcane cropping systems in the different countries. Inorganic fertilizer applications ranged from 60 to 200 kg N/ha/year. Fertilizer recovery under sugarcane in Latin America was about 50%, which is higher than in Mauritius, where a fertilizer N recovery of 20–40% was found (Ng Kee Kwong and Deville, 1994). Recovery fractions in Australia were 20–40% (Vallis *et al.*, 1996a) or 50% (Weier, 1994). If a 50% fertilizer

Table 8.9. Absolute change (unit/year) in soil chemical properties (topsoils only) on sugarcane plantations. Type I and II data.[a]

Soil type	Data type	Period[b] (years)	Depth (m)	pH	C (g/kg)	N (g/kg)	P (mg/kg)	CEC and cations ($mmol_c$/kg)			
								CEC	Ca	Mg	K
Alfisols	II	14	0–0.15	−0.1	−0.1	<−0.1	+1.4	−0.8	nd	nd	−0.2
Alfisols	II	14	0.15–0.30	−0.2	−0.0	<−0.1	+0.7	−1.4	nd	nd	−0.3
Alfisols	II	14	0.45–0.60	−0.4	<+0.1	<−0.1	+0.1	−1.4	nd	nd	−0.2
Alfisols	II	20	0–0.05	nd	−0.1	nd	nd	nd	nd	nd	nd
Alfisols	II	20	0.50–0.60	nd	<+0.1	nd	nd	nd	nd	nd	nd
Alfisols	II	20	0–0.15	−0.1	−0.9	nd	nd	nd	nd	nd	nd
Alfisols	II	20	0.15–0.50	<−0.1	−0.4	nd	nd	nd	nd	nd	nd
Fluvents	II	90	0–0.05	nd	<+0.1	nd	nd	nd	nd	nd	nd
Fluvents	II	90	0.50–0.60	nd	<+0.1	nd	nd	nd	nd	nd	nd
Inceptisols	II	35	0–0.05	nd	−0.2	nd	nd	nd	nd	nd	nd
Inceptisols	II	35	0.50–0.60	nd	+0.3	nd	nd	nd	nd	nd	nd
Oxisols	I	6	0–0.12	−0.1	−4.5	−0.4	+10.3	−16.0	−3.3	−0.2	−0.3
Oxisols	I	6	0.3–0.4	−0.1	+0.6	<+0.1	+0.5	+0.1	+0.4	<+0.1	<−0.1
Oxisols	I	6	0.7–0.8	−0.1	−0.1	<+0.1	−0.2	−3.0	<+0.1	<+0.1	<+0.1
Oxisols	I	6	0–0.12	−0.1	−2.4	−0.4	−0.3	−6.3	−4.5	−1.8	−0.2
Oxisols	I	6	0.3–0.4	<+0.1	+0.2	<+0.1	+0.2	+0.8	−0.6	−0.2	<−0.1
Oxisols	I	6	0.7–0.8	<−0.1	+0.1	<−0.1	+0.3	+1.2	−0.5	−0.4	0
Oxisols	I	6	0–0.12	<+0.1	−2.8	−0.1	+10.7	−0.5	−4.9	+0.3	+0.1
Oxisols	I	6	0.3–0.4	<+0.1	+1.3	+0.1	+1.0	+5.8	−0.2	−0.1	−0.2
Oxisols	I	6	0.7–0.8	<+0.1	<−0.1	<+0.1	−0.7	+4.7	<+0.1	−0.1	<−0.1

Continued

Table 8.9. *Continued.*

Soil type	Data type	Period[b] (years)	Depth (m)	pH	C (g/kg)	N (g/kg)	P (mg/kg)	CEC and cations ($mmol_c$/kg)			
								CEC	Ca	Mg	K
Oxisols	II	17	0–0.15	<–0.1	–0.1	<–0.1	+0.2	–0.2	nd	nd	–0.3
Oxisols	II	17	0.15–0.30	<–0.1	<+0.1	<–0.1	+0.4	+0.1	nd	nd	–0.3
Oxisols	II	17	0.45–0.60	<+0.1	<–0.1	<–0.1	+0.1	–0.0	nd	nd	–0.3
Oxisols	II	20	0–0.20	<–0.1	–0.8	nd	nd	nd	nd	nd	nd
Oxisols	II	20	0.20–0.40	<+0.1	–0.6	nd	nd	nd	nd	nd	nd

[a] Calculated from the references listed in Table 8.1.
[b] Period of cultivation (Type I); years difference between cropped and uncultivated soils (Type II).
nd, No data.

Table 8.10. Relative change (% per year) in soil chemical properties (topsoils only) on sugarcane plantations. Type I and II data[a].

Soil type	Data type	Period[b] (years)	Depth (m)	pH	C (g/kg)	N (g/kg)	P (mg/kg)	CEC and cations ($mmol_c/kg$) CEC	Ca	Mg	K
Alfisols	II	14	0–0.15	<–0.5	–1	–3	+23	–1	nd	nd	–4
Alfisols	II	14	0.15–0.30	–2	<–0.5	–3	+12	–1	nd	nd	–4
Alfisols	II	14	0.45–0.60	–4	<+0.5	–4	+4	–1	nd	nd	–4
Alfisols	II	20	0–0.05	nd	–1	nd	nd	nd	nd	nd	nd
Alfisols	II	20	0.50–0.60	nd	+1	nd	nd	nd	nd	nd	nd
Alfisols	II	20	0–0.15	–1	–3	nd	nd	nd	nd	nd	nd
Alfisols	II	20	0.15–0.50	<–0.5	–2	nd	nd	nd	nd	nd	nd
Fluvents	II	90	0–0.05	nd	<+0.5	nd	nd	nd	nd	nd	nd
Fluvents	II	90	0.50–0.60	nd	+1	nd	nd	nd	nd	nd	nd
Inceptisols	II	35	0–0.05	nd	–1	nd	nd	nd	nd	nd	nd
Inceptisols	II	35	0.50–0.60	nd	+5	nd	nd	nd	nd	nd	nd
Oxisols	I	6	0–0.12	–2	–7	–9	+86	–6	–11	–7	–10
Oxisols	I	6	0.3–0.4	–2	+9	–3	+3	+1	+25	–3	–3
Oxisols	I	6	0.7–0.8	–2	–3	<+0.5	–1	–6	+2	–8	+11
Oxisols	I	6	0–0.12	–1	–5	–10	–3	–4	–13	–14	–9
Oxisols	I	6	0.3–0.4	<+0.5	+1	+2	+2	+1	–8	–11	–6
Oxisols	I	6	0.7–0.8	<–0.5	+1	–4	+7	+2	–11	–15	0
Oxisols	I	6	0–0.12	+1	–7	–3	+152	<–0.5	–11	+8	+3
Oxisols	I	6	0.3–0.4	<+0.5	+13	+6	+11	+8	–2	–4	–12
Oxisols	I	6	0.7–0.8	<+0.5	<–0.5	<+0.5	–13	+8	<+0.5	–6	–8

Continued

Table 8.10. *Continued.*

Soil type	Data type	Period[b] (years)	Depth (m)	pH	C (g/kg)	N (g/kg)	P (mg/kg)	CEC and cations ($mmol_c$/kg)			
								CEC	Ca	Mg	K
Oxisols	II	17	0–0.15	<–0.5	–1	–1	+5	<–0.5	nd	nd	–3
Oxisols	II	17	0.15–0.30	<–0.5	<+0.5	–1	+12	+1	nd	nd	–4
Oxisols	II	17	0.45–0.60	<+0.5	<–0.5	–1	+2	<–0.5	nd	nd	–5
Oxisols	II	20	0–0.20	<–0.5	–2	nd	nd	nd	nd	nd	nd
Oxisols	II	20	0.20–0.40	+1	–2	nd	nd	nd	nd	nd	nd

[a] Calculated from the references listed in Table 8.1.

[b] Period of cultivation (Type I), years difference between cropped and uncultivated soils (Type II).

nd, No data.

Table 8.11. Partial N balance (kg/ha/year) for sugarcane cropping systems in Latin America and the Caribbean. Modified from Ruschel *et al.* (1982).

	Brazil	Ecuador	Dominican Republic	Peru	Trinidad	Range of values
Inputs						
N fixation	15–25	nd	nd	nd	nd	nd
Inorganic fertilizer	60–100[b]	150	200	200	80	60–200
Manure[a]	5	5–10	5–10	5–10	5–10	5–10
Deposition	5	nd	5–10	nd	nd	5–10
Total	100	nd	nd	282	nd	nd
Outputs						
Harvest	50–60	50–60	50–60	150	50–60	50–150
NH_3-volatilization	nd	nd	nd	nd	nd	nd
Leaching	nd	nd	nd	20	nd	nd
Burning	nd	nd	nd	45	nd	30
Forage	nd	nd	nd	20	nd	nd
Total	100	100	100	230	100	100–230
Within system						
Fertilizer recovery (%)	50	50	50	70	50	50–70

[a] Filter cake mud except for Brazil.
[b] Higher values are for ratoon crops.
nd, No data.

N recovery is assumed, it implies that the N balances presented in Table 8.11 are negative in most countries. In the Peruvian case, total N input, after correcting for the fertilizer recovery, is 182 kg N/ha whereas total output was estimated to be 230 kg N/ha. Losses of N by burning were estimated to be 45 kg N/ha and in another study in Peru it was shown that burning losses could account for 30% of the N output in a sugarcane cropping system (Valdivia, 1982). Irrigated sugarcane at the Peruvian coast may be expected to require 300 kg N/ha for economically optimum yields (Valdivia, 1982).

In Coimbatore, India, a partial nutrient balance for sugarcane was calculated by Sundara and Subramanian (1990). The sugarcane was grown on Alfisols and NPK fertilizers were applied annually. The data for the plant cane and first ratoon (2 years) are given in Table 8.12. Much more N and slightly more P were applied with the inorganic fertilizers than removed with the yield. The balance for K is negative and the difference between K applied and removed was 137 kg/ha. If these values are compared with the changes in the topsoil nutrient content, it appears that the soil NPK content declined with an average rate of −13 kg N, −1.5 kg P and −90 kg K/ha/year. So despite the 'positive balance' of N and P, the soil levels of N and P declined. This is because not all outputs were measured. Much of the applied N may have been

Table 8.12. NPK balance and soil changes in a sugarcane field at Coimbatore, India. Calculated from Sundara and Subramanian (1990).

	Balance (kg/ha/year)			Soil changes (0–0.20 m) (kg/ha)		
	Applied with inorganic fertiilzer	Removed at harvest	Difference	Level at the beginning	Content after 2 years	Difference
N	225	107	+118	182	157	–25
P	33	29	+4	35	32	–3
K	100	237	–137	521	341	–180

lost. In other sugarcane-producing countries fertilizer N recovery was found to range from 20% to 50% of the applied N. If a 50% recovery is assumed the balance becomes negative and is more comparable to the net negative changes in the topsoil N content. It shows again that partial nutrient balances should be interpreted with caution.

No study has been conducted quantifying all nutrient inputs and outputs in sugarcane cultivation systems. Although the removal of nutrients with sugarcane is well quantified (de Geus, 1973; Srivastava, 1992; Malavolta, 1994), other outputs and inputs of the nutrient balance have not been studied in sufficient detail, with the exception of biological nitrogen fixation.

Input by biological N fixation (BNF)

In the late 1950s, it was discovered that N_2-fixing bacteria of the genus *Beijerinckia* were present in the rhizosphere of sugarcane (Döbereiner, 1961). Most of the work on BNF in the rhizosphere of sugarcane has been conducted in Brazil following the seminal work of J. Döbereiner. Evidence for substantial inputs of N via N_2 fixation associated with sugarcane, which has been provided by isotope dilution measurements, is consistent with observations made in the field (Chalk, 1991). In Brazil, a conservative estimate of BNF holds that about 17% of total plant N is fixed by the sugarcane, which is equivalent to 17 kg/ha at yields of 70 Mg/ha (Ruschel and Vose, 1982). More recent results of ^{15}N dilution/N balance studies showed that some sugarcane varieties can obtain large contributions of plant-associated BNF, ranging from 60% to 80% of total plant N, equivalent to over 200 kg N/ha/year (Boddey *et al.*, 1991). Cultivar differences are large (Urquiaga *et al.*, 1992) but N fixation is high for most Brazilian cultivars as they have been systematically bred for high yields with low N inputs (Boddey *et al.*, 1995).

Although 100–200 kg N/ha is removed with the harvest of the sugarcane, it is assumed that the N levels in the soil are usually maintained despite continuous cropping with low inputs of inorganic N fertilizer N (Lima *et al.*, 1987). However, in many soils under sugarcane total N levels had decreased (see Tables 8.2, 8.4 and 8.9). Ecosystems with high rates of N fixation often have high loss rates through leaching or possibly denitrification. The precise relationship is not fully understood but is related to trade-off in energetic limitations when switching from uptake of atmospheric N to soil mineral N (Pastor and Binkley, 1998). So the benefits of high N fixation in sugarcane may only be recorded in soils low in mineral N, when no inorganic N fertilizers are applied and when the soil P status is adequate.

Soil-process-oriented Studies

There have been many studies investigating the effects of inorganic fertilizer on sugarcane yield, sugar and leaf nutrient content, and the overall response to inorganic fertilizers (e.g. Stanford and Ayres, 1964; Fox *et al.*, 1967, 1969; Qayyum and Rauf, 1972; Gascho and Snyder, 1976; Yadav and Sharma, 1980; Bowen, 1981; Thomas, 1984; Abayomi, 1987; Medina-Gonzales *et al.*, 1988; Ghosh *et al.*, 1990; Wood, 1990; de Armas *et al.*, 1992; Singh *et al.*, 1995; Aumont and Salas, 1996; Ranjith and Meinzer, 1997). The Diagnosis and Recommendation Integrated System (DRIS), which was originally designed for rubber by E.R. Beaufils, has also been adopted for sugarcane in Florida and South Africa (Sumner and Beaufils, 1975; El Wali and Gascho, 1984).

Also the effects of lime have been fairly well documented, which is important because sugarcane cultivation is prone to acidify the soil when ammonia fertilizers are being used (Ham, 1983; Choudry and Corpuz, 1984; Meyer *et al.*, 1989; Meyer *et al.*, 1991; Turner *et al.*, 1992; Coale, 1993; Coale and Schueneman, 1993; Schroeder *et al.*, 1993; Kingston *et al.*, 1996; Schroeder and Aitken, 1998; Noble and Hurney, 2000). The effects of organic amendments have received less attention in sugarcane cropping systems but several studies are available (e.g. Prasad, 1976a,b; Ng Kee Kwong *et al.*, 1987; Ng Kee Kwong and Deville, 1988; Rodella *et al.*, 1990; Orlando Filho *et al.*, 1991; Sutton *et al.*, 1996; Braunbeck *et al.*, 1999). In recent years growth models have been developed for sugarcane – particularly for sugarcane in Australia (e.g. Wood *et al.*, 1996; Higgins *et al.*, 1998; Keating *et al.*, 1999; Cheeroo-Nayamuth *et al.*, 2000; Lisson *et al.*, 2000)

There are excellent textbooks that summarize soil and fertilizer requirements of sugarcane (Humbert, 1968; de Geus, 1973; Blackburn, 1984; Hunsigi, 1993; Malavolta, 1994). All this literature correctly gives the impression that much is known about sugarcane and its nutrition. Soil-process-oriented studies, in which the fate of nutrients is followed, quantified, explained or better understood, have also been conducted in sugarcane. Most of the soil-process-oriented work has focused on N, possibly because sugarcane is a large N consumer (Malavolta, 1994) and because tropical soils are frequently deficient or low in N (Sanchez, 1976). Less attention is given to K as sugarcane is often grown on alluvial soils in which the K status may be sufficient for sugarcane despite its heavy K requirements and removal (de Geus, 1973). Soil-process-oriented research on P has been minimal, possibly as sugarcane has a low P requirement (Malavolta, 1994).

Leaching, denitrification and inorganic fertilizers

Comprehensive N work has been conducted on the Sugar Industry Research Institute in Reduit, Mauritius. Sugarcane is an important crop in Mauritius and covers more than 90% of the arable land (Ng Kee Kwong *et al.*, 1999). A lysimeter study was initiated in which ^{15}N-labelled N was given as $(NH_4)_2SO_2$ or as $NaNO_3$ at the rate of 100 kg N/ha (Ng Kee Kwong and Deville, 1984). Soils were Ustic Eutropepts (annual rainfall 1550 mm) and Dystropeptic Gibbsiorthox (3700 mm rain per year). The amount of N leached depended more on the duration and intensity of the drying step preceding the rainfall than on the leachate volume. Leaching was greater at the drier site (Inceptisols), but cumulative N loss over 1 year was similar for both soils. This was because frequent, shorter and less intense drying and wetting cycles were as effective in mobilizing soil N than less frequent but longer and more intense drying at the Inceptisols site. More N was leached from soils with higher organic matter content. The Oxisols were able to retain NO_3 by absorption, which reduced N leaching, but K and Ca were more readily leached than N. It was concluded that losses of cations might be more acute than the need for measures to minimize N leaching in both Oxisols and Inceptisols (Ng Kee Kwong and Deville, 1984).

In a subsequent study, Ng Kee Kwong and Deville (1987) measured residual fertilizer N as influenced by timing and N form. The experiment was conducted at an Ustic Eutropepts (62% clay, neutral pH, 31 g C/kg) with an annual rainfall of about 800 mm. Again, 100 kg N/ha was applied as ^{15}N-labelled $(NH_4)_2SO_2$ or $NaNO_3$. Vertical and lateral distribution of residual fertilizer N was not influenced by the type of fertilizer or by the time of application. More than 50% of the residual N remained in the soil surface and less than 30% of fertilizer N had moved laterally more than 0.30 m away from the zone of fertilization. The study provided further evidence that N leaching is not of significant concern in soils under sugarcane in tropical environments similar to that of Mauritius (Ng Kee Kwong and Deville, 1987).

Many studies have shown that an appreciable fraction of fertilizer N invariably remains unaccounted for (Allison, 1966). It is generally assumed that denitrification and volatilization of NH_3 are the major components of this unaccounted-for N. Fertilizer N uptake and N-use efficiency formed the base for another experiment by Ng Kee Kwong and Deville (1994). The pattern of N uptake and dry matter synthesis was studied using ^{15}N-labelled $(NH_4)_2SO_2$ at a rate of 100 kg N/ha. It was found that approximately 10–20 kg/ha of the ^{15}N-labelled N was lost from the green tops (i.e. the aerial parts of the sugarcane). These were gaseous N losses from the plant itself, and fertilizer N-use efficiency derived from recovered N at harvest grossly underestimates the ability of sugarcane to use fertilizer N. The study showed that the uptake of fertilizer N by sugarcane varied from 20% to 40% whereas it was 13% to 18% when the measurements were made at the harvest. So denitrification and volatilization were grossly overestimated because losses of N from the aerial parts constitute a significant proportion of the unaccounted-for N of the sugarcane in Mauritius (Ng Kee Kwong and Deville, 1994).

Various soil-process-oriented studies have been conducted in the sugarcane areas of Australia. Most of these studies focused on the fate and losses of N following N fertilizer applications on trash-harvested fields. Such losses could be high under relatively dry conditions when urea is applied. The essence of the problem is that sugarcane, like all plants, contains the enzyme urease and the urease in trash breaks down urea into CO_2 and NH_3. The trash blanket cannot bind this NH_3 (Ralph, 1992). Volatilization of NH_3 was controlled by the availability of water in the trash and its evaporation. Small additions of water to the trash from dew, light rain and condensation maintained a slow but steady pattern of NH_3 loss, which varied from 32% to 39% of the applied N. In heavy rainfall areas, urea was washed from the trash and losses were 17% of the applied N. Ammonia losses from $(NH_4)_2SO_2$ were less than 2% of the applied N (Freney *et al.*, 1992). Wei-Ping *et al.* (1993) used micrometeorological and microplot experiments in the freshly harvested fields to assess ammonia loss of fertilizer N. Similar results were obtained: high losses under relatively dry conditions. The losses were 20–30% of the applied N within 30 days after applications. The study also showed that leaching of N was absent (Wei-Ping *et al.*, 1993).

Chapman *et al.* (1994) investigated the efficiency of fertilizer N uptake using urea, labelled with ^{15}N, which was either broadcast or buried in different trash-management systems. The proportion of the applied fertilizer N recovered was 33% when the urea was buried and 18% when the urea was broadcast. Again, leaching did not appear to be a significant loss process but it was suspected that denitrification accounted for the majority of the fertilizer N loss (Chapman *et al.*, 1994).

Weier *et al.* (1996) studied the potential for biological denitrification of fertilizer N in soils under sugarcane. Field studies were conducted on Alfisols and Ultisols in northern New South Wales and southern Queensland. Denitrification ranged from 1% to 20% of the applied N and differences between soil orders were considerable. In a glasshouse study, denitrification losses ranged from 13% to 39% of the N applied and the majority of the gaseous N loss occurred as N_2. It was concluded that denitrification is an important cause of fertilizer N loss from fine-textured soils, with N_2O the gaseous N product when soil NO_3 concentrations are high (Weier *et al.*, 1996).

It is interesting that in the Australian research, gaseous N losses from the plant itself were not taken into account, while this was found to be important for the sugarcane in Mauritius (Ng Kee Kwong and Deville, 1994). Fertilizer N recovery, however, was about the same (20–40%) for the sugarcane in both countries. Lower recovery values were found for N applied on the sugarcane grown on Vertisols in Guadeloupe, which is due to the higher rates of volatilization on these soils (Courtaillac *et al.*, 1998).

A study on Grossarenic Paleudults in Florida (USA) showed that leaching losses varied from 6% to 24% of applied N depending on the fertilizer type and irrigation level (El Wali *et al.*, 1980). Leaching of the applied N was mainly in the NO_3 form but when irrigation took place before the N hydrolysed from urea was completely nitrified, leaching in the NH_4 form was considerable. Losses with sulphur-coated urea were lowest, and increased with irrigation. Amounts of N loss ranged from 17% to 24% of the applied N. Some 11–15% of the applied N was not accounted for by the plant, leachate or soil (El Wali *et al.*, 1980). A study on Vertic Haplaquepts in Louisiana (USA) showed that N losses by leaching could be substantial (Southwick *et al.*, 1995). Average NO_3 leaching ranged from 15% to 60% depending on the leaching period and season.

Environmental Impact

Commercial sugarcane is usually grown with high levels of inorganic fertilizers and pesticide inputs, with herbicides representing about 50% of all pesticides applied to sugarcane in many countries (Lanchote *et al.*, 2000). Contamination of ground and surface waters is a major concern, particularly in areas with shallow water-tables. There have been some studies on the environmental impact of sugarcane cultivation and they have mostly focused on the off-site effects including deterioration of surface water and air quality. Nearly all environmental-impact studies have been conducted in the sugarcane areas of the USA and Australia. No studies are available on the effects of sugarcane cropping on wildlife and biodiversity.

Biocides and soil contamination

Southwick *et al.* (1992) measured leaching of the herbicide atrazine on Vertic Haplaquepts under sugarcane in Louisiana (annual rainfall 1500 mm). Maximum concentrations were found within 11 days after application and ranged from 82 to 403 mg/l. The lifetime health advisory limit for drinking water in the USA is 3 mg/l and this concentration was reached 20–30 days after the herbicide was applied (Southwick *et al.*, 1992). In a subsequent study, Southwick *et al.* (1995) measured leaching of atrazine and metribuzin, which are commonly used herbicides in sugarcane. Leaching of both herbicides was high directly after application, but gradually decreased after some weeks. Total losses ranged from 0.4% to 2.0% for atrazine and from 0.4% to 1.7% for metribuzin. Atrazine concentrations in the drainage water were again above the USA health advisory levels, whereas the lifetime health advisory limit for metribuzin was not exceeded (Southwick *et al.*, 1995). Losses of atrazine and metribuzin were also reported by Bengtson *et al.* (1998) and they showed that the method of application was the main factor determining the rate of herbicide loss.

Lanchote *et al.* (2000) measured residues of the herbicides atrazine, simazine and ametryne in surface and groundwater collected in a major sugarcane area near São Paulo, Brazil. Ten water-sampling points were selected in a watershed, of which nine were taken from surface water and one from groundwater. In total 250 samples were collected but ametryne residues were only detected in 17 samples. The concentrations were below those recommended as safe by international agencies of environmental control (Lanchote *et al.*, 2000).

In the Everglades of Florida (USA) an area of about 200,000 ha has been put under sugarcane (Muchovej *et al.*, 2000), and the cultivation takes place in a fragile environment with Histosols as the dominant soil order. In the past century, more than half of the wetlands have been drained (Schrope, 2001) and since 1900 many areas have lost 2.7–4.0 m of surface elevation due to subsidence upon cultivation. Policies that once favoured development are now being reversed by policies and regulating efforts to restore natural ecosystems (Anderson and Rosendahl, 1997). Following serious mercury contamination of freshwater fish in 1989, mercury emission from preharvest burning was investigated at nine locations in the Everglades (Patrick *et al.*, 1994). This contamination was unexpected since the area is non-industrial and very sparsely settled. The average mercury content of the Histosols was only 0.15 mg/kg and also Hg concentrations in the sugarcane stalks were low. The study concluded that direct emission of Hg from sugarcane fields during preharvesting burning was only a minor source (2%) of atmospheric Hg. That was possibly good news for the sugarcane farmers but left the question on the origin of the Hg contamination open. A much more pressing problem in the Everglades was that of the P eutrophication – part of which seemed to originate from inorganic P fertilizers (Anderson and Rosendahl, 1997). Overall, the environmental impact of sugarcane cultivation in the Everglades remains largely unquantified.

In the sugarcane areas of southern China, Chua *et al.* (1998) investigated the accumulation of rare earth elements (REE) by sugarcane, and the study focused on cerium (Ce). It was shown that REEs entered the sugarcane plants in three different ways: via the leaves exposed to atmospheric contaminants, via the roots in soils contaminated by REEs or applied with inorganic fertilizers (Rodriguez-Barrueco, 1996). Official limits to residual concentrations are not available, but high REE concentrations in the soils under sugarcane could lead to harmful effects for humans consuming sugarcane products (Chua *et al.*, 1998).

As mentioned, few studies have looked at the effects of sugarcane cultivation on soil biological properties. Holt and Mayer (1998) quantified microbial biomass in new and old sugarcane fields in Australia. They found significantly lower microbial biomass in soils that had been permanently cultivated with sugarcane. It could not be ascertained whether this decrease was due to the heavy use of agrochemicals (inorganic fertilizers, biocides).

Inorganic fertilizers and eutrophication

Nitrogen concentrations are often not considered to be an important criterion of surface-water quality because of denitrification and biological cycling in an open system (Anderson and Rosendahl, 1997). Nevertheless, some studies have looked at the enrichment of surface water owing to the use of inorganic N fertilizers in sugarcane. In forested wetlands of Louisiana (USA), surface-water quality is being reduced by nutrient input from adjacent agricultural production areas. A ^{15}N study was undertaken to assess the input of fertilizer N applied to sugarcane fields and to forested wetlands (Lindau *et al.*, 1997). The major soils were Vertic Haplaquepts and Aeric Fluvaquents – both were poorly drained. Runoff and surface-water samples were collected from sugarcane fields bordering wetlands over a 16-month period. Fertilizer N draining into adjacent forested wetlands was estimated to be only a small fraction of the amount applied and concentrations of NH_4 and NO_3 were low. About 3–4% of the applied N was removed by runoff. Even after anhydrous NH_3 application, no increase was observed in the NH_4 and NO_3 concentration. This was explained by the high clay contents of the soil and the injection of the anhydrous NH_3 at 0.10–0.15 m below the soil surface.

In order to reduce N losses on sugarcane plantations in Mauritius recent research has focused on the use of drip-fertigation (Ng Kee Kwong *et al.*, 1999). Fertilizer N could be reduced by 30% from 120 to 80 kg N/ha/year without a reduction in growth pattern or sugarcane yields. Obviously, this is applicable to irrigated sugarcane and has major benefits for the farmer and the environment, but investments in drip-fertigation are large and it may be neither economically nor technically feasible for other sugar-producing areas in the tropics.

Discussion and Conclusions

Sugarcane is a major cash crop on the world market. This chapter focused on soil changes on sugarcane plantations in the tropics. Despite the erodibility of many soils under sugarcane, there are very few studies with hard data on soil erosion losses. Erosion can be high after the harvest and with replanting, or when sugarcane is planted on sloping land (Blackburn, 1984). Sugarcane is more prone to soil erosion than other perennial (woody) crops because the periodic harvesting removes almost all vegetation from the field. The use of heavy machinery and the associated soil physical deterioration also increases the risk of soil erosion. Studies in Australia indicated that soil erosion losses can be high, but the fact that sugarcane covers the soil in most parts of the year reduces the risk. In Australia, there seems to be little evidence to support claims that sediment deposition resulting from sugarcane cultivation has had a major impact on the characteristics of the rivers and sugar catchments over the last 50–100 years (Johnson *et al.*, 1997). Likewise, there seems to be no evidence to support claims that sediment deposition following soil erosion under sugarcane is having adverse downstream effects. There is, however, concern about the effects of rising NO_3 levels in groundwater resulting from intensive agricultural cropping in relation to environmentally sensitive areas like the Great Barrier Reef (Thorburn *et al.*, 2003).

Soil chemical changes under sugarcane show that soil acidification is important in most soils under sugarcane. The major cause of soil acidification is the use of N fertilizers containing or producing NH_4^+. All ammoniacal N fertilizers release protons when NH_4^+ is oxidized to NO_3^- by nitrifying microorganisms. Mineralization of organic matter can contribute to soil acidity by the oxidation of N and S to HNO_3 and H_2SO_4 in a similar manner (Sumner, 1997). Since organic matter declined in most studies, it may have contributed to the increase in soil acidity. Acidity is largely a reversible process, particularly where liming readily restores productivity. If acidification has also taken place in the subsoil, amelioration is much more difficult, expensive and time-consuming (Sumner, 1997). Research has shown that there is only a small response of sugarcane to lime on moderately acid soils (Turner *et al.*, 1992) whereas in other studies a decrease in the sugar content was found after lime applications (Kingston *et al.*, 1996). Avoiding strong soil acidification might be a better option than the use of lime to correct for high-acidity inputs.

Soil organic C dynamics have received considerable attention in sugarcane cropping systems, but some conflicting reports have appeared in the literature. Part of the problem is that total soil organic C, as determined by the Walkley & Black or the dry combustion method, is not very sensitive to short-term changes in land use, and long-term observations are required to pick up statistically significant differences in soil organic C levels. It is also related to the spatial variability in total soil organic C, which is linked to variation in texture (Sitompul *et al.*, 2000). Total soil organic C decreased in most topsoils and in most soil types. In a number of soils, it was found that levels of soil organic C increased in the subsoil, which also seems to occur in the subsoils under sugarcane in Papua New Guinea (see Chapter 9). While most reports indicate a decline in soil organic C under long-term sugarcane cultivation, Bramley *et al.* (1996) found little or no difference in soil organic C between cropped and uncultivated land. This may be explained by the existence of trash retention in north Queensland, which has the potential to maintain or increase soil organic carbon contents (Wood, 1991; Vallis *et al.*, 1996b).

Several process-oriented studies have been conducted under sugarcane and most studies looked at the fate of nutrients from inorganic fertilizers. Sugarcane is grown either under rainfed conditions with high annual rainfall or under irrigated conditions, which may enhance leaching losses of N. Leaching losses of fertilizer N were low or absent in studies conducted in the sugarcane areas of Australia and Mauritius. Research in the USA showed that leaching losses (up to 60% of the applied N) were high. Gaseous losses are important in all sugarcane areas. There are indications that sugarcane may lose some of its N through the aerial parts of the plant. Fertilizer N recovery fractions were higher if this form of gaseous loss is taken into account. In most studies N applications are high and the recovery of fertilizer N ranges from 20% to 50%. The fact that N rarely accumulates in the soils and the likelihood that the remaining N is lost to the air or surface waters implies that sugarcane cultivation may have environmental effects.

Most environmental-impact studies in sugarcane areas have been conducted in the USA and Australia. The studies focused on the effects on water quality and little attention has been paid to the effects on soil quality. Sugarcane is cultivated with relatively high levels of agrochemical inputs (herbicides, pesticides, inorganic fertilizers, etc). The heavy use of agrochemicals in a fragile environment should be a reason for concern, but these inputs guarantee that high sugarcane productivity is achieved (Fig. 8.3). Precision agriculture has great potential in sugarcane monocropping systems as it may result in increased crop yield, savings in inorganic fertilizers and reduced potential for off-farm environmental damage (Wood *et al.* 1996).

Fig. 8.3. Application of herbicides (left) and sulphate of ammonia at the sugarcane plantation in Papua New Guinea: good for the crop and good for the environment?

References

Abayomi, A.Y. (1987) Growth, yield and crop quality performance of sugarcane cultivar Co 957 under different rates of application of nitrogen and potassium fertilizers. *Journal of Agricultural Science (Cambridge)* 109, 285–292.

Abruña-Rodriguez, F. and Vicente-Chandler, J. (1967) Sugarcane yields as related to acidity of a humid tropical Ultisol. *Agronomy Journal* 59, 330–332.

Ahmad, N. (1996) Management of Vertisols in rainfed conditions. In: Ahmad, N. and Mermut, A. (eds) *Vertisols and Technologies for their Management. Developments in Soil Science 24.* Elsevier, Amsterdam, pp. 363–428.

Alaban, R.A., Barredo, F.C. and Aguirre, A.L. (1990) An assessment of some indicators and determinants of farm productivity and soil fertility in the VMC district: trends, associations, interactions 1969–1990. *Proceedings of the Philippine Sugar Technologists* 64–83.

Allison, F.E. (1966) The fate of nitrogen applied to soils. *Advances in Agronomy* 18, 219–258.

Anderson, D.L. and Rosendahl, P. (1997) The development and application of environmental policy: the Everglades and Florida sugar industry. In: Keating, B.A. and Wilson, J.R. (eds) *Intensive Sugarcane Production – Meeting the Challenges Beyond 2000.* CAB International, Wallingford, UK, pp. 381–402.

Aumont, G. and Salas, M. (1996) Effect of stage of maturity, cultivar, nitrate fertilization, soil type on mineral content of sugarcane. *Journal of the Agricultural University of Puerto Rico* 80, 37–46.

Bengtson, R.L., Selim, H.M. and Ricaud, R. (1998) Water quality from sugarcane production on alluvial soils. *Transactions of the ASAE* 41, 1331–1336.

Blackburn, F. (1984) *Sugar-cane.* Longman, Harlow.

Boddey, R.M., Urquiaga, S., Reis, V. and Döbereiner, J. (1991) Biological nitrogen fixation associated with sugar cane. *Plant and Soil* 137, 111–117.

Boddey, R.M., Deoliveira, O.C., Urquiaga, S., Reis, V.M., Deolivares, F.L., Baldani, V.L.D. and Döbereiner, J. (1995) Biological nitrogen fixation associated with sugar cane and rice – contributions and prospects for improvement. *Plant and Soil* 174, 195–209.

Bowen, J.E. (1981) Micro-element nutrition of sugar-cane. 1 Effects of micro-elements on growth and yield. *Tropical Agriculture* 58, 13–21.

Bramley, R.G.V., Ellis, N., Nable, R.O. and Garside, A.L. (1996) Changes in soil chemical properties under long-term sugar cane monoculture and their possible role in sugar yield decline. *Australian Journal of Soil Research* 34, 967–984.

Braunack, M., Wood, A.W., Dick, R.G. and Glimour, J.M. (1993) The extent of soil compaction in sugarcane soils and a technique to minimise it. *Sugar Cane* 5, 12–18.

Braunbeck, O., Bauen, A., Rosillo-Calle, F. and Cortez, L. (1999) Prospects for green cane harvesting and cane residue use in Brazil. *Biomass and Bioenergy* 17, 495–506.

Bunn, S.E., Davies, P.M. and Kellaway, D.M. (1997) Contributions of sugar cane and invasive pasture grass to the aquatic food web of a tropical low-land stream. *Marine and Freshwater Research* 48, 173–179.

Caron, J., Espindola, C.R. and Angers, D.A. (1996) Soil structural stability during rapid wetting – influence of land use on some aggregate properties. *Soil Science Society of America Journal* 60, 901–908.

Cerri, C.C. and Andreux, F. (1990) Changes in organic carbon content in Oxisols cultivated with sugar cane and pasture, based on ^{13}C natural abundance measurement. In: *14th International Congress of Soil Science.* Kyoto, vol. IV, pp. 98–103.

Chalk, P.M. (1991) The contribution of associative and symbiotic nitrogen fixation to the nitrogen nutrition of non-legumes. *Plant and Soil* 132, 29–39.

Chapman, L.S., Haysom, M.B.C. and Saffigna, P.G. (1994) The recovery of ^{15}N from labelled urea fertilizer in crop components of sugarcane and in soil profiles. *Australian Journal of Agricultural Research* 45, 1577–1585.

Cheeroo-Nayamuth, F.C., Robertson, M.J., Wegener, M.K. and Nayamuth, A.R.H. (2000) Using a simulation model to assess potential and attainable sugar cane yield in Mauritius. *Field Crops Research* 66, 225–243.

Choudry, B.A. and Corpuz, I.T. (1984) Profitability of fertilizers and dolomite application to sugar cane in the Philippines. *Sugar y Azucar* 79, 29–44.

Chua, H., Zhao, Y.G., Kwang, Y.H., Liu, Q.Y. and Liu, X.H. (1998) Accumulation of environmental residues of rare earth elements in sugar-cane. *Environment International* 24, 287–291.

Coale, F.J. (1993) Sugarcane cultivar response to dolomite amendment of an acidic sandy soil. *Journal of American Society of Sugar Cane Technologists* 13, 94–97.

Coale, F.J. and Schueneman, T.J. (1993) Sugarcane response to limestone and gypsum application on acidic sandy soils. *Journal of American Society of Sugarcane Technologists* 13, 73–86.

Courtaillac, N., Baran, R., Oliver, R., Casabianca, H. and Ganry, E. (1998) Efficiency of nitrogen fertilizer in the sugarcane-vertical system in Guadeloupe according to growth and ratoon age of the cane. *Nutrient Cycling in Agroecosystems* 52, 9–17.

de Armas, R., Valadier, M.H., Champigny, M.L. and Lamaze, T. (1992) Influence of ammonium and nitrate on the growth and photosynthesis of sugarcane. *Journal of Plant Physiology* 140, 531–535.

de Boer, H.G. (1997) Strip tillage in sugar cane agriculture: techniques and challenges. *Zuckerindustrie* 5, 371–375.

de Geus, J.G. (1973) *Fertilizer Guide for the Tropics and Subtropics.* Centre d'Etude de l'Azote, Zürich.

Dick, R.G. and Hurney, A.P. (1986) Technology change in sugarcane agriculture. *Proceedings of Australian Society of Sugar Cane Technologists* 69–75.

Döbereiner, J. (1961) Nitrogen-fixing bacteria of the genus *Beijerinckia* Derx in the rhizosphere of sugar cane. *Plant and Soil* 15, 211–216.

El Wali, A.M.O. and Gascho, G.K. (1984) Soil testing, foliar analysis, and DRIS as guides for sugarcane fertilization. *Agronomy Journal* 76, 466–470.

El Wali, A.M.O., Le Grand, F. and Gascho, G.J. (1980) Nitrogen leaching from soil and uptake by sugarcane from various urea-based fertilizers. *Soil Science Society of America Journal* 44, 119–122.

Fox, R.L., Silva, J.A., Younge, O.R., Plucknett, D.L. and Sherman, G.D. (1967) Soil and plant silicon and silicate response by sugar cane. *Soil Science Society of America Proceedings* 31, 775–779.

Fox, R.L., Silva, J.A., Plucknett, D.L. and Teranishi, D.Y. (1969) Soluble and total silicon in sugar cane. *Plant and Soil* 30, 81–92.

Freney, J.R., Denmead, O.T., Wood, A.W., Saffigna, P.G., Chapman, L.S., Ham, G.J., Hurney, A.P. and Stewart, R.L. (1992) Factors controlling ammonia loss from trash covered sugarcane fields fertilized with urea. *Fertilizer Research* 31, 341–349.

Garside, A.L., Bramley, R.G.V., Bristow, K.L., Holt, J.A., Magarey, R.G., Nable, R.O., Pankhurst, C.E. and Skjemstad, J.O. (1997) Comparisons between paired old and new land sites for sugarcane growth and yield and soil chemical, physical, and biological properties. *Proceedings of Australian Society of Sugarcane Technologists*, 60–66.

Gascho, G.J. and Snyder, G.H. (1976) Sulfur-coated fertilizers for sugarcane: I. Plant response to sulfur-coated urea. *Agronomy Journal* 40, 119–122.

Georges, J.E.W., Mohamed, M.S. and Harvey, W.O. (1985) Effects on soil water content and compactive effort on soil compaction and sugar cane regrowth. *Proceedings West Indies Sugar Technologies* 635–653.

Ghosh, A.K., Rai, R.K., Saxena, Y.R. and Shrivastava, A.K. (1990) Effect of sulphur application on the nutritional status, yield and quality of sugar cane. *Journal of the Indian Society of Soil Science* 38, 73–76.

Ham, G. (1983) Responses to gypsum and lime in the Burdekin. *BSES Bulletin* 4, 16.

Hassink, J. (1992) Effects of soil texture and structure on carbon and nitrogen mineralization in grassland soils. *Biology and Fertility of Soils* 14, 126–134.

Henry, P.C. and Ellis, R.D. (1995) Soil as a factor in sugarcane ratoon yield decline on an irrigated estate in Swaziland. *Proceedings International Society of Sugarcane Technologists* 236–245.

Higgins, A.J., Muchow, R.C., Rudd, A.V. and Ford, A.W. (1998) Optimising harvest date in sugar production: a case study for the Mossman mill region in Australia – I. Development of operations research model and solution. *Field Crops Research* 57, 153–162.

Holt, J.A. and Mayer, R.J. (1998) Changes in microbial biomass and protease activities of soil associated with long-term sugar cane monoculture. *Biology and Fertility of Soils* 27, 127–131.

Humbert, R.P. (1959) Soil as a factor in varietal yield decline. *International Society of Sugarcane Technologists* 51–62.

Humbert, R.P. (1968) *The Growing of Sugar Cane*. Elsevier, Amsterdam.

Hunsigi, G. (1993) *Production of Sugarcane, Theory and Practice*. Springer-Verlag, Berlin.

Johnson, A.K.L., McDonald, G.T., Shrubsole, D.A. and Walker, D.H. (1997) Sharing the land – the sugar industry as part of the wider landscape. In: Keating, B.A. and Wilson, J.R. (eds) *Intensive Sugarcane Production – Meeting the Challenges Beyond 2000*. CAB International, Wallingford, UK, pp. 361–380.

Juang, T.C. and Uehara, G. (1971) Effects of ground-water table and soil compaction on nutrient element uptake and growth of sugarcane. *International Society of Sugarcane Technologists* 679–687.

Keating, B.A. and Wilson, J.R. (eds) (1997) *Intensive Sugarcane Production: Meeting the Challenges Beyond 2000*. CAB International, Wallingford, UK.

Keating, B.A., Robertson, M.J., Muchow, R.C. and Huth, N.I. (1999) Modelling sugarcane production systems I. Development and performance of the sugarcane module. *Field Crops Research* 61, 253–271.

King, N.J., Mungomery, R.W. and Hughes, C.G. (1953) *Manual of Cane Growing. Australian Agricultural and Livestock Series*. Angus and Robertson, Sydney.

Kingston, G., Hurney, A.P. and Aitken, R.L. (1996) Depression of CCS in sugarcane after use of liming products: economic significance and associations with soil chemistry and leaf nutrient composition. *Proceedings of Australian Society of Sugar Cane Technologists* 164–173.

Lanchote, V., Bonato, P.S., Cerdeira, A.L., Santos, N.A.G., de Carvallho, D. and Gomes, M.A. (2000) HPLC screening and GC-MS confirmation of triazine herbicides residues in drinking water from sugar cane area in Brazil. *Water, Air, and Soil Pollution* 118, 329–337.

Lawes, R.A., Wegener, M.K., Basford, K.E. and Lawn, R.J. (2000) Commercial cane sugar trends in the Tully sugar district. *Australian Journal of Experimental Agriculture* 40, 969–973.

Lima, E., Boddey, R.M. and Döbereiner, J. (1987) Quantification of biological nitrogen fixation associated with sugar cane using a ^{15}N aided nitrogen balance. *Soil Biology and Biochemistry* 19, 165–170.

Lindau, C.W., Delaune, R.D. and Alford, D.P. (1997) Monitoring nitrogen pollution from sugarcane runoff using ^{15}N analysis. *Water, Air, and Soil Pollution* 89, 389–399.

Lisson, S.N., Robertson, M.J., Keating, B.A. and Muchow, R.C. (2000) Modelling sugarcane production systems – II: analysis of system performance and methodology issues. *Field Crops Research* 68, 31–48.

Maclean, N.R. (1975) Long-term effects of sugar cane production on some physical and chemical properties of soils in the Goondi mill area. *Proceedings Queensland Society of Sugarcane Technologists* pp. 123–126.

Malavolta, E. (1994) *Fertilizing for High Yield Sugarcane*. International Potash Institute, Basel.

Masilaca, A.S., Prasad, R.A. and Morrison, R.J. (1985) The impact of sugarcane cultivation on three Oxisols from Vanua Levu, Fiji. *Tropical Agriculture* 63, 325–330.

McGarry, D., Bigwood, R.C., Pillay-McGarry, U.P., Bray, S.G. and Moody, P.W. (1996) Is cane damaging the soil? A comparison of a ten-year-old block with the adjoining treeline. In: Egan, B.T. (ed.) *Proceedings of the 1996 Conference of the Australian Society of Sugarcane Technologists*, Mackay, pp. 189–194.

McGarry, D., Braunack, M.V., Cunningham, G., Halpin, N., Sallaway, M. and Waters, D. (1997) Comparison of soil physical properties of row and inter-row: a basis for control traffic in cane. In: Egan, B.T. (ed.) *Proceedings of the 1997 Conference of the Australian Society of Sugar Cane Technologists*, Cairns, pp. 263–269.

Medina-Gonzales, O.A., Fox, R.L. and Bosshart, R.P. (1988) Solubility and availability to sugarcane (*Saccharum* spp.) of two silicate materials. *Fertilizer Research* 16, 3–13.

Meyer, E., Boevey, T.C. and Meyer, J.H. (1989) The incorporation of lime under minimum tillage conditions. *Proceedings of the South African Sugar Technologists' Association*, pp. 142–146.

Meyer, J.H., Turner, P.E.T. and Fey, M.V. (1991) Interim evaluation of phosphogypsum as an ameliorant for soil acidity in sugarcane. *Proceedings of the South African Sugar Technologists' Association*, pp. 41–46.

Moody, P.W. and Aitken, R.L. (1995) Soil acidification and lime use in some agricultural systems in Queensland, Australia. In: Date, R.A., Grundon, R.J., Rayment, G.E. and Probert, M.E. (eds) *Plant Soil Interactions at Low pH.* Kluwer Academic Publishers, Dordrecht, pp. 749–752.

Moody, P.W. and Aitken, R.L. (1997) Soil acidification under some tropical agricultural systems. 1. Rates of acidification and contributing factors. *Australian Journal of Soil Research* 35, 163–173.

Muchovej, R.M.C., Luo, Y., Shine, J.M. and Jones, J.C. (2000) Nutritional problems associated with low yields of sugarcane on mineral soils. *Soil and Crop Society of Florida Proceedings* 59, 146–150.

Mulkins, L. (2000) From sugarcane monoculture to agro-ecological village. *ILEIA Newsletter* 16, 22–23.

Ng Kee Kwong, K.F. and Deville, J. (1984) Nitrogen leaching from soils cropped with sugarcane under the humid tropical climate of Mauritius, Indian Ocean. *Journal of Environmental Quality* 13, 471–474.

Ng Kee Kwong, K.F. and Deville, J. (1987) Residual fertilizer nitrogen as influenced by timing and nitrogen forms in a silty clay soil under sugarcane in Mauritius. *Fertilizer Research* 14, 219–226.

Ng Kee Kwong, K.F. and Deville, J. (1988) Filter cake as a nitrogen fertilizer for plant cane in Mauritius. *Tropical Agriculture* 65, 364–366.

Ng Kee Kwong, K.F. and Deville, J. (1994) The course of fertilizer nitrogen uptake by rainfed sugarcane in Mauritius. *Journal of Agricultural Science* 122, 385–391.

Ng Kee Kwong, K.F., Deville, J., Cavalot, P.C. and Riviere, V. (1987) Value of cane trash in nitrogen nutrition of sugarcane. *Plant and Soil* 102, 79–83.

Ng Kee Kwong, K.F., Paul, J.P. and Deville, J. (1999) Drip-fertigation – a means for reducing fertilizer nitrogen to sugarcane. *Experimental Agriculture* 35, 31–37.

Noble, A.D. and Hurney, A.P. (2000) Long-term effects of lime additions on sugarcane yield and soil chemical properties in north Queensland. *Experimental Agriculture* 36, 397–413.

Orlando Filho, J., Silva, L.F.C. and Rodella, A.A. (1991) Effects of filter cake applications on sugarcane yields in Brazil. *Sugar Journal* 54, 22–24.

Parton, W.J., Sanford, R.L., Sanchez, P.A. and Stewart, W.B. (1989) Modeling soil organic matter dynamics in tropical soils. In: Coleman, D.C., Oades, J.M. and Uehara, G. (eds) *Dynamics of Soil Organic Matter in Tropical Ecosystems*. NiFtal Project, University of Hawaii, Hawaii, pp. 153–171.

Pastor, J. and Binkley, D. (1998) Nitrogen fixation and the mass balances of carbon and nitrogen in ecosystems. *Biogeochemistry* 43, 63–78.

Patrick, W.H., Gambrell, R.P., Parkpian, P. and Tan, F. (1994) Mercury in soils and plants in the Florida Everglades sugarcane area. In: Watras, C.J. and Huckabee, J.W. (eds) *Mercury Pollution – Integration and Synthesis*. Lewis Publishers, Boca Raton, Florida, pp. 609–614.

Prasad, M. (1976a) Response of sugar cane to filter press mud and N, P and K fertilizers. I. Effect on sugarcane yield and sucrose content. *Agronomy Journal* 68, 539–543.

Prasad, M. (1976b) Response of sugar cane to filter press mud and N, P and K fertilizers. II. Effect on plant composition and soil chemical properties. *Agronomy Journal* 68, 543–547.

Prove, B.G., Doogan, V.J. and Truong, P.N.V. (1995) Nature and magnitude of soil erosion in sugarcane land on the wet tropical coast of north-eastern Queensland. *Australian Journal of Experimental Agriculture* 35, 641–649.

Qayyum, F. and Rauf, A. (1972) Sugarcane yield and quality as affected by NPK. *Pakistan Journal of Scientific Research* 24, 210–213.

Ralph, W. (1992) Restricting nitrogen loss in cane-fields. *Rural Research* 157, 16–19.

Ranjith, S.A. and Meinzer, F.C. (1997) Physiological correlates of variation in nitrogen-use efficiency in two contrasting sugarcane cultivars. *Crop Science* 37, 818–825.

Rao, S.R. and Narasimham, R.L. (1988) Effect of soil compaction in sugarcane: part III. On nutrient indices and uptake. *International Journal of Tropical Agriculture* 6, 232–236.

Rodella, A.A., da Silva, L.C.F. and Filho, J.O. (1990) Effects of filter cake application on sugarcane yields. *Turrialba* 40, 323–326.

Rodriguez-Barrueco, C. (ed.) (1996) *Fertilizers and Environment: Developments in Plant and Soil Sciences*. Kluwer Academic Publishers, Dordrecht.

Ruschel, A.P. and Vose, P.B. (1982) Nitrogen cycling in sugarcane. *Plant and Soil* 67, 139–146.

Ruschel, A.P., Tamas, F.R., Ahmad, N., Mendoza-Palacios, H.H. and Valdivia-Liga, S. (1982) Report of the work group on sugarcane. *Plant and Soil* 67, 395–397.

Sanchez, P.A. (1976) *Properties and Management of Soils in the Tropics*. John Wiley & Sons, New York.

Schroeder, B.L. and Aitken, R.L. (1998) Effect of lime on green cane trash decomposition. In: *National Soil Acidification Conference*, Queensland, Australia.

Schroeder, B.L., Meyer, J.H., Wood, R.A. and Turner, P.E.T. (1993) Modifying lime requirements for sandy, to sandy clay loam soils in the Natal Midlands. *Proceedings of the South African Sugar Technologists' Association* 49–52.

Schroeder, B.L., Robinson, J.B., Wallace, M. and Turner, P.E.T. (1994) Soil acidification: occurrence and effects in the South African sugar industry. *Proceedings of the South African Sugar Technologists' Association* 70–74.

Schrope, M. (2001) Save our swamp. *Nature* 409, 128–130.

Singh, S.N., Singh, S.B., Singh, N.P. and Chaudhary, S.C. (1995) Studies on the effect of plant cane harvesting and time of nitrogen application on yield and quality of sugarcane ratoon. *Cooperative Sugar* 26, 359–361.

Sitompul, S.M., Hairiah, K., Cadisch, G. and van Noordwijk, M. (2000) Dynamics of density fractions of macro-organic matter after forest conversion to sugarcane and woodlots, accounted for in a modified Century model. *Netherlands Journal of Agricultural Science* 48, 61–73.

Skjemstad, J.O., Taylor, J.A., Janik, L.J. and Marvanek, S.P. (1999) Soil organic carbon dynamics under long-term sugarcane monoculture. *Australian Journal of Soil Research* 37, 151–164.

Southwick, L.M., Willis, G.H. and Salim, H.M. (1992) Leaching of Atrazine from sugarcane in Southern Louisiana. *Journal of Agricultural and Food Chemistry* 40, 1264–1268.

Southwick, L.M., Willis, G.H., Johnson, D.C. and Selim, H.M. (1995) Leaching of nitrate, atrazine and metribuzin from sugarcane in southern Louisiana. *Journal of Environmental Quality* 24, 684–690.

Srivastava, A.K. (1984) Soil compaction: a problem in sugarcane culture. *Indian Sugar* 34, 503–504.

Srivastava, S.C. (1992) Sugarcane. In: Halliday, D.J. and Trenkel, M.E. (eds) *IFA World Fertilizer Use Manual.* International Fertilizer Industry Association, Paris, pp. 257–266.

Stanford, G. and Ayres, A.S. (1964) The internal nitrogen requirement of sugarcane. *Soil Science* 98, 338–344.

Sumner, M.E. (1997) Opportunities for amelioration of soil physical and chemical constraints under intensive cropping. In: Keating, B.A. and Wilson, J.R. (eds) *Intensive Sugarcane Production – Meeting the Challenges Beyond 2000.* CAB International, Wallingford, UK, pp. 305–326.

Sumner, M.E. and Beaufils, E.R. (1975) Diagnosis of the NPK requirements of sugarcane irrespective of plant age and season using Beaufils system (DRIS) – preliminary observations. *Proceedings of the South African Sugar Technologists' Association*, pp. 137–141.

Sundara, B. and Subramanian, S. (1990) Changes in soil available NPK in multiple cropping systems based on short duration sugarcane crops relative to a conventional sugarcane cropping system. *Fertilizer Research* 24, 67–76.

Sutton, M.R., Wood, A.W. and Saffigna, P.G. (1996) Long term effects of green cane trash retention on Herbert river soils. In: Wilson, J.R., Hogarth, D.M., Campbell, J.A. and Garside, A.L. (eds) *Sugarcane: Research Towards Efficient and Sustainable Production.* CSIRO Division of Tropical Crops and Pastures, Brisbane, pp. 178–180.

Swinford, J.M. and Boevey, T.M.C. (1984) The effects of soil compaction due to infield transport on ratoon cane yields and soil physical characteristics. *Proceedings of the South African Sugar Technologists' Association*, pp. 198–203.

Thomas, J.R. (1984) Calibration of leaf N percentages for predicting the N fertilizer needs of sugarcane. *Sugar Journal* 47, 19–22.

Thorburn, P.J., Biggs, J.S., Weier, K.L. and Keating, B.A. (2003) Nitrate in groundwaters of intensive agricultural areas in coastal Northeastern Australia. *Agriculture Ecosystems and Environment* 94, 49–58.

Trouse, A.C. and Humbert, R.P. (1961) Some effects of soil compaction on the development of sugar cane roots. *Soil Science* 91, 208–217.

Turner, P.E.T., Meyer, J.H. and Wood, R.A. (1992) Some results of field experiments to test lime and phosphogypsum for aluminium amelioration. *Proceedings of the South African Sugar Technologists' Association* 44–49.

Urquiaga, S., Cruz, K.H.S. and Boddey, R.M. (1992) Contribution of nitrogen fixation to sugar cane: nitrogen-15 and nitrogen balance estimates. *Soil Science Society of America Journal* 56, 105–114.

Valdivia, S. (1982) Nitrogen gains and losses in sugarcane (*Saccharum* sp.) agro-ecosystems on the coast of Peru. *Plant and Soil* 67, 147–156.

Vallis, I., Catchpoole, V.R., Hughes, R.M., Myers, R.J.K., Ridge, D.R. and Weier, K.L. (1996a) Recovery in plants and soils to ^{15}N applied as surface bands of urea to sugarcane. *Australian Journal of Agricultural Research* 47, 355–370.

Vallis, I., Parton, W.J., Keating, B.A. and Wood, A.W. (1996b) Simulation of the effects of trash and N fertilizer management on soil organic matter levels and yields of sugarcane. *Soil and Tillage Research* 38, 115–132.

van Noordwijk, M., Cerri, C., Woomer, P.L., Nugroho, K. and Bernoux, M. (1997) Soil carbon dynamics in the humid tropical forest zone. *Geoderma* 79, 187–225.

Vitorello, V.A., Cerri, C.C., Andreux, F., Feller, C. and Victoria, R.L. (1989) Organic matter and natural carbon-13 distribution in forested and cultivated Oxisols. *Soil Science Society of America Journal* 53, 773–778.

Weier, K.L. (1994) Nitrogen use and losses in agriculture in subtropical Australia. *Fertilizer Research* 39, 245–257.

Weier, K.L. (1998) Sugarcane fields – sources or sinks for greenhouse gas emissions? *Australian Journal of Agricultural Research* 49, 1–9.

Weier, K.L., McEwan, C.W., Vallis, I., Catchpoole, V.R. and Myers, R.J.K. (1996) Potential for biological denitrification of fertilizer nitrogen in sugarcane soils. *Australian Journal of Agricultural Research* 47, 67–79.

Wei-Ping, K., Saffigna, P.G., Wood, A.W. and Freney, J.R. (1993) Evaluation of methods for control of ammonia volatilization from surface-applied nitrogen fertilizers to sugarcane trash in North Queensland. *Pedosphere* 3, 321–330.

Wood, A.W. (1985) Soil degradation and management under intensive sugarcane cultivation in North Queensland. *Soil Use and Management* 1, 120–124.

Wood, A.W. (1991) Management of crop residues following green harvesting of sugarcane in north Queensland. *Soil and Tillage Research* 20, 69–85.

Wood, A.W., Muchow, R.C. and Robertson, M.J. (1996) Growth of sugarcane under high input conditions in tropical Australia. 3. Accumulation, partitioning and use of nitrogen. *Field Crops Research* 48, 223–233.

Wood, R.A. (1990) The roles of nitrogen, phosphorus and potassium in the production of sugarcane in South Africa. *Fertilizer Research* 26, 89–98.
Yadav, R.L. and Sharma, R.K. (1980) Effect of nitrogen levels and harvesting dates on quality characters and yield of four sugarcane genotypes. *Indian Journal of Agricultural Science* 50, 581–589.
Yadav, R.L. and Singh, K. (1986) Long term experiments with sugarcane and intensive cropping systems and variation in soil fertility. *Indian Journal of Agronomy* 31, 322–325.

9

Case 1 – Sugarcane Plantation, Papua New Guinea

> Commercial sugarcane production would be a risky undertaking in Papua New Guinea because this is its centre of origin and therefore has a broad range of pests and diseases.
>
> J.J.H. Szent-Ivany and J.H. Ardley (1962)

In the late 1970s a sugarcane plantation was established in Papua New Guinea, where sugarcane is indigenous. The plantation is owned by the government of Papua New Guinea and managed by a British plantation company. The sugarcane is not irrigated and as the availability of manual labour is limited, the plantation was designed for relatively high rates of mechanization. In 1996 and 1997, studies were conducted to investigate whether soil chemical and physical properties had changed due to the cultivation of sugarcane. This case study focuses on the changes in soil chemical properties and the information is partly derived from earlier publications. The data presented in these publications have been re-evaluated and updated where appropriate. As was done in the previous sections, the soil data are presented based on their mode of collection: Type I data, Type II data and a nutrient balance.

Introduction

Plans for establishing a sugarcane plantation in Papua New Guinea date back to the 1930s, when Singara Sugar Estates Ltd proposed to establish a plantation near Buna in the Northern Province (van der Veer, 1937). The plantation was never established. In the decades that followed, various reports suggested that commercial sugarcane production was technically feasible. It was emphasized that it would be a great risk because of a broad range of pests and diseases. In the mid-1970s the demand for sugar increased and world market prices fluctuated strongly, and the government decided to establish a national sugar industry. Initial investigations were carried out in the Markham valley to identify a suitable site. Several sites were found, but the Gusap-Dumpu area on the north bank of the Ramu River was favoured because it did not need irrigation or flood-protection works and land preparation costs were lower (Chartres, 1981). In 1979, a detailed soil survey was undertaken and about 7000 ha of suitable or moderately suitable land was identified. The first sugarcane was planted in 1979 and the plantation was named Ramu Sugar.

Initially, most attention was paid to the establishment of the plantation and factory, but in 1987 a soil management plan was developed based on expert knowledge (Booker Agriculture International, 1987). The plan has received only lip-service by the plantation management because they mainly focused on the control of insect pests and diseases, which severely affected production.

Physical Environment

Ramu Sugar Plantation (6°S, 146°E) is located in Madang Province. Before sugarcane was planted in 1979, the site was natural grassland with some forest and swamp vegetation in poorly drained and low-lying areas. The grassland was dominated by *Imperata cylindrica*, which was found on the deeper and fine-textured soils (Booker Agriculture International, 1979). Its dominance was probably due to annual burning, as *Imperata* regenerates rapidly compared with other species. On shallower and stony soils *Themeda australis* (kangaroo grass) dominated the natural vegetation, whereas *Saccharum spontaneum* and *Ophiuros* sp. occurred in the wetter sites along streams and rivers (Chartres, 1981).

Climate

The plantation is in an area that is directly affected by the passage of the 'Inter-Tropical Convergence Zone', which occurs twice yearly. Consequently, there is a seasonal rainfall pattern (unimodal) with a dry season from May to November and a rainy season from December to April. The average rainfall at the plantation is 1954 mm per year, but between 1980 and 2000 annual rainfall varied from 1480 to 2557 mm (Fig. 9.1). June to September are the driest months, with an average of less than 90 mm of rain per month. March is the wettest month, with an average rainfall of 284 mm. Evaporation (class A open pan) is about 2281 mm per year and exceeds rainfall from May to November. Mean annual temperatures are 26.7°C, with only minor fluctuations through the year. The climate is classified as Am (Köppen): tropical rainy climate with a short dry season.

Geomorphology and soils

The Ramu Valley is drained by the perennial Ramu River and several tributaries with erratic flow characteristics. The valley covers an area of about 7500 km^2 and forms, together with the Markham Valley, a large *graben*, which has been a zone of subsidence since the Late Tertiary period (Löffler, 1977). At the site of the plantation, the valley is about 10 km wide (Fig. 9.2). The Ramu Valley contains unconsolidated and poorly consolidated Quaternary marine and terrestrial clastic sediments overlying Tertiary sedimentary rocks. The valley comprises a series of alluvial fans with slopes of up to 5% on the higher parts of the fans but less than 0.5% in the lower parts. The plantation is about 400 m above sea level.

The parent material of the soils at the plantation is alluvium. The soils have been developed in clayey, silty and sandy sediments, and from the weathering products of the water-worn stones and boulders

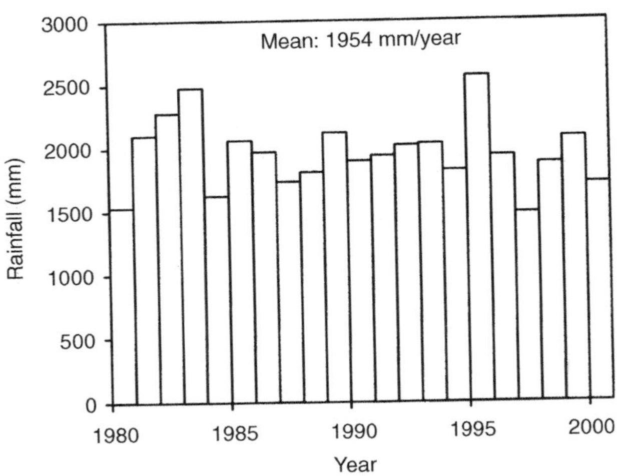

Fig. 9.1. Annual rainfall between 1980 and 2000 at the sugarcane plantation in Papua New Guinea.

Fig. 9.2. The Markham-Ramu *graben* and the sugarcane plantation.

of varying lithology. The stones and boulders originate from sandstone, siltstone and limestone, but also from basalt and igneous rocks with coarser textures. The soil pH indicates no apparent danger from exchangeable Al, or excess $CaCO_3$. Soil salinity is not a problem in the topsoils. The deeper subsoils are slightly alkaline. Sheet and gully erosion are a threat in some areas and terraces have been dug across the contour to control surface water.

Fluvents are the dominant major soils at the plantation, and most soils are classified as Tropofluvents. Some Entisols are classified as Tropopsamments (Bleeker, 1983). The soil temperature regime is isohyperthermic and the soil moisture regime udic, indicating that, in most years, the soils are dry for less than 90 cumulative days per year. Vertisols cover about one-quarter of the plantation. These soils are dominated by montmorillonite or some other smectite mineral. The soils contain no calcareous concretions, which are commonly absent in Vertisols under high rainfall (Blokhuis, 1980). At the great group level the soils are classified as Hapluderts.

Land management under sugarcane

The first 3 ha of sugarcane were planted in 1979 but the total area grew rapidly from 1592 ha in 1981, to 5011 ha in 1983, and to 6030 ha in 1995. The plantation was established for rainfed sugarcane production.

Feasibility studies for irrigation have been conducted in the past but it was soon realized that other constraints were more important. About 1800 ha of sugarcane are planted mechanically each year. Up to 1994, planting took place at the beginning of the wet season (September to November) but currently most of the cane is planted from late February to May as this reduces the risk of certain insect pests. The harvesting season lasts from May to October and cutter–chopper–loader harvesters are used, with 20-Mg tractors and trailers transporting the cane to the factory. Most of this equipment has conventional tyres. About half of the sugarcane is trash-harvested (no burning before harvesting). Up to five crops (i.e. plant cane plus four ratoons) are sometimes obtained, after which the land is replanted; in some blocks cowpea (*Vigna unguiculata*) may be sown and ploughed under after 1 year. Before 1989, inorganic N fertilizer was applied as urea (46% N), but when trash-harvesting replaced preharvest burning it was suggested that considerable amounts of urea-N would be lost through volatilization (see page 249). Therefore N fertilizer supplied after 1989 was in the form of sulphate of ammonia (21% N); on average, 90 kg N/ha/year was applied during the period 1991 to 1995. It is not known whether there was a yield increase because of sulphate of ammonia. Nitrogen applications are usually broadcast between August and November. Phosphorus and K fertilizers are not applied.

Research Methods

To investigate changes in soil chemical properties and leaf nutrient concentrations, all available analytical data from 1978 to 1995 were collected. With the establishment of the plantation in 1979, the area was divided into blocks of 10–20 ha. Between 1982 and 1994, soil samples were taken in most sugarcane blocks for routine analysis, and the analytical data of 487 topsoil (0–0.15 m) and some 50 subsoil samples were available (Type I data). Also, the chemical data of 21 soil profiles (15 Fluvents, six Vertisols) from the initial soil survey report of 1979 were available. The existing data were used for a preliminary assessment of soil chemical changes. The data were grouped according to soil orders using the 1:25,000 soil map of the plantation. Based on the preliminary assessment, additional sampling sites were selected for the collection of extra Type I and II data. Figure 9.3 summarizes the work plan.

The overall aim of the investigations was an analysis of change at the reconnaissance level. The objective of the study was not to investigate changes in each individual sugarcane block, but to quantify in the two major soil orders (Fluvents and Vertisols) as a result of sugarcane cultivation.

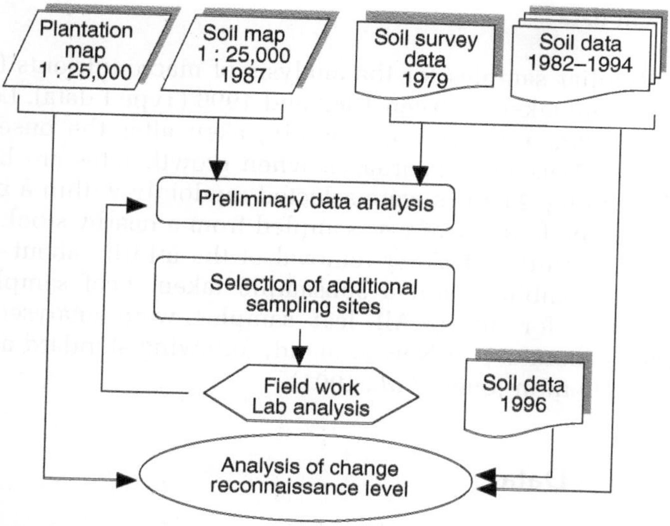

Fig. 9.3. Work plan for the evaluation of soil chemical changes at the sugarcane plantation in Papua New Guinea.

Soil sampling and analysis

Between 1982 and 1994, topsoil samples were commonly taken after the last ratoon when the sugarcane was ploughed-out. Samples were bulked from 20–50 locations in a sugarcane block using an Edelman auger. The samples taken in 1996 were composites from 10–15 locations in a sugarcane block, and mini-pits were used for the 0–0.15 m soil horizons. All soil samples of 1996 were taken in the inter-row of the sugarcane. In addition, soil samples were taken in natural grassland bordering sugarcane fields (Type II data).

Air-dried, ground and sieved samples (2 mm) were analysed at the Cambridge Laboratory, Cambridge, New Zealand or at the National Analytical Chemistry Laboratory, Port Moresby, PNG. The procedures for soil analysis were identical at both laboratories, and were as follows: pH H_2O in 1:2.5 or 1:5 w/v; organic C by $K_2Cr_2O_7$ and H_2SO_4 oxidation (Walkley & Black); total N by Kjeldahl; available P by $NaHCO_3$ extraction (Olsen); exchangeable cations and CEC percolation by 1 M NH_4OAc followed by spectrophotometry (K, Na), AAS (Ca, Mg) and titration (CEC); particle size analysis by hydrometer. The soil samples of the initial soil survey in 1979 were analysed at the laboratories of Hunting Technical Services Ltd in the UK. Except for available P, all methods were identical to the ones described above.

Leaf nutrient data

About 600 foliar samples for the analysis of macronutrients (N, P, K, Ca, Mg, S) were taken between 1982 and 1996 (Type I data). Leaf samples at the plantation were commonly taken after the onset of the rainy season (December–February), when growth rates are high. For the leaf sampling, 21 rows were selected randomly within a block. At 30–40 paces the fourth leaf was sampled from a nearby stool; the first leaf was the unfurl leaf. After removal of the midrib, about 400–600 leaves were combined and a subsample taken. Leaf samples were dried at 80°C for 48 h. All leaf samples were analysed at the Cambridge Laboratory in New Zealand, following standard analytical procedures (Benton-Jones *et al.*, 1991).

Type I and II Data

Type I data

All soil chemical data from sugarcane blocks with Fluvents or Vertisols as soil order were grouped and averages were calculated for each year (Table 9.1). Between 1979 and 1996, the topsoil pH decreased from about 6.5 to 5.8 in both Fluvents and Vertisols. The soil acidification was accompanied by a change in the levels of exchangeable cations. Organic C levels declined by about 40% between 1979 and 1996. Levels of available P declined but variation was large.

Soil chemical data were available from the early 1980s and early 1990s from seven sugarcane blocks on Fluvents and five sugarcane blocks on Vertisols (Table 9.2). A *t*-test was conducted and a significant decrease was found in the pH, available P and CEC of the Fluvents. In the Vertisols only the decline in soil pH was significant and the rate was about similar to the pH change of the Fluvents.

Soil acidification

The data in Tables 9.1 and 9.2 showed that under sugarcane the topsoils had significantly acidified. Changes in the pH of the subsoils were also investigated. In 1986 several sugarcane blocks were sampled up to 0.60-m or 0.90-m depth and these blocks were re-sampled in 1996. Table 9.3 presents the pH data for the different depths. A significant decrease of 0.2 to 0.4 pH unit was found to a depth of 0.60 m after 10 years of continuous sugarcane cultivation. In addition, soil samples were taken from five sugarcane blocks and adjacent grass-

Table 9.1. Topsoil (0–0.15m) chemical properties of Fluvents and Vertisols between 1979 and 1996. Values are the arithmetic mean ±1 SD. Type I data, modified from Hartemink (1998c).

Soil order	Year	Number of samples[a]	pH H_2O (1:2.5)	Organic C (g/kg)	Available P (mg/kg)	CEC ($mmol_c$/kg)	Exchangeable cations ($mmol_c$/kg)		
							Ca	Mg	K
Fluvents	1979[b]	15	6.5 ± 0.4	58 ± 15	nd	389 ± 43	228 ± 78	93 ± 41	13.0 ± 5.0
	1982	14	6.2 ± 0.1	nd	36 ± 4	459 ± 55	275 ± 35	113 ± 24	12.9 ± 2.0
	1983	44	6.3 ± 0.1	nd	37 ± 10	435 ± 48	256 ± 35	100 ± 16	12.4 ± 2.8
	1984	9	6.1 ± 0.1	nd	42 ± 10	437 ± 52	266 ± 45	102 ± 21	12.9 ± 3.8
	1994	12	5.9 ± 0.1	35 ± 6	28 ± 9	384 ± 65	232 ± 47	101 ± 22	10.8 ± 2.3
	1996	8	5.8 ± 0.2	31 ± 7	28 ± 12	374 ± 33	220 ± 30	99 ± 13	8.0 ± 2.0
Vertisols	1979[b]	6	6.6 ± 0.1	52 ± 9	nd	421 ± 21	293 ± 69	123 ± 39	15.5 ± 2.7
	1982	17	6.2 ± 0.1	nd	43 ± 5	490 ± 29	286 ± 22	131 ± 16	16.1 ± 2.9
	1983	40	6.3 ± 0.2	nd	40 ± 13	477 ± 94	290 ± 83	114 ± 33	12.9 ± 2.3
	1986	7	6.2 ± 0.2	nd	37 ± 18	490 ± 108	307 ± 77	112 ± 37	12.3 ± 5.6
	1994	12	5.9 ± 0.1	32 ± 3	32 ± 11	452 ± 79	273 ± 50	129 ± 34	13.4 ± 3.9
	1996	12	5.8 ± 0.2	32 ± 6	28 ± 11	421 ± 102	276 ± 73	115 ± 38	9.0 ± 3.0

[a] Composite topsoil samples of continuously cultivated fields, except for 1979.
[b] Soil samples taken in grassland prior to the establishment of the plantation.
nd, No data.

Table 9.2. Changes in soil chemical properties (0–0.15 m) of Fluvents and Vertisols under sugarcane between the 1980s and 1990s. Type I data.

	Fluvents (n = 7 pairs)			Vertisols (n = 5 pairs)		
	1982–1983	1991–1994	Difference	1982–1984	1991–1994	Difference
pH H_2O (1:2.5 w/v)	6.3	5.9	P <0.001	6.4	6.0	P <0.001
Available P (mg/kg)	37.2	29.0	P = 0.04	35.4	24.6	ns
CEC ($mmol_c$/kg)	412	354	P <0.001	450	403	ns
Exchangeable Ca ($mmol_c$/kg)	229	213	ns	269	250	ns
Exchangeable Mg ($mmol_c$/kg)	100	94	ns	109	95	ns
Exchangeable K ($mmol_c$/kg)	11.0	9.5	ns	13.0	10.1	ns

ns, Not significant.

Table 9.3. Change in pH H_2O with depth based on samples from the same site at different times, and from different land use sampled at the same time. Type I and II data, modified from Hartemink (1998a).

Type I data					Type II data				
Sampling depth (m)	Sample pairs	1986	1996	Difference	Sampling depth (m)	Sample pairs	Natural grassland	Continuous sugarcane[a]	Difference
0–0.15	9	6.2	5.8	P <0.001	0–0.15	5	6.3	5.8	P = 0.02
0.15–0.30	9	6.2	5.9	P <0.001	0.15–0.30	5	6.3	6.1	P = 0.02
0.30–0.45	7	6.5	6.1	P = 0.02	0.30–0.50	5	6.6	6.4	P = 0.05
0.45–0.60	7	6.6	6.4	P = 0.01	0.50–0.70	5	6.7	6.6	ns
					0.70–0.90	5	6.9	6.8	ns

[a] Soils were continuously cultivated with sugarcane for at least 10 years.
ns, Not significant.

lands that were never cultivated (Type II data). These data also showed a significant decrease of 0.5 pH units in the topsoil, and a decrease of 0.2 pH units to a depth of 0.5 m. The data show that acidification under sugarcane was not restricted to the topsoil.

pH data were available from different sampling times from 80 sugarcane blocks. The difference in years between the initial sample at t_1 and the second sample at t_2 was plotted against the difference in the measured pH values. It appeared that the decline in pH was related to the initial pH value (Fig. 9.4). Although the data are scattered, they suggest that a larger decline occurred when the initial pH was high. This relationship does not, however, take into account the time elapsed between the pH measurements. Based on the 80 sample pairs, the decline in pH with time was calculated where t_1 was the initial value and t_2 the pH value of the second sampling. In only a few samples the pH increased or had not changed (i.e. pH at t_2 minus pH at $t_1 \geqslant 0$) but in the majority of the sample pairs, there was a decline in pH (i.e. pH at t_2 minus pH at

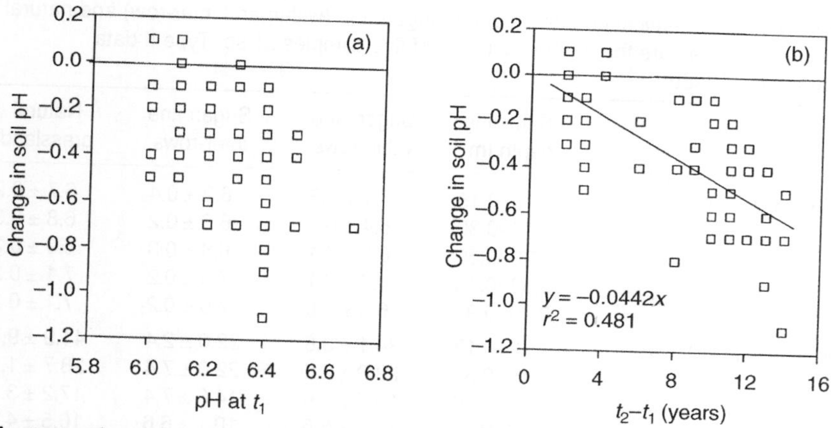

Fig. 9.4. Changes in topsoil pH (0–0.15 m) in relation to the initial pH at t_1 (a), and the change in topsoil pH with time (b). Based on 80 sample pairs. Type I data. Modified from Hartemink (1998a).

$t_1 < 0$). The largest decrease in pH occurred after 10 years ($t_2 - t_1 \geqslant 10$), and nearly 50% of the variation was explained by the linear function: $\Delta pH = -0.044 * (t_2 - t_1)$.

An important cause for the soil acidification trend is the yearly application of N fertilizers. Since these fertilizers contain N in the ammonium form, nitrification reactions result in acidification. The acidity produced by 1 kg N in urea is 71 g H^+, which is equivalent to about 3.6 kg $CaCO_3$ (Adams, 1984). Most of the N fertilizers in the mid-1990s were applied as sulphate of ammonia. For sulphate of ammonia, the potential soil acidity produced through nitrification is equivalent to 143 g H^+/kg applied N.

Type II data

Samples were taken in sugarcane fields in the inter-row and within the rows, and in adjoining natural grassland areas that had never been cultivated (Table 9.4). A pH difference of 0.6 units was observed between topsoils (0–0.15 m) under natural grassland and within the sugarcane rows. The pH values of the inter-row were slightly higher than within the sugarcane rows. Below 0.3-m depth, there were only slight differences between the soils under sugarcane and natural grassland.

Organic C levels in the topsoils within the sugarcane rows were about 8 g/kg lower than under natural grassland but on average 2.1 g/kg higher than in soils of the sugarcane inter-row. The inter-row had a lower organic C content in the subsoil, whereas organic C in natural grassland and within the row was comparable with depth. For total N content, a

Table 9.4. Soil chemical properties under sugarcane (within and inter-row) and natural grassland. Values are the arithmetic mean of five samples ±1 SD. Type II data (Hartemink, 1998b).

	Sampling depth (m)	Sugarcane within rows	Sugarcane inter-rows	Natural grassland
pH H_2O (1:2.5)	0–0.15	6.1 ± 0.3	6.2 ± 0.4	6.7 ± 0.2
	0.15–0.30	6.4 ± 0.2	6.6 ± 0.2	6.8 ± 0.3
	0.30–0.50	6.8 ± 0.1	6.8 ± 0.3	6.9 ± 0.2
	0.50–0.70	6.9 ± 0.1	7.0 ± 0.2	7.1 ± 0.2
	0.70–0.90	6.9 ± 0.6	7.0 ± 0.2	7.1 ± 0.2
Organic C (g/kg)	0–0.15	34.1 ± 3.6	32.0 ± 2.4	41.9 ± 9.1
	0.15–0.30	29.0 ± 2.8	22.0 ± 7.4	28.7 ± 1.9
	0.30–0.50	18.7 ± 4.6	14.6 ± 7.4	17.2 ± 3.3
	0.50–0.70	12.7 ± 6.6	10.1 ± 6.6	10.5 ± 4.2
	0.70–0.90	9.0 ± 5.1	8.1 ± 4.2	7.9 ± 4.2
Total N (g/kg)	0–0.15	2.3 ± 1.6	1.8 ± 0.3	2.4 ± 0.7
	0.15–0.30	1.4 ± 0.2	1.2 ± 0.5	1.6 ± 0.1
	0.30–0.50	0.9 ± 0.3	0.7 ± 0.4	0.9 ± 0.2
	0.50–0.70	0.6 ± 0.3	0.4 ± 0.4	0.6 ± 0.2
	0.70–0.90	0.3 ± 0.3	0.3 ± 0.1	0.3 ± 0.2
Available P (mg/kg)	0–0.15	22 ± 10	22 ± 11	27 ± 10
	0.15–0.30	17 ± 10	11 ± 7	16 ± 11
	0.30–0.50	7 ± 5	6 ± 4	7 ± 6
	0.50–0.70	6 ± 4	6 ± 4	5 ± 3
	0.70–0.90	4 ± 2	4 ± 1	5 ± 2
Exchangeable Ca ($mmol_c$/kg)	0–0.15	278 ± 73	280 ± 49	283 ± 48
	0.15–0.30	280 ± 61	249 ± 74	246 ± 34
	0.30–0.50	283 ± 71	262 ± 70	257 ± 33
	0.50–0.70	275 ± 52	274 ± 57	263 ± 24
	0.70–0.90	251 ± 21	250 ± 17	270 ± 66
Exchangeable Mg ($mmol_c$/kg)	0–0.15	104 ± 16	91 ± 12	92 ± 15
	0.15–0.30	104 ± 19	93 ± 26	83 ± 18
	0.30–0.50	116 ± 13	94 ± 21	97 ± 18
	0.50–0.70	119 ± 28	101 ± 19	109 ± 21
	0.70–0.90	103 ± 9	93 ± 14	106 ± 40
Exchangeable K ($mmol_c$/kg)	0–0.15	10.8 ± 4.9	10.3 ± 5.5	12.8 ± 6.3
	0.15–0.30	6.4 ± 5.8	4.1 ± 1.8	5.8 ± 4.4
	0.30–0.50	2.5 ± 1.2	2.9 ± 1.2	4.5 ± 4.6
	0.50–0.70	2.5 ± 0.7	2.5 ± 0.9	4.3 ± 4.8
	0.70–0.90	2.0 ± 0.4	2.5 ± 0.6	4.6 ± 5.1

similar pattern was followed, with lower content in the topsoils of the sugarcane inter-row as compared to within the rows. Levels of available P in the topsoils were similar in the inter-row and within the rows, but P levels were lower in the subsoil of the inter-rows. Soils under natural grassland had higher levels of available P in the topsoil (+5 mg/kg), but only small differences were found with depth between soils under grassland and within the sugarcane rows. The considerably higher exchangeable Mg content within the sugarcane rows as compared to the inter-row and natural grassland was striking. Also, exchangeable Ca levels appeared to be slightly higher within the sugarcane rows. Exchangeable K was highest under natural grassland. Potassium levels were similar in the topsoils within and between the rows of sugarcane.

Rates of Change

From the Type I and II data presented in the previous sections the rates of change in soil chemical properties are calculated. Absolute changes in unit per year are presented in Table 9.5. A consistent decline in all soil chemical properties was found in both soil orders. When the decline was linked to the period between the two soil samplings, it seems that a lower annual decline is found when the period between the two soil sampling is larger, which suggests that the decline in soil chemical properties is initially larger and levels off with time.

The relative change, based on the initial level, is presented in Table 9.6. The decline in soil organic C is 1–3% per year and the rate of decline in soil pH was 1% in both soil orders. The decline in CEC and exchangeable Ca and Mg ranged from 0% to 1% per year, but the annual decline in exchangeable K was larger.

Regression analysis of paired samples

From 30 sugarcane blocks (13 Fluvents, 17 Vertisols), soil chemical data were available from different sampling times. These data were used to calculate the rates of change. The difference in years between the initial soil sample at t_1 and the second sample at t_2 was plotted against the difference in the measured value for each soil chemical property (similar to Fig. 9.4b). Different functions (linear, logarithmic, polynomial) were fitted through the data, and the function with the highest correlation coefficient (r^2) was used to calculate the rates of change. Clear trends with time were found for pH, available P, CEC and exchangeable K.

In Fluvents and Vertisols, the pH decreased with time (pH at t_2 minus pH at t_1 <0), and from a linear equation fitted through the data it was calculated that the pH decrease was about 0.5 and 0.3 units after 10 years ($t_2 - t_1 = 10$), and 0.7 and 0.4 units after 15 years ($t_2 - t_1 = 15$) in the topsoils of Fluvents and Vertisols, respectively (Table 9.7). Rates of change for P were higher in Fluvents (–20 mg/kg in 15 years) than in Vertisols (–11 mg/kg in 15 years) and also the decline in CEC

Table 9.5. Absolute change (unit/year) in soil chemical properties (topsoils only) on the sugarcane plantation in Papua New Guinea. Type I and II data.

Soil order	Data type	Period[a] (years)	pH H_2O	C (g/kg)	N (g/kg)	P (mg/kg)	CEC and exchangeable cations ($mmol_c$/kg) CEC	Ca	Mg	K
Fluvents	I	11	−0.1	nd	nd	−0.8	−5.3	−1.5	−0.6	−0.1
Fluvents	II	15	<−0.1	−0.6	<0.1	−0.3	nd	−0.3	+0.4	−0.2
Fluvents	I	17	−0.1	−1.6	nd	−0.5	−0.9	−0.5	+0.4	−0.3
Vertisols	I	11	−0.1	nd	nd	−1.0	−4.3	−1.7	−1.3	−0.3
Vertisols	I	17	−0.1	−1.2	nd	−0.9	0	−1.0	−0.5	−0.4

[a] Period of cultivation (Type I data); years difference between cropped and uncultivated soils (Type II data).
nd, No data.

Table 9.6. Relative change (rounded figures in % per year) in soil chemical properties (topsoils only) on the sugarcane plantation in Papua New Guinea. Type I and II data.

Soil order	Data type	Period[a] (years)	pH H_2O	C (g/kg)	N (g/kg)	P (mg/kg)	CEC and exchangeable cations ($mmol_c$/kg) CEC	Ca	Mg	K
Fluvents	I	11	−1	nd	nd	−2	−1	−1	−1	−1
Fluvents	II	15	−1	−1	−1	−1	nd	−0	0	−1
Fluvents	I	17	−1	−3	nd	−1	0	0	0	−2
Vertisols	I	11	−1	nd	nd	−3	−1	−1	−1	−2
Vertisols	I	17	−1	−2	nd	−2	0	0	0	−2

[a] Period of cultivation (Type I data); years difference between cropped and uncultivated soils (Type II data).
nd, No data

Table 9.7. Approximate rates of changes in topsoil (0–0.15 m) chemical properties of Fluvents and Vertisols under sugarcane. Type I data.

Soil order	Soil chemical property	Line fitted[a]	Approximate change in			
			r^2	5 years	10 years	15 years
Fluvents (13 pairs)	pH (1:2.5)	pH = $-0.048x$	0.301	–0.2	–0.5	–0.7
	Available P (mg/kg)	P = $-0.098x^2 + 0.159x$	0.607	–2	–8	–20
	CEC ($mmol_c$/kg)	CEC = $-0.531x^2 + 0.374x$	0.212	–11	–49	–114
	Exchangeable K ($mmol_c$/kg)	K = $-0.172x$	0.182	–0.9	–1.7	–2.6
Vertisols (17 pairs)	pH 1:2.5	pH = $-0.029x$	0.471	–0.1	–0.3	–0.4
	Available P (mg/kg)	P = $-0.734x$	0.914	–4	–7	–11
	CEC ($mmol_c$/kg)	CEC = –4.553lx	0.265	–23	–46	–68
	Exchangeable K ($mmol_c$/kg)	K = $-0.439x$	0.224	–2.2	–4.4	–6.6

[a] Line fitted through: t_2 minus t_1 (x–axis) vs. value at t_2 minus value at t_1 (y–axis); t_1 = initial sample and t_2 = second sample.

was larger in Fluvents. Changes in exchangeable K were, however, larger in Vertisols (–6.6 $mmol_c$/kg in 15 years) than in Fluvents (–2.6 $mmol_c$/kg in 15 years).

Semiquantitative Data

In this section an estimate is made of the major inputs and outputs of nutrients at the sugarcane plantation. Yield data from 1991 to 1995 (Fig. 9.5) were multiplied with a range of nutrient-removal data (kg nutrient per Mg cane/ha) given by Hunsigi (1993). These were compared with the nutrients applied by the inorganic fertilizers and the difference was calculated for the low and high range in nutrient removal (Table 9.8). It appeared that the difference between N removal and application was positive, whereas for P and K a negative difference between removal and fertilizer application was found.

Table 9.9 shows the difference between the highest and lowest nutrient removal and the annual inorganic fertilizer applications. This assumes a 100% recovery of the nutrients in the inorganic fertilizers, which never occurs. In reality, the balance is therefore much more negative. The negative K balance of the sugarcane is due to lack of inorganic K applications and the relatively large removal by the crop.

Changes in Leaf Nutrient Contents

In addition to the soil chemical data, changes in leaf nutrient concentrations were investigated. Median N concentrations in the sugarcane

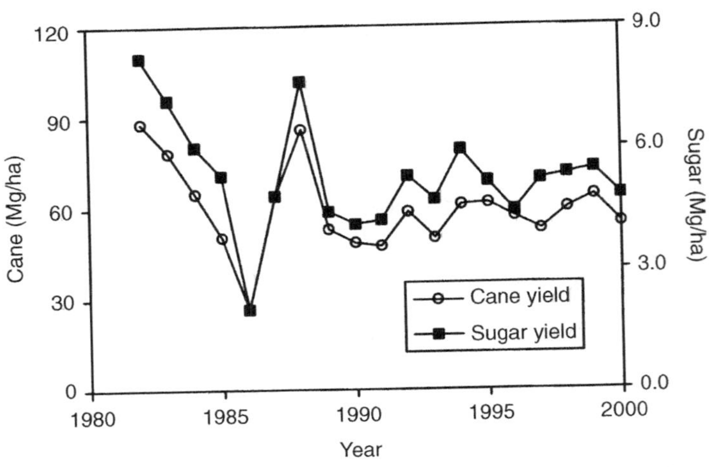

Fig. 9.5. Annual sugarcane and sugar yield at the plantation between 1982 and 2000.

Table 9.8. Nutrient removal (range) and nutrient input with inorganic fertilizers between 1991 and 1995.

	Nutrient removal (kg/ha)[a]						Inorganic fertilizer applications (kg/ha)		
	N		P		K				
	Low	High	Low	High	Low	High	N	P	K
1991	27	57	8	17	48	119	34	12	0
1992	33	71	9	21	59	148	115	4	0
1993	28	60	8	17	50	124	105	1	0
1994	35	75	10	22	62	156	81	0	0
1995	35	75	10	22	63	156	83	3	0

[a] Highest and lowest removal values as given by Hunsigi (1993) multiplied by the sugarcane yield from the plantation.

Table 9.9. Difference between nutrient removal (range in kg/ha) and nutrient input with inorganic fertilizers between 1991 and 1995. Low and high refers to removal range of Table 9.8.

	N		P		K	
	Low	High	Low	High	Low	High
1991	7	−23	4	−5	−48	−119
1992	82	44	−6	−17	−59	−148
1993	77	46	−7	−16	−50	−124
1994	47	7	−10	−22	−62	−156
1995	48	8	−7	−19	−63	−156

leaves varied from 19.3 to 22.0 g N/kg between 1983 and 1994 (Table 9.10). The lowest figure was the median of the 27 leaf samples taken in 1994. There appears to be a declining trend in the P concentration of sugarcane leaves but the median value of 2.4 g/kg in 1994 is above the optimum concentration of 1.8 g/kg as given by Anderson and Bowen (1990) and 2.1 g/kg as given by de Geus (1973). The trend in leaf P coincides with the decrease observed in the levels of available P in the topsoils (Table 9.1). Leaf K concentrations were in the optimum range in the 1980s, but the median value in 1994 was at the lower end of the optimum range at 12.5 g/kg. Levels of S, Ca and Mg show no apparent trend and all median values are in the optimum range.

All major nutrients in the sugarcane leaves decreased significantly between the mid-1980s and 1990s (Table 9.11). The largest decrease was found in the Ca and Mg concentrations, which had decreased by 36% and 33%, respectively. Small but highly significant differences were found in the leaf P concentrations of the 1980s and 1990s.

Table 9.10. Macronutrient concentrations (g/kg) of sugarcane leaves between 1983 and 1994. Median values with CV% in parentheses. Type I data, modified from Hartemink and Kuniata (1996).

Year	Number of samples	Nutrient concentration in g/kg											
		N		P		K		S		Ca		Mg	
1983	481[a]	22.0	(11%)	3.5	(16%)	15.0	(14%)	1.3	(12%)	2.9	(16%)	1.8	(12%)
1987	69	20.0	(13%)	2.7	(9%)	16.0	(15%)	1.8	(14%)	4.4	(16%)	2.5	(14%)
1989	24	21.0	(12%)	2.9	(17%)	16.1	(15%)	1.8	(30%)	3.5	(21%)	1.7	(13%)
1994	27	19.3	(10%)	2.4	(7%)	12.5	(11%)	nd		3.1	(20%)	1.3	(17%)

[a] There were only 11 samples of S, Ca and Mg in 1983.

nd, No data.

Table 9.11. Macronutrient concentrations (g/kg) of sugarcane leaves in the 1980s and 1990s. Type I data, modified from Hartemink (1998b).

Period	Number of samples	Nutrient concentration in g/kg				
		N	P	K	Ca	Mg
1985–1987	93	20.3	2.8	14.7	4.4	2.4
1994–1996	160	19.4	2.6	13.8	2.8	1.6
Difference		$P<0.001$	$P<0.01$	$P<0.001$	$P<0.001$	$P<0.001$

There are several keys to the interpretation of leaf nutrient concentration for sugarcane, but much depends on the age of the plant at sampling, the sugarcane variety, the plant part sampled, soil conditions and inorganic fertilizer applications. The first row in Table 9.12 summarizes the critical nutrient concentration for the fourth leaf as compiled from the literature (de Geus, 1973; Nelson, 1980; Orlando Filho, 1985; Anderson and Bowen, 1990; Srivastava, 1992; Malavolta, 1994). Average nutrient concentration in the 1980s and the 1990s exceeded critical levels. However, the percentage of samples below the critical level increased dramatically between the 1980s and 1990s (Table 9.12). The increase was particularly high for N and K and in the 1990s, about 40% of the samples were below the critical N concentration, whereas 47% of the samples were below the critical K concentration. Although Ca and Mg concentrations had decreased dramatically, there were few samples below the critical level in the 1990s.

For a few blocks (fields) at the sugarcane plantation, soil and leaf analytical data were available from different periods (Table 9.13). In the blocks on Fluvents a decline in the soil pH was found. The decrease in available P and exchangeable K coincided with a decrease in leaf P and K levels in both Fluvents and Vertisols. Available P and exchangeable Mg and K decreased in the two sugarcane blocks on Vertisols. Although the data are few there seems to be fair agreement in the trend in soil chemical data and leaf tissue concentrations.

Table 9.12. Critical nutrient concentration (g/kg) in sugarcane leaves, and percentage samples below this level in the mid-1980s and mid-1990s. Modified from Hartemink (1998b).

	N	P	K	Ca	Mg
Critical nutrient concentration	19.0	2.0	13.0	1.5	1.0
% samples below the critical level in mid-1980s	17	1	23	0	0
% samples below the critical level in mid-1990s	40	17	47	1	3

Table 9.13. Soil and leaf analytical data of four sugarcane blocks at different sampling times. Type I data, modified from Hartemink and Kuniata (1996).

	Soil data (0–0.15 m)							Leaf data			
					Exchangeable cations ($mmol_c$/kg)				Nutrient (g/kg)		
Soil order	Year	pH H_2O (1:2.5)	Available P (mg/kg)	CEC ($mmol_c$/kg)	Ca	Mg	K	Year	N	P	K
Fluvents	1983	6.4	39	420	235	101	11.9	1982	17	3.2	15
	1991	5.6	33	nd	243	94	8.4	1994	21	2.3	10
	1983	6.4	35	370	213	89	10.7	1982	19	3.4	16
	1994	6.0	44	367	217	99	10.2	1984	22	3.3	16
Vertisols	1983	6.4	40	550	341	113	12.7	1983	20	4.0	16
	1986	6.4	26	410	307	57	8.7	1987	19	2.8	14
	1982	6.5	41	510	286	146	16.7	1983	20	3.7	15
	1991	6.1	26	442	297	40	9.2	1988	24	2.7	15

nd, No data.

The Effects on Yield

Mean sugarcane yields at the sugarcane plantation have varied in the past decades from 27 to 106 Mg/ha/year; sugar yield varied from 2.0 to 8.2 Mg/ha/year (Fig. 9.5). There appears to be a declining trend towards the mid-1980s but this can be largely explained by pests and diseases, some of which can have a high impact on yield if poorly controlled (Hartemink and Kuniata, 1996).

The disease Ramu stunt was first observed in 1985; in 1986 the disease was widespread in the sugarcane variety Ragnar that occupied most of the plantation. The disease was so severe that it could have caused the closure of the plantation. Also, the white grub was present in 1984 and 1985 but its effects were not very obvious, although potential losses of up to 36 Mg cane/ha/year can be expected. As a result of the Ramu stunt infestation, most of the sugarcane was replanted in 1986 with the resistant variety Cadmus. However, Cadmus appeared to be very susceptible to the moth stem borer, and in 1987 a severe outbreak was observed, damaging up to 60% of the crop. Average cane yields in 1988 were substantially higher because of the prolonged droughts in 1987, which significantly reduced the number of stem borers. Yields dropped again in 1989 due to an outbreak of cicadas, which reduced yields to about 50 Mg cane/ha. The cicadas were controlled by ploughing-out, followed by a fallow period of 2–4 months. This was effectively practised from 1989 onwards.

The relative high sugar yield in 1997 coincided with a severe drought due to El Niño, which hit the Pacific severely (Fig. 9.1). Since the late 1980s, yields have stabilized at around 60 Mg cane/ha/year. Yield levels at the plantations are relatively low and can be explained through the planting of varieties resistant to pests and diseases, but these varieties have generally a low yield potential. In Australia yield levels are around 75 Mg/ha but inorganic N applications are on average 200 kg/ha (Moody and Aitken, 1995).

Highly productive varieties were planted in 1993, resulting in higher yields but also a higher population of moth stem borers in 1995 and 1996. Yields are also limited by the competition between sugarcane and weeds. Dominant weeds at Ramu Sugar Plantation are itchgrass (*Rottboellia* sp.) and nutgrass (*Cyperus rotundus*); weed competition trials have shown that itchgrass can give yield reductions of up to 54 Mg cane/ha. In commercial fields, an average loss of 26 Mg cane/ha was observed in 1993, but losses were reduced to 5 Mg cane/ha in 1995 as a result of improved weed-control measures. It confirms the general belief that sugarcane does not tolerate competition for water and nutrients.

Discussion and Conclusions

In the previous section, evidence was presented for changes in young alluvial soils under sugarcane cultivation since 1979. Figure 9.6 summarizes the major changes brought about by permanent sugarcane cultivation at the plantation. Both soil physical and soil chemical changes occurred. Soil erosion and surface sealing were not measured but the observed soil compaction and reduced water infiltration may imply that such processes take place at the sugarcane plantations (Hartemink, 1998c).

A similar degree of soil acidification and fertility decline was found when Type I and Type II data were compared. The acidification in both topsoil and subsoil is probably one of the most significant soil changes to have occurred at the plantation. The initial decrease in pH from grassland to sugarcane (1979–1982) may have been due to the increased mineralization of organic matter, which is a common cause of soil acidification. There were no organic C data available from the early 1980s, but the levels declined from about 56 g C/kg in 1979 to 30 g C/kg in 1994 (Table 9.1).

The significant pH decline observed in the 1990s coincided with a change in fertilizer policy resulting from a change in harvesting technique. Since 1989, Australian cutter–chopper–loader harvesters were used instead of preharvest burning (Fig. 9.7). These harvesters may leave

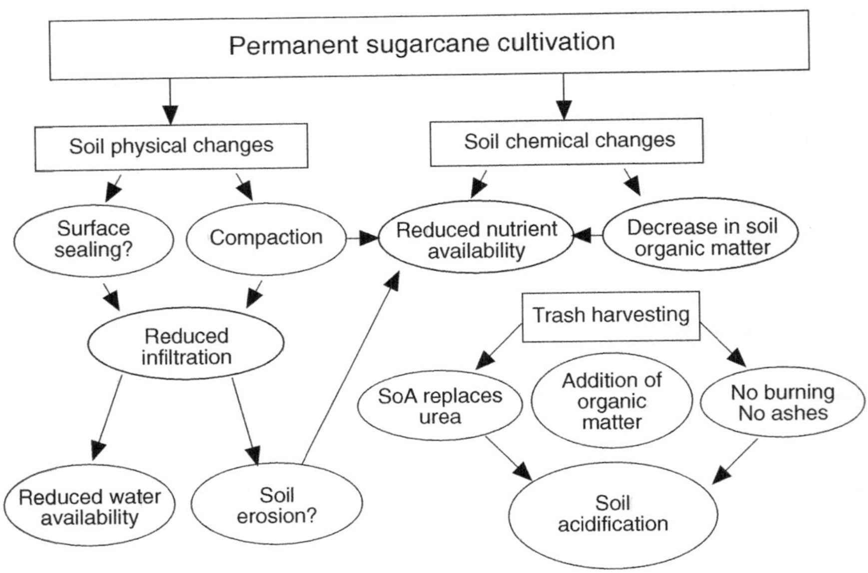

Fig. 9.6. Summary diagram showing major changes under permanent sugarcane cultivation.

Fig. 9.7. Trash harvester replaced preharvest burning. About 10–20 Mg trash/ha remains after the harvest and the trash slowly decomposes.

up to 10 Mg/ha of crop residues or trash (Ng Kee Kwong *et al.*, 1987). In Australia, it has been found that, when urea is applied to trash-covered fields, ammonia losses of 40% can be expected (Freney *et al.*, 1992). Accordingly, in the early 1990s, urea was replaced by sulphate of ammonia, which is less vulnerable to NH_4-volatilization. The theoretical acidity produced by sulphate of ammonia is twice that from urea-N, and that may explain the significant increase in soil acidity observed in the 1990s. Contributing causes are possibly the end of burning, by which no more pH-increasing ashes are returned to the soil, and the yearly addition of sugarcane trash, which increases the organic matter content, resulting generally in a lowering of the pH (Moody and Aitken, 1997). Although the trash-harvesting method may favour the organic matter content (Vallis *et al.*, 1996), in the young alluvial soils of this sugarcane plantation, it is likely to have resulted in significant acidification.

The decline in topsoil and subsoil pH has a number of unwanted consequences. Although sugarcane tolerates a pH range of 5–8 (Blackburn, 1984), some studies have shown that optimum yields are obtained when the pH is about 6.0 (Coale and Schueneman, 1993). In some of the soils under sugarcane, the pH had decreased below 5.5, the point at which Al starts to become soluble. Although sugarcane is relatively more tolerant of Al in solution than, for example, maize (Hetherington *et al.*, 1988), a decline in productivity may be expected. Cation availability is decreased at a lower pH because the increase in protons displaces cations from the exchange sites, which are subsequently leached (Haynes and Swift, 1986). Indeed levels of exchangeable cations declined over time. In particular, the levels of exchangeable K declined, possibly due to a combination of the large K removal by the sugarcane (Yates, 1978) and leaching losses.

Another important change was the change in soil organic C. Since 1979, soil organic C levels have decreased by about 40%, but the current practice of trash harvesting is likely to increase soil organic matter (Wood, 1991). Such an increase would affect many soil properties. For example, the pH-buffering capacity may increase, reducing the acidifying effects of sulphate of ammonia. An increase in soil organic C may also reduce the compactibility of the soil by increasing resistance to deformation (Soane, 1990). Although the soil organic C levels in the topsoils had declined, soil organic C data from natural grasslands compared with soils under sugarcane suggested higher C contents in the subsoils under sugarcane. Higher C levels in the subsoil under sugarcane were also found by Skjemstad *et al.* (1999) and McGarry *et al.* (1996) in Australia. This may be related to the deeper rooting of the sugarcane compared with natural grassland.

Given the fact that yields at the plantation were largely controlled by the sudden and catastrophic outbreak of pests and diseases, it is difficult to assume a prominent role for the changes in soil chemical properties. The question remaining is: what do the changes in soil chemical properties indicate for the sustainability of sugarcane cultivation at the plantation? Discussion of this question involves a review of indicators of sustainable land management, threshold values in soil chemical properties, and requirements for sustainable land management under sugarcane. This issue is addressed in Chapter 11.

References

Adams, F. (1984) Crop responses to lime in the southern United States. In: Adams, F. (ed.) *Soil Acidity and Liming*. ASA, Madison, Wisconsin, pp. 211–265.

Anderson, D.L. and Bowen, J.E. (1990) *Sugarcane Nutrition*. Potash and Phosphate Institute, Norcross, Georgia.

Benton-Jones, J., Wolf, B. and Mills, H.A. (1991) *Plant Analysis Handbook, a Practical Sampling, Preparation, Analysis, and Interpretation Guide*. Micro-Macro Publishing, Athens, Georgia.

Blackburn, F. (1984) *Sugar-cane*. Longman, Harlow, UK.

Bleeker, P. (1983) *Soils of Papua New Guinea*. Australian National University Press, Canberra.

Blokhuis, W.A. (1980) Vertisols. In: *Land Reclamation and Water Management – Developments, Problems and Challenges*. ILRI Publication No. 27. International Institute for Land Reclamation and Improvement, Wageningen, pp. 43–72.

Booker Agriculture International (1979) *Ramu Sugar Ltd, Soil Survey Report.* BAI Ltd, London.

Booker Agriculture International (1987) *Final Report Soil Management at Ramu Sugar Ltd.* BAI Ltd, London.

Chartres, C.J. (1981) Land resource assessment for sugar-cane cultivation in Papua New Guinea. *Applied Geography* 1, 259–271.

Coale, F.J. and Schueneman, T.J. (1993) Sugarcane response to limestone and gypsum application on acidic sandy soils. *Journal of American Society of Sugarcane Technologists* 13, 73–86.

de Geus, J.G. (1973) *Fertilizer Guide for the Tropics and Subtropics.* Centre d'Etude de l'Azote, Zürich.

Freney, J.R., Denmead, O.T., Wood, A.W., Saffigna, P.G., Chapman, L.S., Ham, G.J., Hurney, A.P. and Stewart, R.L. (1992) Factors controlling ammonia loss from trash covered sugarcane fields fertilized with urea. *Fertilizer Research* 31, 341–349.

Hartemink, A.E. (1998a) Acidification and pH buffering capacity of alluvial soils under sugarcane. *Experimental Agriculture* 34, 231–243.

Hartemink, A.E. (1998b) Changes in soil fertility and leaf nutrient concentration at a sugar cane plantation in Papua New Guinea. *Communications in Soil Science and Plant Analysis* 29, 1045–1060.

Hartemink, A.E. (1998c) Soil chemical and physical properties as indicators of sustainable land management under sugar cane in Papua New Guinea. *Geoderma* 85, 283–306.

Hartemink, A.E. and Kuniata, L.S. (1996) Some factors influencing yield trends of sugarcane in Papua New Guinea. *Outlook on Agriculture* 25, 227–234.

Haynes, R.J. and Swift, R.S. (1986) Effects of soil acidification and subsequent leaching on levels of extractable nutrients in a soil. *Plant and Soil* 95, 327–336.

Hetherington, S.J.C., Asher, C.J. and Blamey, F.P.C. (1988) Comparative tolerance of sugarcane, navybean, soybean and maize to aluminum toxicity. *Australian Journal of Agricultural Research* 38, 171–176.

Hunsigi, G. (1993) *Production of Sugarcane, Theory and Practice.* Springer-Verlag, Berlin.

Löffler, E. (1977) *Geomorphology of Papua New Guinea.* Australian National University Press, Canberra.

Malavolta, E. (1994) *Fertilizing for High Yield Sugarcane.* International Potash Institute, Basel.

McGarry, D., Bigwood, R.C., Pillay-McGarry, U.P., Bray, S.G. and Moody, P.W. (1996) Is cane damaging the soil? A comparison of a ten-year-old block with the adjoining treeline. In: Egan, B.T. (ed.) *Proceedings of the 1996 Conference of the Australian Society of Sugarcane Technologists.* Mackay, pp. 189–194.

Moody, P.W. and Aitken, R.L. (1995) Soil acidification and lime use in some agricultural systems in Queensland, Australia. In: Date, R.A., Grundon, R.J., Rayment, G.E. and Probert, M.E. (eds) *Plant Soil Interactions at Low pH.* Kluwer Academic, Dordrecht, The Netherlands, pp. 749–752.

Moody, P.W. and Aitken, R.L. (1997) Soil acidification under some tropical agricultural systems. 1. Rates of acidification and contributing factors. *Australian Journal of Soil Research* 35, 163–173.

Nelson, L.E. (1980) Phosphorus nutrition of cotton, peanuts, rice, sugarcane, and tobacco. In: Khasawneh, F.E., Sample, E.C. and Kamprath, E.J. (eds) *The Role of Phosphorus in Agriculture.* ASA-CSSA-SSSA, Madison, Wisconsin, pp. 693–736.

Ng Kee Kwong, K.F., Deville, J., Cavalot, P.C. and Riviere, V. (1987) Value of cane trash in nitrogen nutrition of sugarcane. *Plant and Soil* 102, 79–83.

Orlando Filho, J. (1985) Potassium nutrition of sugarcane. In: Munson, R.D. (ed.) *Potassium in Agriculture.* ASA-CSSA-SSSA, Madison, Wisconsin, pp. 1045–1062.

Skjemstad, J.O., Taylor, J.A., Janik, L.J. and Marvanek, S.P. (1999) Soil organic carbon dynamics under long-term sugarcane monoculture. *Australian Journal of Soil Research* 37, 151–164.

Soane, B.D. (1990) The role of organic matter in soil compactibility: a review of some practical aspects. *Soil and Tillage Research* 16, 179–201.

Srivastava, S.C. (1992) Sugarcane. In: Halliday, D.J. and Trenkel, M.E. (eds) *IFA World Fertilizer Use Manual.* International Fertilizer Industry Association, Paris, pp. 257–266.

Szent-Ivany, J.J.H. and Ardley, J.H. (1962) Insects of *Saccharum* sp. in the territory of Papua New Guinea. *Proceedings of the International Society of Sugar Cane Technologists,* 11.

Vallis, I., Parton, W.J., Keating, B.A. and Wood, A.W. (1996) Simulation of the effects of trash and N fertilizer management on soil organic matter levels and yields of sugarcane. *Soil and Tillage Research* 38, 115–132.

van der Veer, K. (1937) Land- en Tuinbouw. In: Klein, W.C. (ed.) *Nieuw Guinee – Deel II.* J.H. de Bussy, Amsterdam.

Wood, A.W. (1991) Management of crop residues following green harvesting of sugarcane in north Queensland. *Soil and Tillage Research* 20, 69–85.

Yates, R.A. (1978) The environment for sugarcane. In: *Land Evaluation for Standards for Rainfed Agriculture.* World Soil Resources Report No 49. FAO, Rome, pp. 58–72.

Case 2 – Sisal Plantations, Tanzania

10

> The most suitable lands for the cultivation of the sisal plant are poor, rocky dry, gravelly soils, rich in lime, worn-out sugar-cane and arrowroot fields.
>
> H.H. Smith (1929)

> Sisal will grow on relatively infertile soils.
>
> P.M. Ahn (1993)

In this chapter, a case study is presented based on experimental work conducted in the Tanga Region of Tanzania in the late 1980s and early 1990s. Tanzania was a world market leader in sisal but production has declined since the mid-1960s. In the 1980s a large rehabilitation programme was set up to revive the Tanzania sisal industry. As part of that programme detailed soil surveys were conducted of many sisal plantations and these aimed to provide information about the soil conditions and where to plant sisal or alternative crops. During the soil surveys, soil chemical properties received special attention because no inorganic and organic nutrient inputs were made on sisal plantations. This case study focuses on the soil fertility status of sisal plantations and the information presented is partly derived from previously published work. The data have been re-evaluated and updated where appropriate. Similar to the approach in the previous chapters, data are presented based on their mode of collection: Type I data, Type II data and the nutrient balance. Rates of change in soil chemical properties are calculated for various soil orders based on different data types.

Introduction

Sisal (*Agave sisalana* Perrine) was introduced in Tanzania in 1893 by the German Dr Richard Hindorf (Fig. 10.1). The first 62 sisal plants were planted near Pangani in the Tanga region, and these plants were the foundation of the sisal industry in East Africa (Lock, 1969). The German settlers had not found a crop that would thrive in the coastal plain, where it was too dry for *Hevea* rubber and other plantation crops. Sisal, being fairly drought resistant, appeared well adapted to the environmental conditions and the first plantations were established in 1900.

Sisal research was started after the First World War at the East African Agricultural Research Station in Amani in the East Usambara Mountains with G. Milne as soil chemist. The site was ecologically not representative for the main sisal-growing areas, and in 1934 the Sisal Research Station Mlingano in Tanga region was founded. One of the achievements of this station was the breeding of a long-fibre agave hybrid, No. 11648 (bred on 11th June 1948), which replaced the lower-yielding *A. sisalana*. The hybrid, however, is more susceptible

Fig. 10.1. 'Sisal – a German pioneer effort. The German pioneer Dr Richard Hindorf imported the hard-fibre sisal from Mexico into German East Africa.' From Ettighoffer (1943). The first plants were planted at Mwera estate near the Pangani River.

to diseases than *A. sisalana*, and cannot withstand temporary water-logging. The fibre from the hybrid is similar to that of *A. sisalana* and commercially no distinction is made.

In Tanzania, sisal is grown on plantations and large sisal areas can be found between Tanga and Moshi, between Dar es Salaam and Morogoro and in the south near Mtwara and Lindi. In Tanzania, sisal has never been grown extensively by smallholders. For extraction of the fibre, heavy machinery is required, which is only economical if there is 1000 ha or more planted with sisal.

Sisal production

The first exports of sisal from Tanzania occurred in 1898, and reached 7.5 Mg in 1900. Exports rose to 1400 Mg in 1905 increasing to 20,000 Mg by 1913 (Fig. 10.2). Sisal growing was then firmly established in Tanzania (Lock, 1969). With the construction of railways, sisal was planted away from the coast and production rapidly increased. As in many countries in the tropics, agricultural development followed the construction of railways. The First World War brought a halt to the expansion and production of sisal growing in Tanzania. During the British administration that followed, production was revived and annual exports rose to nearly 50,000 Mg in 1930. During the global economic crisis of the 1930s, there was no decline in sisal production and annual exports reached 100,000 Mg by 1938.

After the Japanese invasion of the Philippines and Indonesia towards the end of 1941, East Africa became the world's main hard-fibre producer. After the Second World War, sisal production increased steadily and reached its highest production in 1964, when

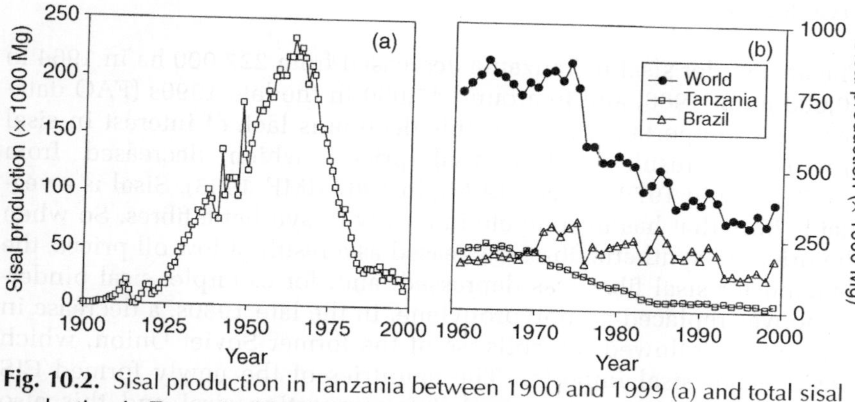

Fig. 10.2. Sisal production in Tanzania between 1900 and 1999 (a) and total sisal production in Tanzania, Brazil, and the world between 1960 and 1999 (b). Data from Lock (1969) and FAOSTAT (http://apps.fao.org).

Tanzania produced 234,000 Mg of sisal fibre. In 1963 and 1964, sisal export earnings exceeded US$60 million, or one-third of the total agricultural export earnings of Tanzania. In those years, agricultural export earnings were more than 80% of the national income. Over half of the sisal was produced at plantations in the Tanga Region.

From 1964 onwards, production began to decline and in the mid-1980s sisal production had decreased to 38,000 Mg/year. It further decreased to 32,000 Mg in 1989, which was the same production level as in 1927. In order to revive sisal production, large rehabilitation programmes were launched. Foreign companies were allowed to buy shares in the nationalized estates and many abandoned fields were replanted. Although sisal prices remained low, the results of the liberalization of politics in the country have enabled many of the rehabilitation programmes to be profitable.

In the 1960s, Tanzania produced nearly one-third of the world's sisal output. Currently, Tanzania produces about 6% of the world sisal production and ranks after Brazil and Kenya. Sisal contributes presently less than 1% of Tanzania's total export earnings. In the early 1970s, Brazil became the main sisal-producing country, and the current production of Brazil is more than half of the world's sisal output. Unlike Tanzania, sisal in Brazil is grown mainly by smallholders and the quality of the fibre is usually lower. Total world annual sisal production decreased from about 800,000 Mg in the 1960s to 400,000 Mg in the 1990s. A very sharp increase occurred between 1974 and 1976 when the oil crisis temporarily stimulated a high demand for sisal. Afterwards, when sisal prices dropped, production declined again. The decline in sisal production is attributed to a decline in both the acreage under sisal and the sisal yields per hectare.

Area decline

The area under sisal in Tanzania decreased from 227,000 ha in 1964 to 50,000 ha in 1986, and to around 45,000 in the late 1990s (FAO databases). An important cause for this decline is lack of interest in sisal growing as result of low sisal prices, which decreased from US$713/Mg in 1979 to US$519/Mg in 1987 (IMF, 1988). Sisal is a vegetable fibre that has to compete in cost with synthetic fibres. So when the prices of synthetic fibres decreased as a result of low oil prices, the demand for sisal fibre was depressed, and, for example, sisal binder-twine was replaced by polypropylene. In the late 1980s, a decrease in sisal prices followed the collapse of the former Soviet Union, which was a major sisal importer. The countries of the newly formed CIS lacked the hard currency needed for importing sisal and this also decreased the prices.

Besides sisal prices, the area decline was also the result of the nationalization of sisal plantations during the late 1960s and early 1970s. Lack of sufficient management capacity following the nationalization, resulted in the abandonment of many sisal plantations. Another factor that contributed to the area decline was the shortage of labour owing to low wages and the harshness of the work. Furthermore, mechanization problems resulting from the unavailability of spare parts and a lack of investment in tractors, lorries and decorticators resulted in a reduction of the area under sisal.

Physical Environment

The research was conducted on eight sisal plantations located in Tanga Region in northeast Tanzania. The region covers about 26,900 km^2 and is the main area of sisal production. Four major physiographic units can be recognized in the region: (i) mountains; (ii) uplands; (iii) coastal area; and (iv) alluvial plains. Sisal is only grown in the uplands and the coastal area, which cover about 76% of the region.

Climate

The seasonal pattern of rainfall in Tanga Region is greatly influenced by the Indian Ocean. Throughout the region, rainfall is bi-modal with the main rains falling in April and May (southeast monsoon), and the small or short rains falling between October and December (northeast monsoon). Rainfall at the plantations is on average between 1000 and 1200 mm/year. Potential evaporation (Penman) is about 1500–1700 mm/year, and is exceeded by rainfall in April and May only. There is considerable interannual variation in rainfall, and throughout Tanga Region rainfall is highly unpredictable. Mean annual temperature is 26°C with only minor fluctuations. Most of the coastal area has an ustic soil moisture regime with an isohyperthermic soil temperature regime but in the western part of the Tanga region, soil moisture regimes are aridic.

Soils

The sisal plantations are located in the uplands and coastal area. Main rock types in the undulating to rolling uplands are Precambrian rocks of the Basement Complex and include schist, granulite, quartzite and gneiss of acid and intermediate composition. Gneiss of intermediate composition is a common soil parent material in Tanga Region.

Soils derived *in situ* from gneiss are generally red, very deep (>4 m), with clayey textures and sesquioxides and kaolinite as predominant minerals. Many of the soils have a oxic B horizon (CEC_{clay} <160 $mmol_c$/kg), and are classified as Oxisols. The Oxisols have typically 350–500 g clay/kg in the topsoil, and 550–650 g clay/kg in the subsoil. Silt contents are lower than 150 g/kg in the topsoil and less than 100 g/kg in the subsoils. The Oxisols must have formed in a wetter climate as the present rainfall levels are too low for the formation of such highly weathered and leached soils. Other common soils derived from gneiss are Ultisols and Alfisols. These soils have a textural B horizon with a CEC_{clay} exceeding 160 or 240 $mmol_c$/kg. They may have lower clay contents in both topsoils and subsoils than the Oxisols. Silt contents are similar to the Oxisols.

In large areas of the uplands the soils constitute a recurrent topographic sequence (catena), which was first recognized by G. Milne in the 1930s (Milne, 1935). On the hill crests and slopes, the soils are dusky red and well drained (Oxisols, Ultisols). On the footslopes, the soils are more yellow, gravelly and moderately well drained (Ultisols, Plinthic and aquic subgroups) and in the valleys the soils are imperfectly to poorly drained and have brownish to black colours (Fluvents). The colour changes usually correspond to various haematite to goethite ratios and coincide with changes in texture.

In the coastal area, soils are very heterogeneous and relations between landform and soils are hard to establish. This heterogeneity is caused by the irregular deposition of fluvial sediments from the hinterland and the uneven surface of the underlying coral rock and limestone by which the depth of soil varies over short distances. Differences in soils can mainly be explained by the varying textural composition of the fluvial deposits and the depth to bedrock. Most soils developed in a mixture of weathering products from Neogene limestone and heterogeneous Quaternary sediments. They are generally less deep (<2 m), much younger and less weathered than the soils in the uplands. The soils are classified as Inceptisols, Alfisols, and Mollisols. Some of the soils developed in Quaternary sediments have a poor soil chemical fertility and are classified as Psamments and Oxisols.

During the soil surveys of the plantations it became apparent that there was a large difference in the soil chemical status between soil orders. Table 10.1 shows the soil chemical status of Alfisols, Oxisols, Ultisols and Inceptisols that had been under sisal for more than 50 years. Although there is a range of values, the following picture emerged: Oxisols had an extremely acid soil reaction with low levels of exchangeable cations, and Al saturation was extremely high in the subsoil. Ultisols under sisal had acid topsoils and extremely acid subsoils, but exchangeable Al was lower than in the Oxisols. Soil organic C contents of the Alfisols, Oxisols and Ultisols were similar, but

Table 10.1. Soil chemical properties of Oxisols, Ultisols and Inceptisols under sisal cropping (>50 years). Values are the arithmetic mean with range in parentheses. Modified from Hartemink (1997b).

	Sampling depth (m)	Alfisols ($n = 3$)	Oxisols ($n = 5$)	Ultisols ($n = 5$)	Inceptisols ($n = 3$)
pH H_2O (1:2.5)	0–0.20	5.9 (5.5–6.6)	4.5 (4.3–4.7)	5.0 (4.3–5.3)	7.7 (7.4–7.9)
	0.30–0.50	5.4 (5.1–5.8)	4.3 (4.2–4.5)	4.4 (4.0–4.9)	7.5 (6.6–8.1)
Organic C (g/kg)	0–0.20	17 (8–28)	18 (16–21)	18 (17–19)	25 (20–34)
	0.30–0.50	7 (5–8)	9 (8–10)	12 (11–13)	11 (9–15)
Avail. P (mg/kg)[a]	0–0.20	2 (2–4)	3 (<0.5–4)	3 (2–4)	2 (1–4)
	0.30–0.50	1 (1–2)	1 (<0.5–1)	1 (1–2)	2 (1–3)
CEC ($mmol_c$/kg)	0–0.20	136 (108–178)	nd	nd	291 (242–320)
	0.30–0.50	128 (90–151)	nd	nd	353 (221–386)
Exchangeable Ca ($mmol_c$/kg)	0–0.20	50 (40–67)	10 (4–14)	23 (13–28)	213 (140–272)
	0.30–0.50	32 (23–37)	3 (1–6)	16 (6–21)	195 (107–282)
Exchangeable Mg ($mmol_c$/kg)	0–0.20	19 (8–37)	7 (3–10)	15 (8–19)	33 (20–42)
	0.30–0.50	20 (12–31)	2 (1–7)	12 (2–20)	25 (8–44)
Exchangeable K ($mmol_c$/kg)	0–0.20	4 (2–9)	2 (1–3)	2 (<0.5–3)	4 (2–8)
	0.30–0.50	1 (1–1)	1 (<0.5–1)	1 (<0.5–2)	2 (1–3)
Base saturation (%)	0–0.20	54 (42–64)	nd	nd	86 (58–100)
	0.30–0.50	44 (27–57)	nd	nd	73 (60–85)
Exchangeable Al ($mmol_c$/kg)	0–0.20	0	6 (5–11)	3 (0–9)	0
	0.30–0.50	0	11 (4–16)	6 (0–9)	0
Al saturation (% ECEC)[b]	0–0.20	0	25 (17–39)	7 (0–28)	0
	0.30–0.50	0	60 (33–80)	17 (0–33)	0

[a] Soil pH <7: P-Bray I; soil pH >7: P-Olsen.
[b] Aluminium saturation of the ECEC is calculated as: Al/Ca+Mg+K+Na+H+Al*100.
nd, No data.

higher in Inceptisols, which also had a very high base saturation and a slightly alkaline soil reaction despite being cropped with sisal for over 60 years. Common to all these soils under sisal are the very low levels of available P, which are on average below 4 mg/kg in the topsoil and less than 3 mg P/kg in the subsoil. Exchangeable K was low in most soil orders, but levels over 8 $mmol_c$ K/kg were recorded in some of the topsoils of the Alfisols and Inceptisols.

Land management under sisal

Sisal is propagated from bulbils, i.e. small plantlets that appear in large quantities in the inflorescence after anthesis (particularly in *A. sisalana*). The bulbils are raised in nurseries for about 2 years. In general, 2–3 years after transplanting the bulbils to the field, the first leaves can be cut and cutting may continue for up to 8 years. Thereafter the plants produce a pole, start flowering and leaf production ceases. From planting to flowering lasts about 10 years and this is termed a cycle. The length of a cycle depends on the growth rate of the sisal plants, which is influenced by soil conditions, temperature and rainfall (Lock, 1969). On good soils and with proper management, one cycle of Hybrid 11648 may yield 25 Mg/ha of fibre. After each cycle, the vegetation is cleared with heavy machinery and the debris is burned, whereafter the land is harrowed and replanted. Some sisal growers use a rotational system by which the land is left fallow for 10–20 years after each cycle of sisal. This bush-fallow is usually cleared with the use of heavy machinery. Manuring or the applications of inorganic fertilizers have never been widely adopted by sisal growers and on the plantations, sisal is grown with low levels of agrochemical inputs. Also the waste from the sisal processing factory is not used by sisal growers despite the fact that it contains large amounts of nutrients (Lock, 1969).

Soil requirements of sisal

In Tanzania, sisal has been planted on a wide range of soils and although this might give the impression that sisal is not critical in its soil requirements, for high yields the crop needs to be grown on the most suitable soil. A key requirement for sisal is that the soil is well drained as even the shortest period of waterlogging seriously affects growth and production. Soil chemical requirements are related to the fact that it originates from the limestone areas in Yucatan, Mexico. Sisal prefers soils with a neutral soil reaction and has high requirements for Ca and K (Table 10.2). Nitrogen and Mg requirements are lower, and P is only needed in small quantities (National Soil Service, 1988).

Table 10.2. Key to soil fertility assessment for sisal. Modified from Rijkebusch and Osborne (1965) and National Soil Service (1988).

	Very low	Low	Moderate	High	Very high
pH H_2O	<5.0	5.0–5.9	6.0–6.9	7.0–8.0	>8.0
Organic C (g/kg)	<10	10–20	21–40	41–60	>60
Available P (mg/kg)		<7	7–20	>20	
CEC ($mmol_c$/kg)	<50	50–80	81–150	151–300	>300
Exchangeable Ca ($mmol_c$/kg)	<30	30–50	51–99	100–200	>200
Exchangeable Mg ($mmol_c$/kg)	<5	5.1–15.0	15.1–30.0	>30	
Exchangeable K ($mmol_c$/kg)	<1.5	1.5–2.9	3.0–5.0	5.1–10.0	>10.0
Base saturation (%)	<20	21–40	41–60	61–100	

Research Methods

This section briefly describes the methods used to investigate the soil chemical changes at the sisal plantations, including soil sampling and analysis, statistical analysis of the data, and the construction of nutrient balances. Full details are given in the original publications (Hartemink and Bridges, 1995; Hartemink *et al.*, 1996; Hartemink, 1997a,b)

Soil sampling and analysis

The selection of soil sampling sites on the sisal plantations was based on detailed soil maps (1:20,000 or 1:30,000) prepared at the end of the 1980s and early 1990s. Sample sites were selected from records of field history provided by the sisal plantation management and both Type I and Type II data were collected. Permanently cropped fields sampled in the 1950s or 1960s were re-sampled in the late 1980s and early 1990s (Type I data). Samples were taken in five different soil orders and on three sisal plantations. The sampled soils had been under sisal cropping since the 1930s or 1940s.

Topsoil samples were taken in sisal fields and in similar soils immediately outside the plantation that had never been cropped (Type II data). Sampled sites were within a distance of 100 m of each other. The uncultivated land was usually covered with thick woodland (bush vegetation). Such samples were taken in Oxisols, Ultisols, and Inceptisols on three plantations. All soil samples were taken at 10–15 locations in an area of about 0.5 ha. At each location a mini-pit was dug with hoes, and soil samples were taken with a small spade at two depths: 0–0.20 m and 0.30–0.50 m. The samples were taken in the middle of the sisal rows. Sampling methods in the 1950s and 1960s were similar (J.F. Osborne, personal communication, 1995). Soil analyses were carried out at the National Soil Service Laboratories in Mlingano following standard procedures: pH H_2O in 1:2.5 suspension of soil and water; organic carbon by $K_2Cr_2O_7$ and H_2SO_4 oxidation (Walkley & Black); exchangeable cations Ca, Mg, K, Na and CEC percolation by 1 M NH_4OAc followed by spectrophotometry (K, Na), AAS (Ca, Mg) and titration (CEC); available P by NH_4F and HCl extraction (Bray I) for soils with pH <7, and $NaHCO_3$ extraction (Olsen) for soils with pH >7; exchangeable acidity (H, Al) extraction by 1 M KCl, and particle size analysis by hydrometer. In the comparison between the data of the 1950s–1960s and 1980s–1990s, pH H_2O, organic C, and exchangeable cations data could be used because the analytical methods had not changed.

Statistical analysis – Type I and II data

A statistical analysis was carried out for the soil chemical data of the Oxisols, Ultisols and Inceptisols. For other soil orders, insufficient data was available for statistical analysis. Statistical analysis was carried out on the historical soil data after the data had been log-transformed because the data were skewly distributed. The number of soil samples from the 1950s–1960s was not equal to the number of samples from the 1980s–1990s for both Oxisols (31 vs. 25) and Inceptisols (28 vs. 29). The statistical analysis for groups of unequal sizes follows, however, almost exactly the pattern for groups of equal sizes (Snedecor and Cochrane, 1989). The pooled variance for the data was calculated from the sum of squared deviations within the population. This was followed by the calculation of the *t*-value taking into consideration the unequal population size. The difference in the geometric means between the two sampling periods is reported.

Differences in soil chemical properties between bush vegetation and permanent sisal (Type II data) were analysed by an ANOVA. The data were tested for normality using ANOVA with untransformed data, after which the residuals were examined. The residuals of some soil chemical parameters were not normally distributed and showed a variance that increased with the mean. Log-transformation of data was used to overcome the skewed distribution and non-constant variance of these parameters. For data including small values, 1 was added before log-transformation. Student's *t*-tests were applied for comparing means of the two sampling depths. Back transformation was used for the skewly distributed data and differences in geometric means in soil properties between bush vegetation and sisal cropping are presented. For the normally distributed data differences in the arithmetic mean is given.

Nutrient balances

A nutrient balance was calculated for a sisal field (block) for which yield and soil data were available from 1966 to 1990. The sisal field was located on an elongated ridge at an altitude of 180 m a.s.l., and the soils were very deep (>4m), uniform and highly weathered. They were classified as very fine, isohyperthermic, Rhodic Haplustox. The natural forest at the site was cleared in 1956 and the field planted with sisal (*A. sisalana*) in 1957. After the sisal had poled, the land was cleared and replanted in 1966. In 1976, a third crop was planted, which was harvested up to the late 1980s. The planted crop in 1976 was Hybrid 11648. From 1966 to 1990, sisal yield (Mg/ha/year) and rainfall (mm/day) were recorded. The field had never been fertilized or manured.

In this field, a composite topsoil sample was taken by workers of the Sisal Research Station Mlingano in 1966 before the second planting. The sample was taken from *c.* 0.5 ha and consisted of 15 subsamples (J.F. Osborne, personal communicaton, 1995). In the same field, a new composite topsoil sample was taken in 1990 by the present writer. Soil analytical methods for the 1966 and 1990 samples were the same.

Ring samples (100 ml) were taken in soil pits in 1990 for the determination of the bulk density. Topsoil bulk density was 1.31 Mg/m^3 (mean of three replicates) and this value was used to calculate soil nutrient contents in kg/ha. The measured bulk densities of 1990 were also used for the 1966 soil chemical data. As the soils had already been under cultivation for 10 years in 1966, it is assumed that their bulk densities had not significantly changed between 1966 and 1990.

The following nutrient inputs were included in the balance: wet deposition, non-symbiotic N fixation, and nutrients added with the planting material. Nutrients in the wet deposition were calculated with the regression equation of Stoorvogel and Smaling (1990). At the research site, annual rainfall varied from 599 to 1669 mm between 1966 and 1990. Nitrogen deposited with the rainfall varied from 3.4 to 5.8 kg/ha/year (mean 4.6 kg N/ha/year), P deposits ranged from 0.6 to 0.9 kg/ha/year (mean 0.8 kg P/ha/year), and K deposition was on average 3.0 kg/ha/year (range 2.3–3.8 kg K/ha/year). A small source of N input occurs with non-symbiotic fixation. It occurs in most soils but its contribution to the N balance is difficult to quantify. Stoorvogel and Smaling (1990) used an equation in which non-symbiotic N fixation was considered to be rainfall-dependent: $A = 2 + (P - 1350) \times 0.005$, in which A is kg N/ha/year and P is mm/year. For the sisal field, non-symbiotic fixation hence ranged from 0 to 3.6 kg N/ha/year (mean <0.5 kg N/ha/year).

An important source of nutrient inputs for a sisal field is the planting material. Input of nutrients with seeds or planting material is rarely included in nutrient balances, but for sisal such inclusion is required, as at the beginning of a cycle thousands of small sisal plants (*c.* 2 kg each) are brought to the field. These plants are raised in nurseries at a density of 80,000/ha. Nutrient removal from such nurseries is 257 kg N, 78 kg P, 283 kg K, 699 kg Ca and 102 kg Mg/ha (Osborne, 1967). At the research site, plant densities of field sisal were 5000/ha. Input of nutrients with the planting material in 1966 and 1976 was hence calculated as 5000/80,000 multiplied by the nutrient removal of sisal nurseries.

The only output of nutrients that could be fairly well quantified was the removal with the harvested products. Crop residues were not removed from the field and are not included in the balance. It has been frequently observed that erosion is negligible on Oxisols under sisal cultivation in Tanga Region when there is a grass cover, and this has also been reported by Ngatunga *et al.* (1984). Moreover, sisal is a perennial crop with a grass cover between the rows, which gives better soil protection than annual crops. Losses of nutrients by erosion are therefore considered negligible. Soil erosion losses can be substantial in nurseries on sloping land (Lock, 1969), but at the site slopes were less than 5% and soil erosion was not observed.

Yield data were multiplied by nutrient removal data (kg/ha) to arrive at kg nutrient per Mg fibre/ha. Nutrient removal data, which vary between cultivars (Table 4.9), were taken from Osborne (1967), who conducted research on leaf analysis and nutrient removal in Mlingano in the 1960s. For *A. sisalana*, Osborne (1967) found a removal per Mg fibre of: 27 kg N/ha, 7 kg P/ha, 69 kg K/ha, 70 kg Ca/ha, and 34 kg Mg/ha. Nutrient removal for sisal Hybrid 11648 was estimated to be: 26 kg N/ha, 3.5 kg P/ha, 44 kg K/ha, 82 kg Ca/ha, and 31 kg Mg/ha per Mg of fibre. For the nutrient balance, the *A. sisalana* removal data were multiplied by the yields from 1966 to 1976, and the sisal Hybrid 11648 data by the yields from 1976 to 1990.

Type I and II Data

Type I data

An Oxisol sampled in 1966 had an acid soil reaction with low amounts of exchangeable cations. In 1987 the same field had a strongly acid soil reaction (pH 5.0) and very low levels of exchangeable Ca and Mg. Exchangeable K was already low in 1966 but it further declined by 3 $mmol_c$/kg between 1966 and 1987 (Table 10.3).

Table 10.3. Changes in soil chemical properties (0–0.20 m) of sisal fields at different sampling times. Type I data, modified from Hartemink (1997b).

Soil order	Years between sampling	pH H_2O (1:2.5)	Organic C (g/kg)	Exchangeable cations ($mmol_c$/kg) Ca	Mg	K
Alfisols	27	+0.1	nd	+3	+3	0
Inceptisols (lithic)	28	+0.9	nd	+6	+44	−3
Mollisols	28	−0.2	nd	−82	+10	−8
Oxisols	21	−0.5	−10	−13	−8	−3
Ultisols	24	−1.0	−3	−34	−11	−2

nd, No data.

The topsoil pH of an Ultisol had decreased by one pH unit in 25 years. Levels of exchangeable cations in 1990 were about 60% of their 1966 levels. The pH and exchangeable cations of an Alfisol had changed little between 1960 and 1987. Soil chemical properties in a shallow Inceptisol overlying coral limestone had hardly changed between 1959 and 1987, but exchangeable K decreased from 5 to 2 $mmol_c$/kg. In a Mollisol exchangeable K declined by 8 $mmol_c$/kg in 28 years. A considerable number of topsoil chemical data were available for Oxisols in the uplands and Inceptisols in the coastal area. The data were grouped for the 1950s–1960s and from the late 1980s (Table 10.4). The soil pH of the Oxisols had declined by 1.2 units and organic C as well as exchangeable cations had declined. Statistical analysis revealed a significant decline in pH and levels of exchangeable cations in Oxisols. In Inceptisols, only exchangeable K levels had declined significantly; other changes were not significant.

Table 10.4. Soil chemical properties (0–0.20 m) of sisal fields on Oxisols and Inceptisols at different sampling times (median with range of values in parentheses). Type I data, modified from Hartemink and Bridges (1995).

	Oxisols				Inceptisols (overlying limestone)			
Years of sampling:	1956–1966		1987–1990		1958–1960		1987–1989	
Number of samples:	31		25		28		29	
pH H_2O (1:2.5)	6.1	(5.2–7.3)	4.9	(4.3–5.6)	7.0	(5.5–8.0)	6.6	(5.4–8.1)
Organic C (g/kg)	20	(17–25)	17	(14–21)	nd	–	18	(6–30)
Exchangeable Ca ($mmol_c$/kg)	39	(9–113)	14	(3–16)	116	(2–755)	89	(23–633)
Exchangeable Mg ($mmol_c$/kg)	25	(1–45)	10	(3–23)	19	(6–43)	15	(2–45)
Exchangeable K ($mmol_c$/kg)	2	(1–17)	1	(< 0.5–8)	5	(1–28)	3	(1–9)
Base saturation (%)	56	(36–88)	33	(14–79)	93	(53–100)	68	(27–100)

nd, No data.

Type II data

The topsoil pH of an Oxisol under bush vegetation was 1.0 unit higher than in a similar soil under sisal, and the higher pH was accompanied by a higher CEC and base saturation (Table 10.5). Exchangeable cations were higher under bush than under sisal but absolute levels were low in Oxisols under both bush vegetation and sisal. Available P was very low regardless of land use. Exchangeable Al was nil under bush but 10 $mmol_c/kg$ (42% ECEC) in the subsoils under sisal. Ultisols under bush vegetation had a topsoil pH that was 1.5 units higher than under sisal. Exchangeable Ca and Mg levels were also much lower under sisal. In Inceptisols, the exchangeable K, Mg and available P levels were lower under sisal cropping than in the soils under bush vegetation (Table 10.5).

Table 10.5. Soil chemical properties of Oxisols, Ultisols and Inceptisols under bush vegetation and sisal. Type II data, modified from Hartemink (1997b).

	Sampling depth (m)	Oxisols		Ultisols		Inceptisols	
		Bush	Sisal	Bush	Sisal	Bush	Sisal
pH H_2O (1:2.5)	0–0.20	6.2	5.2	6.1	4.6	7.5	7.4
	0.30–0.50	5.7	5.1	5.6	4.7	7.5	6.6
Organic C (g/kg)	0–0.20	21	17	15	11	19	34
	0.30–0.50	9	6	4	3	13	15
Available P (mg/kg)[a]	0–0.20	3	3	3	< 0.5	9	4
	0.30–0.50	1	1	< 0.5	< 0.5	4	1
CEC ($mmol_c$/kg)	0–0.20	125	88	157	110	310	310
	0.30–0.50	105	60	127	97	481	221
Exchangeable Ca ($mmol_c$/kg)	0–0.20	68	13	38	11	161	140
	0.30–0.50	23	9	17	11	97	107
Exchangeable Mg ($mmol_c$/kg)	0–0.20	26	5	23	5	70	36
	0.30–0.50	21	3	9	5	40	23
Exchangeable K ($mmol_c$/kg)	0–0.20	5	1	5	3	3	1
	0.30–0.50	4	< 0.5	5	2	3	1
Base saturation (%)	0–0.20	80	21	45	24	76	58
	0.30–0.50	54	20	31	27	29	60
Exchangeable Al ($mmol_c$/kg)	0–0.20	0	9	0	nd	0	0
	0.30–0.50	0	10	0	nd	0	0
Al saturation (% ECEC)[b]	0–0.20	0	32	0		0	0
	0.30–0.50	0	42	0		0	0

[a] Soil pH <7: P-Bray I; soil pH >7: P-Olsen.
[b] Aluminium saturation of the ECEC is calculated as: Al/Ca+Mg+K+Na+H+Al*100.
nd, No data.

At one plantation soil samples were taken in natural forest and in fields that had been under two and three cycles of sisal cultivation. Soil fertility was highest under forest and decreased with increasing periods of sisal cultivation (Table 10.6). An exception is the soil P content, which was very low in all three soils. The pH decreased by 0.5 of a unit in the topsoil with an extra cycle of sisal but soil organic C contents were not influenced by the number of sisal cycles. Exchangeable Ca, Mg and K contents decreased considerably with an extra cycle of sisal, and K was exhausted after three sisal cycles. The decline of exchangeable cations is greater in the subsoil, and this is accompanied by moderate to high levels of exchangeable Al.

From five sisal fields on Oxisols and three fields on Ultisols, soil samples were taken under bush vegetation and sisal. Differences in soil chemical properties were statistically analysed for both soil orders (Table 10.7). The pH had significantly decreased in both soil orders, and the decrease was largest in the topsoils of the Ultisols. Organic C and available P were significantly lower in the topsoils of the Oxisols, but did not differ between Ultisols under bush vegetation or sisal. Levels of exchangeable cations had decreased significantly in both soils under sisal, except for the exchangeable K level in the subsoils of the Ultisols. Likewise, base saturation had decreased significantly and the decrease was largest in the subsoils of the Oxisols. The CEC did not alter significantly in both soil orders. Exchangeable Al had increased significantly in the subsoils of the Oxisols and Ultisols.

Table 10.6. Soil chemical properties of Oxisols under natural forest compared to soils after 2 or 3 cycles of sisal cultivation. Type II data, modified from Hartemink (1991).

	Natural forest[a]		Two cycles of sisal[b]		Three cycles of sisal[b]	
Sampling depth (m)	0–0.20	0.30–0.50	0–0.20	0.30–0.50	0–0.20	0.30–0.50
pH H_2O (1:2.5)	6.2	5.7	5.7	5.2	5.2	5.1
Organic C (g/kg)	21	9	17	6	17	6
Total N (g/kg)	1.9	0.8	1.3	0.5	1.4	0.6
Available P (Bray I) (mg/kg)	3	1	2	1	3	1
Exchangeable Ca ($mmol_c$/kg)	68	23	32	18	13	9
Exchangeable Mg ($mmol_c$/kg)	26	21	16	10	5	3
Exchangeable K ($mmol_c$/kg)	5	4	6	3	1	<0.05
CEC ($mmol_c$/kg)	125	105	117	98	88	60
Base saturation (%)	80	54	48	32	21	20
Exchangeable Al ($mmol_c$/kg)	0	0	0	5	9	10
Al saturation (% CEC)	0	0	0	5	10	17

[a] Data of 1 composite topsoil sample (= 15 subsamples of about 0.5 ha).
[b] Mean of 2 composite topsoil samples; 1 cycle is 10 years.

Table 10.7. Differences in soil chemical properties of Oxisols and Ultisols under bush vegetation and sisal. Values are calculated as sisal cropping minus bush vegetation. Type II data, modified from Hartemink (1997b).

	Oxisols ($n = 10$)		Ultisols ($n = 6$)	
Sampling depth (m):	0–0.20	0.30–0.50	0–0.20	0.30–0.50
pH H_2O (1:2.5)	–1.2 ***	–1.2 ***	–1.4 **	–1.3 **
Organic C (g/kg)	–5 *	–1 ns	–2 ns	–1 ns
Available P Bray I (mg/kg)	–3 **	–1 ns	–1 ns	0
CEC ($mmol_c$/kg)	–5 ns	+2 ns	–42 ns	–38 ns
Exchangeable Ca ($mmol_c$/kg)	–25 *	–13 **	–29 **	–18 *
Exchangeable Mg ($mmol_c$/kg)	–10 *	–10 *	–23 *	–22 *
Exchangeable K ($mmol_c$/kg)	–2.6 *	–2.4 *	–3 *	–2 ns
Base saturation (%)	–40 **	–44 **	–25 *	–21 *
Exchangeable Al ($mmol_c$/kg)	+2 ns	+7 **	+5 ns	+11 *
Al saturation (% ECEC)	+3 ns	+20 **	+21 ns	+34 ns

***,**,* significant difference at $P < 0.001$, $P < 0.01$, $P < 0.05$, respectively.
ns, No significant difference.

Rates of Change

From the Type I and II data presented in the previous sections the rates of change in soil chemical properties are calculated. Absolute changes in unit per year are presented in Table 10.8. No changes occurred in the Alfisols and in some of the Inceptisols. A consistent decline in all soil fertility properties was found in Oxisols and Ultisols.

The relative change, based on the initial level, is presented in Table 10.9. The decline in soil organic C in the Oxisols is about 2% per year whereas the rate of decline in soil reaction ranges from 0% to 2%. The decline in exchangeable cations is much higher and values of up to 7% were found. Exchangeable Ca decreased in most soils but both the absolute and relative decrease was highest in the Oxisols.

Semiquantitative Data

Differences between nutrient input with rainfall, non-symbiotic N fixation and planting material, and output with the harvested product (yield) were calculated for the five major nutrients. In the absence of fertilizers and manure the balance is negative for each nutrient (Table 10.10).

Total input of N during the period 1966–1990 was 166 kg N/ha, of which the major part was deposited with the rainfall. Output of N with the harvested product was about three times larger than the sum of inputs, and the total differences between 1966 and 1990 was –326

Table 10.8. Absolute change (unit/year) in soil chemical properties (topsoils only) on sisal plantations in northeast Tanzania. Soils were permanently cropped and had received no manure or inorganic fertilizer input. Based on Type I and II data.

Soil order	Data type	Period[a] (years)	pH H_2O	C (g/kg)	N (g/kg)	P (mg/kg)	CEC and exchangeable cations ($mmol_c$/kg)			
							CEC	Ca	Mg	K
Alfisols	I	27	0	nd	nd	nd	nd	+0.11	+0.11	0
Inceptisols	I	28	+0.03	nd	nd	nd	nd	+0.21	+1.57	−0.11
Inceptisols	I	30	−0.01	nd	nd	nd	nd	−0.90	−0.13	−0.07
Mollisols	I	28	−0.01	nd	nd	nd	nd	−2.93	+0.36	−0.29
Oxisols	II	10	−0.05	−0.40	−0.06	−0.10	−0.80	−3.60	−1.00	+0.10
Oxisols	II	20	−0.05	−0.20	−0.03	0	−1.85	−2.75	−1.05	−0.20
Oxisols	I	21	−0.02	−0.48	nd	nd	nd	−0.62	−0.38	−0.14
Oxisol	I	24	−0.15	−0.32	nd	nd	nd	−4.72	−1.84	0
Oxisols	I	30	−0.04	−0.10	nd	nd	nd	−0.83	−0.50	−0.03
Ultisols	I	24	−0.04	−0.13	nd	nd	nd	−1.42	−0.46	−0.08

[a] Period of cultivation (Type I data); years difference between cropped and uncultivated soils (Type II data).
nd, No data.

Table 10.9. Relative change (in % per year) in soil chemical properties (topsoils only) on sisal plantations in northeast Tanzania. Soils were permanently cropped and had received no manure or inorganic fertilizer input. Type I and II data.

							CEC and exchangeable cations ($mmol_c$/kg)			
Soil order	Data type	Period[a] (years)	pH H_2O	C (g/kg)	N (g/kg)	P (mg/kg)	CEC	Ca	Mg	K
Alfisols	I	27	0	nd	nd	nd	nd	0	+1	0
Inceptisols	I	28	0	nd	nd	nd	nd	0	+9	–2
Inceptisols	I	30	0	nd	nd	nd	nd	–1	–1	–1
Mollisols	I	28	0	nd	nd	nd	nd	–1	+1	–3
Oxisols	II	10	–1	–2	–3	–3	–1	–5	–4	+2
Oxisols	II	20	–1	–1	–1	0	–1	–4	–4	–4
Oxisols	I	21	0	–2	nd	nd	nd	–3	–3	–4
Oxisols	I	24	–2	–2	nd	nd	nd	–6	–7	0
Oxisols	I	30	–1	–1	nd	nd	nd	–2	–2	–2
Ultisols	I	24	–1	–1	nd	nd	nd	–2	–2	–2

[a] Period of cultivation (Type I data); years difference between cropped and uncultivated soils (Type II data).
nd, No data.

Table 10.10. Nutrient balance of a sisal field during the period 1966 to 1990 (kg/ha). Modified from Hartemink (1997a).

	N	P	K	Ca	Mg
Input with rainfall	115	19	75	213	105
Input with BNF	19	0	0	0	0
Input with planting material	32	10	35	87	13
Output with yield	491	100	1067	1400	605
Difference	–326	–71	–957	– 1100	–487

kg N/ha. Little P was deposited with the rainfall (<1 kg P/ha/year) but the output of P was also moderate (–100 kg P/ha in 25 years). The total difference was –71 kg P/ha. Although considerable amounts of cations were deposited in the rainfall and supplied with the planting material, there was a large shortfall for each of the three cations. The removal of K with the yield was tenfold greater than the sum of inputs. For Ca and Mg, the nutrient removal with the yield was about five times larger than the sum of inputs.

Soil chemical properties of a sisal field sampled in 1966 and 1990 were used to calculate topsoil nutrient contents. The content of each of the five nutrients had been seriously reduced during this period and Table 10.11 presents the major nutrient content of the topsoil. The largest decrease was found in the soil N content and 2620 kg/ha had disappeared from the topsoil. The largest relative decrease was found in the soil P content and only 8 kg P/ha was available in the topsoils in 1990. Exchangeable cation contents in 1990 were less than 30% of the contents in 1966.

The nutrient balance was compared with the difference in soil nutrient contents (Table 10.12). More N had disappeared from the topsoil than was calculated from the nutrient balance. Differences in topsoil P contents were slightly lower than was calculated from the

Table 10.11. Soil nutrient content of a sisal field sampled in 1966 and in 1990 (kg/ha for 0–0.20 m soil depth). Modified from Hartemink (1997a).

	N	P	K	Ca	Mg
Content in 1966[a]	5764	52	369	996	355
Content in 1990	3144	8	82	271	97
Difference	–2620	– 44	–287	–725	–258

[a] For the nutrient content in 1966 the bulk density measurements of 1990 were used.

Table 10.12. Nutrient balance and differences in soil nutrient contents between 1966 and 1990 (kg/ha/year). Modified from Hartemink (1997a).

	N	P	K	Ca	Mg
Nutrient balance (Table 10.10)	–13	–2.8	–38	–44	–19
Soil changes (Table 10.11)	–104	–1.8	–11	–29	–10

nutrient balance. The annual decrease in exchangeable cations was considerably lower than the shortfall based on the nutrient balance. Noticeably more K had disappeared when calculated with the nutrient balance than was lost from the topsoil.

Nutrient depletion for the whole of Tanzania was estimated by Stoorvogel and Smaling (1990) (see Chapter 5, 'Semiquantitative Studies', on page 155). For the year 2000, the N balance was –32 kg/ha, the P balance –5 kg/ha, and the K balance was estimated to be –21 kg/ha (Table 5.10). Comparing these values with those in Table 10.12 shows that the sisal nutrient balance is less negative for N and P, but worse for K. It shows the divergence in values obtained with calculations from soil chemical properties (Table 10.11) and by semi-quantitative approaches (Table 10.10) and the data from Stoorvogel and Smaling (1990).

The Effects on Yield

Soil fertility had seriously declined in some of main soils under sisal in Tanzania. In the introduction to this chapter, it was shown that sisal production had been drastically reduced since the mid-1960s. The question arises of how much of the production decline can be attributed to the decline in soil fertility and other agronomic factors?

On the plantations in Tanzania sisal is grown in blocks (fields) of 5–10 ha but sisal yields are not collected for individual blocks. So it was not possible to relate soil chemical data, which were available for many blocks, to sisal yield data. For most plantations, annual total fibre production is available, which divided by the area under mature sisal gives average fibre yields per ha. Interannual variation in plantation yields may be considerable because of changing production patterns. Therefore yield data from five plantations were collected and it was found that the average annual yield declined by 0.5 Mg/ha between 1968 and 1990 (Hartemink and Wienk, 1995). The decline in yield was almost 30%. From one plantation in the Tanga Region leaf length data were available from measurements of the percentage fibre grade 3L+1 – representing the longest fibres. The percentage of 3L+1 fibre has decreased from above 80% in the early 1970s to less than 40% in the late 1980s, which indicates a severe reduction in long sisal leaves (Hartemink and Wienk, 1995).

In Tanzania, lower yields and shorter leaves have been attributed to the unintentional planting of less productive *Agave* hybrids. In the late 1980s, some evidence was generated that many fields had not been planted with the high-yielding *Agave* Hybrid No. 11648 (Fig. 10.3) but with another, less-productive hybrid with shorter leaves. This hybrid, which was nicknamed *kaptura* (Swahili for shorts), had probably spread as a result of uncontrolled dissemination of planting material. Some growers believe that *kaptura* is a mutant or a genetically degenerated Hybrid No. 11648. The scale at which *kaptura* is found, and the fact that at Mlingano during the many years following the selection of Hybrid No. 11648 off-types were never encountered, does not make this a likely explanation. The effects of soil fertility decline on sisal yield are further discussed in Chapter 11, 'Effect on crop production', page 326.

Discussion and Conclusions

Large differences were found in soil chemical changes between different soil orders. Although for some soil orders the data were few,

Fig. 10.3. A healthy crop of Hybrid 11648 at the sisal research station in Mlingano. Photograph from the 1960s. Such well-maintained and highly productive fields are nowadays rarely seen in Tanga Region.

Oxisols and Ultisols showed a striking decline in soil fertility under sisal. Soil fertility has hardly changed in shallow Inceptisols, Alfisols and Mollisols as a result of sisal cultivation. These soils are able to resist the process of decline as nutrients removed by the crop are replenished by weathering of the underlying limestone rock and the subsequent capillary rise of cations. This replenishment has maintained high Ca and Mg levels and the pH has remained near neutral. Levels of exchangeable K have decreased in these soils. Apparently K is not replenished by the weathering of primary minerals or the underlying limestone rock whereas sisal removes considerable amounts of K (Lock, 1969). Rates of decline in Oxisols were about the same for Type I and II data and there was good consistency between the two data types. The nutrient balance based on inputs and outputs compared with a soil nutrient balance of a sisal field confirmed the observations made by the Type I and II data.

The decline in soil chemical properties is related to the soil management practices at the sisal plantations. Some sisal growers leave their land fallow after a cycle of sisal and assume that the soil chemical fertility is sufficiently restored for a new cycle of sisal (10 years). In order to verify this, soil samples were taken in Oxisols and Ultisols that had been fallow for 18 years and in adjacent sisal fields that had been cropped for many decades (Hartemink *et al.*, 1996). In Oxisols, a slightly higher pH was found in the soils under bush fallow compared with soils under sisal cultivation, but no differences were found in the levels of soil organic C or exchangeable cations (Table 10.13). The main difference was the reduction in Al saturation in both topsoils and subsoils. In Ultisols, the pH increase under bush fallow was much higher but again no differences were found in the soil organic C levels. Exchangeable Ca and K were higher but levels of exchangeable Mg were similar. In these poor soils it appears that the amount of nutrients gained by the fallow are inadequate for a new cycle of sisal. Apparently fallowing has little effect and this confirmed the observations by Nye and Greenland (1960), who had reported that fallows did not restore the soil chemical fertility when the land was cultivated for prolonged periods.

It is evident that the decline in soil chemical fertility of the Oxisols is caused by the removal of cations owing to sisal cropping and lack of inorganic fertilizers and lime, resulting in a negative nutrient balance, which has to be offset from the soil nutrient pool. Moreover, the regenerating effect of a fallow period is very limited. Since these soils have few weatherable minerals remaining, they are easily depleted and nutrient inputs are required. The increasing acidity of the Oxisols and Ultisols under permanent sisal has a number of unfavourable effects. The uptake of Ca, Mg and K is suppressed in the presence of high proton concentrations, which is

Table 10.13. Soil chemical properties of Oxisols and Ultisols under 18 years of bush fallow and sisal cultivation. Values are the mean of three samples (±1 SD). Type II data, modified from Hartemink *et al.* (1996).

	Sampling depth (m)	Oxisols: 18 years of bush fallow	Oxisols: Permanent sisal cultivation	Ultisols: 18 years of bush fallow	Ultisols: Permanent sisal cultivation
pH H_2O (1:2.5)	0–0.20	4.8 ± 0.1	4.5 ± 0.1	5.9 ± 0.5	5.0 ± 0.4
	0.30–0.50	4.9 ± 0.1	4.3 ± 0.1	5.1 ± 0.6	4.4 ± 0.3
Organic C (g/kg)	0–0.20	15 ± 0.9	18 ± 1.8	19 ± 2.3	18 ± 0.7
	0.30–0.50	9 ± 0.8	9 ± 0.6	11 ± 1.3	12 ± 0.7
Available P Bray I (mg/kg)	0–0.20	1 ± 1.1	3 ± 1.5	4 ± 0.7	3 ± 0.5
	0.30–0.50	1 ± 0.8	1 ± 0.7	1 ± 0.3	1 ± 0.2
Exchangeable Ca ($mmol_c$/kg)	0–0.20	12 ± 1.8	10 ± 3.7	31 ± 12.4	23 ± 5.8
	0.30–0.50	8 ± 1.8	3 ± 1.6	9 ± 7.3	16 ± 6.2
Exchangeable Mg ($mmol_c$/kg)	0–0.20	8 ± 2.2	7 ± 2.2	17 ± 4.9	15 ± 4.2
	0.30–0.50	2 ± 1.5	2 ± 1.9	11 ± 4.7	12 ± 7.1
Exchangeable K ($mmol_c$/kg)	0–0.20	4 ± 2.0	2 ± 1.0	3 ± 1.7	2 ± 1.3
	0.30–0.50	3 ± 2.1	1 ± 0.8	2 ± 2.0	1 ± 0.9
Al saturation (% ECEC)[a]	0–0.20	4 –	25 ± 9.8	1 ± 1.4	7 ± 10.9
	0.30–0.50	10 –	60 ± 16.8	8 ± 6.0	17 ± 10.7

[a] Aluminium saturation of the ECEC is calculated as: Al/Ca+Mg+K+Na+H+Al*100.
–, Insufficient data to calculate a standard deviation.

particularly problematic in soils with very low CECs (Kamprath, 1984). As acidity increases, Al is released from soil particles into the soil solution. Also Mn becomes soluble and may be found at toxic levels. Furthermore, high levels of acidity in the subsoil affect the rooting depth, which limits deep nutrient and water uptake (Ritchey *et al.*, 1980). These factors will seriously affect sisal productivity, as the crop prefers alkaline soil reactions and high levels of Ca and Mg (Table 10.2).

Soil organic C decreased significantly in most topsoils, and it has consequences for sisal production because organic matter supplies most of the N taken up by unfertilized crops (Sanchez, 1976). For a field in Tanga Region cropped with sisal since 1957, it was calculated that owing to dwindling organic matter contents, 115 kg N/ha was released with organic matter mineralization in 1966 and only 63 kg N/ha in 1990 (Hartemink, 1995). In the Oxisols and Ultisols, most of the P is in the organic form and may become available with mineralization. The combination of increasing acidity resulting in P immobilization with the decrease in soil organic matter may severely reduce the P availability.

Figure 10.4 shows the main factors involved in the sisal production decline of Tanzania. Owing to low sisal-fibre prices and inadequate management practices sisal growing became less intensive. Crop husbandry standards fell and yield levels dropped, which further reduced the income at the sisal plantations. This indirectly resulted in the decline in soil fertility and it is likely that it has contributed to the dramatic fall in sisal production in Tanzania.

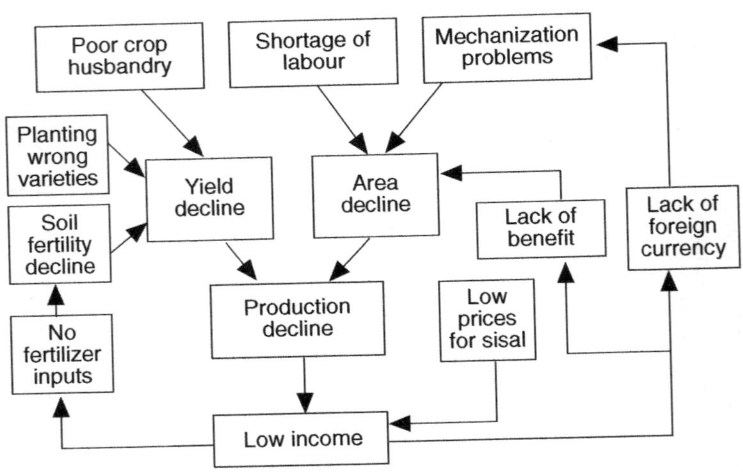

Fig. 10.4. Relationship diagram showing factors affecting sisal production and decline in Tanzania (Hartemink and Wienk, 1995).

References

Ahn, P.M. (1993) *Tropical Soils and Fertiliser Use.* Longman Scientific and Technical, Harlow, UK.

Ettighoffer, P.C. (1943) *Sisal das Blonde Gold Afrikas.* G. Bertelsmann, Gütersloh.

Hartemink, A.E. (1991) *Soils of Bamba Estate and their Suitability for Sisal, Teak and Cashew.* Cillington Corporation, London.

Hartemink, A.E. (1995) *Soil Fertility Decline under Sisal Cultivation in Tanzania.* ISRIC, Wageningen.

Hartemink, A.E. (1997a) Input and output of major nutrients under monocropping sisal in Tanzania. *Land Degradation and Development* 8, 305–310.

Hartemink, A.E. (1997b) Soil fertility decline in some major soil groupings under permanent cropping in Tanga region, Tanzania. *Geoderma* 75, 215–229.

Hartemink, A.E. and Bridges, E.M. (1995) The influence of parent material on soil fertility degradation in the coastal plain of Tanzania. *Land Degradation and Rehabilitation* 6, 215–221.

Hartemink, A.E. and Wienk, J.F. (1995) Sisal production and soil fertility decline in Tanzania. *Outlook on Agriculture* 24, 91–96.

Hartemink, A.E., Osborne, J.F. and Kips, P.A. (1996) Soil fertility decline and fallow effects in Ferralsols and Acrisols of sisal plantations in Tanzania. *Experimental Agriculture* 32, 173–184.

IMF (1988) *Primary Commodities. Market Development and Outlook.* Commodities Division of the Research Department, International Monetary Fund, Washington, DC.

Kamprath, E.J. (1984) Crop response to lime on soils in the tropics. In: Adams, F. (ed.) *Soil Acidity and Liming,* 2nd edn. ASA-CSSA-SSSA, Madison, Wisconsin, pp. 349–368.

Lock, G.W. (1969) *Sisal,* 2nd edn. Longman, Harlow, UK.

Milne, G. (1935) Some suggested units of classification and mapping, particularly for East African soils. *Soil Research* 4, 183–198.

National Soil Service (1988) *Soils of Kwamdulu Estate and their Potential for Hybrid Sisal Cultivation.* NSS, Mlingano, Tanzania.

Nye, P.H. and Greenland, D.J. (1960) *The Soil under Shifting Cultivation.* Commonwealth Bureau of Soils, Harpenden.

Osborne, J.F. (1967) *Some Preliminary Estimates of Nutrient Removal by Agave Crops.* Tanganyika Sisal Growers' Association, Mlingano, Tanzania.

Rijkebusch, P.A.H. and Osborne, J.F. (1965) *Interpretation of Soils Analyses in Relation to the Manuring of Sisal Fields.* Tanganyika Sisal Growers' Association, Mlingano, Tanzania.

Ritchey, K.D., Souza, D.M.G., Lobato, E. and Correa, O. (1980) Calcium leaching to increase rooting depth in a Brazilian savannah Oxisol. *Agronomy Journal* 72, 40–44.

Sanchez, P.A. (1976) *Properties and Management of Soils in the Tropics.* John Wiley & Sons, New York.

Smith, H.H. (1929) *Sisal Production and Preparation.* John Bale, Sons and Danielson, London

Snedecor, G.W. and Cochrane, W.G. (1989) *Statistical Methods*, 8th edn. Iowa State University Press, Ames.

Stoorvogel, J.J. and Smaling, E.M.A. (1990) *Assessment of Soil Nutrient Decline in Sub-Saharan Africa, 1983–2000*. Winand Staring Centre-DLO, Wageningen.

Synthesis 11

> ...people expect scientists to give them clear and unambiguous advice when new and worrying problems appear. Difficulties arise when the uncertainties in scientific advice to policy makers are not caused by probabilistic predictions, but rather derive from a fundamental lack of understanding of new phenomena at or beyond the frontiers of present knowledge.
>
> R. May (2001)

In the previous chapters, evidence was presented for soil changes under annual cropping (Chapter 5), perennial crops and forest plantations (Chapters 6 and 7), sugarcane (Chapter 8 and 9) and sisal (Chapter 10). The information was presented by soil order. In this chapter, the data are brought together to unravel similarities and differences in soil chemical changes. Firstly, it is investigated how soil chemical properties change under different land-use systems and how changes differ between soil orders. This is followed by a general discussion on the trends and study methods. From the investigation, a set of general observations on the effects of cropping on soils in the tropics is derived, followed by a discussion on the implications for crop production and sustainable land management.

Rates of Change in Soil Chemical Properties

Quantifying rates of change in soil properties for different soils and climates is necessary. These rates of change are needed to quantify soil fertility decline but very few studies have been conducted in which these rates were calculated. At the Royal Society Meeting in 1996 this concern was expressed as follows: much more needs to be known of the rate at which soil productivity declines under stress due to cultivation, and the reversibility of this degradation, which differs between soils because of the differences in their resilience to different kinds and intensities of stress (Greenland *et al.*, 1997). The rates can be calculated in different ways. First, the absolute change in a soil chemical property can be calculated. For example, if the total N level of the topsoil was 2.0 g N/kg in 1990 and 1.0 g N/kg in 2000, the absolute change is –1.0 g N/kg over 10 years, or on average –0.1 g N/kg per year. The decrease can also be expressed as a percentage over the whole period (50% decrease over 10 years), or as change in percentage per year (–5% per year).

As was discussed in the section 'Interpretation of the Results' (Chapter 4, page 123), the change in a soil chemical property is related to period of observation and a decrease may be expected when the observational period is long, the soil is permanently cropped and no amendments are made. Changes also depend on the initial level and the time of observation in relation to the first crop (Fig. 4.8(b)). Generally, large changes in soil chemical properties are observed directly after the forest is cut, burned and the first crop planted (Nye and Greenland, 1960; Sanchez *et al.*, 1983; Lal, 1986). If the initial level is high, the rate of change is also high. With time, rates of change decrease and an equilibrium is established.

From the soil chemical data presented in this book, rate of change was calculated for each topsoil chemical property (χ) measured over a given time span (t_1 to t_2), as follows: $((\chi_1 - \chi_2)/\chi_1) / (t_1 - t_2) \times 100$. This gives the change in percentage per year of the initial level, and takes into account initial level (χ_1) and the absolute change ($\chi_1 - \chi_2$) as well as the period between the soil samplings ($t_1 - t_2$). Plotting the rate of change against the period of observation reveals what changes can be expected depending on the period between two soil samplings.

Data on changes in topsoil chemical properties (pH, organic C, total N, available P, CEC and exchangeable Ca, Mg and K) under different land-use systems were plotted (Fig. 11.1). The data from the forest plantations were excluded as they distorted the pattern: much of the data was from degraded soils and an increase in soil chemical properties was common (Chapter 7). For the soil pH, a funnel-shaped pattern was found with some variation when the observational period was less than 10 years. Overall, the rates of change in soil pH were small and ranged from –3 to +3%. For other soil chemical properties there is a clear relationship between the rates of change and the period of observation. The annual decline in soil fertility is high with short-term observations and lower with long-term observations. In some cases a positive relation was found, namely increase in exchangeable Mg under sisal. These data were from

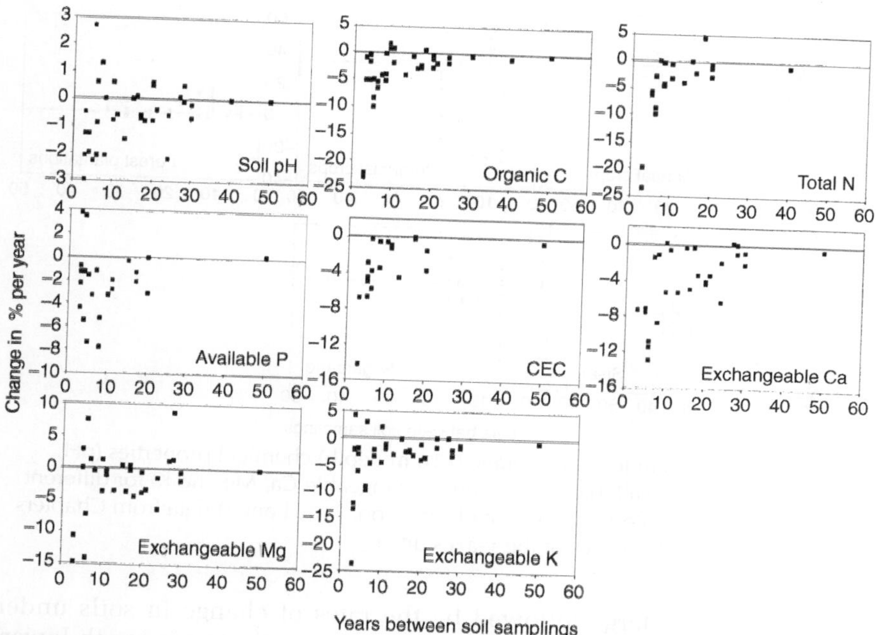

Fig. 11.1. Changes in topsoil (average 0.16 m depth) chemical properties in various land-use systems related to period of observation. Type I and II data presented in Chapters 5 to 10 – excluding data from forest plantations. Note that the scales of the *y*-axes differ.

highly heterogeneous soils overlying limestone and an increase in exchangeable Mg was found despite cropping with sisal for 30 years.

The length of time that the site had been under cultivation was not taken into account when the initial sample (t_1) was taken. So some of the data might be from soils that had been under cultivation for 10 years, whereas other data might be from soils that were cultivated from forest. It may affect the rates of change (i.e. larger in newly cultivated soils) and could partly explain the range of values found in some of the soil chemical properties.

Changes in Different Land-use Systems

Rates of change as calculated for individual soil properties were also calculated for each of the five land-use systems (annual crops, perennial crops, forest plantations, sisal and sugarcane) – Fig. 11.2. The largest rates of change in soil chemical properties were found in soils under annual crops, and much of the data from the annual cropping systems were short term (<10 years). Rates of change under sugarcane

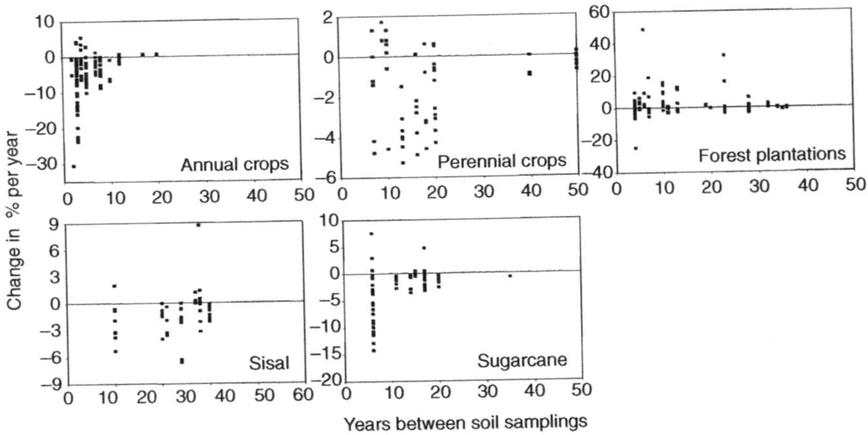

Fig. 11.2. Changes in topsoil (average 0.16 m depth) chemical properties (pH, organic C, total N, available P, CEC and exchangeable Ca, Mg and K) for different land-use systems related to period of observation. Type I and II data from Chapters 5 to 10. Note that the scales of the *y*-axes differ.

were also fairly large followed by the rates of change in soils under sisal. The rates of change under perennial cropping were much lower, but in soils of forest plantations the rates were variable and in various studies, positive rates of change were found, indicating an increase in soil chemical fertility.

For each land-use system, the average rate of change was calculated per soil chemical property (Table 11.1). Although there were differences in the period of observation, rates of change were highest in soils under

Table 11.1. Mean change in soil chemical properties (topsoils only; average 0.16 ± 0.05 m) for different land-use systems, in percentage per year (±1 SD). Based on Type I and Type II data presented in Chapters 5 to 10.

		Woody plantation crops		Herbaceous plantation crops	
	Annual crops	Perennial crops	Forest plantations	Sisal	Sugarcane
pH	−1 ± 2	<−0.5 ± 1	<+0.5 ± 1	−1 ± 1	−1 ± 1
Organic C	−5 ± 6	−1 ± 2	−1 ± 5	−1 ± 1	−2 ± 2
Total N	−5 ± 7	−1 ± 3	+2 ± 3	−2 ± 1	−4 ± 4
Available P	−4 ± 8	−1 ± 2	+2 ± 5	−2 ± 2	−1 ± 3
CEC	−6 ± 4	−3 ± 2	+8 ± 16	−1 ± 1	−1 ± 2
Exchangeable Ca	−7 ± 4	−3 ± 2	+5 ± 14	−2 ± 2	−5 ± 6
Exchangeable Mg	−5 ± 7	−1 ± 3	+1 ± 6	−1 ± 4	−2 ± 6
Exchangeable K	−7 ± 8	−3 ± 2	+3 ± 7	−2 ± 2	−3 ± 4

annual crops, followed by sugarcane. Rates of change in soils under sisal and perennial crops were about similar, and the rates were positive for most soil chemical properties under forest plantations.

Soil organic C declined on average in all land-use systems, but rates were found to be higher under annual crops and sugarcane. Also rates of change in soil N were high under annual crops. Rates of change in exchangeable cations were particularly high under sugarcane, which follows the general soil acidification trend. Part of the variation in the data is explained by the fact that these are the mean data from different soils; in the next section the possible differences between soil orders are investigated.

Changes by Soil Order

In this book, data were presented from highly weathered Oxisols and Ultisols to fertile Inceptisols and Mollisols. These soil orders have a different inherent soil fertility, which is determined by the parent material, the age and degree of weathering and the climate and vegetation; however, soil fauna and topography or relief are also important (Jenny, 1941, 1980). Cropping and nutrient management greatly affect the nutrient status but rates of change are probably related to the rate of pedogenesis and the stage of soil weathering. Using the work of Jackson and Sherman (1953), who based the soil-weathering stage on the minerals present, Richter and Markewitz (2001) distinguished three soil-weathering stages (Table 11.2).

The general distinction given in Table 11.2 was used to group the soil data presented in Chapters 5 to 10 (excluding data from forest plantations). Changes in soil chemical properties in percentage per year were plotted against the years between two soil samplings – similar to Figs 11.1 and 11.2. The results are presented for Alfisols, Entisols and Inceptisols, Ultisols and Oxisols (Fig. 11.3). For other soil orders there were not enough data.

Table 11.2. General weathering stages relating soil orders with common soil minerals (Richter and Markewitz, 2001).

	Soil weathering stage		
	Early	Intermediate	Advanced
Soil order	Entisol, Andisol	Inceptisol, Mollisol, Alfisol	Ultisol, Oxisol
Common soil minerals	Gypsum, calcite, olivine, biotite, feldpsar	Feldspar, muscovite, vermiculite, smectite	Kaolinite, gibbsite, hydrous Fe oxides

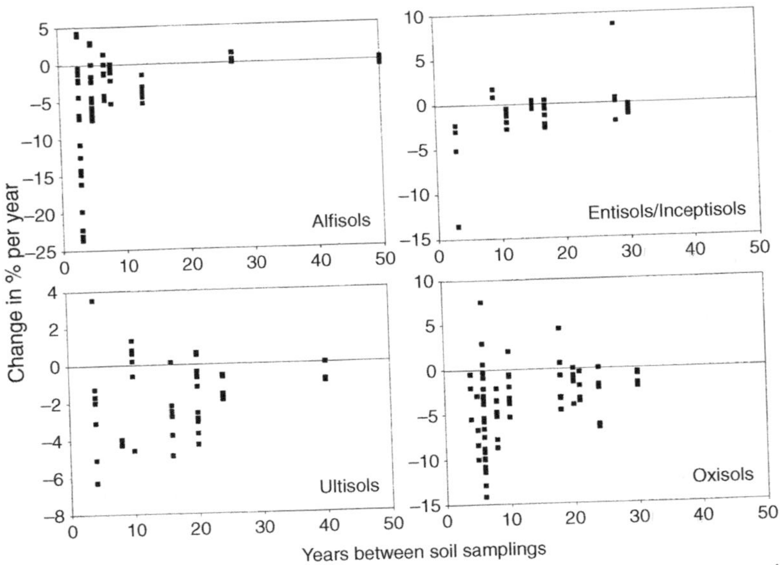

Fig. 11.3. Relative change (in % per year) in topsoil (average 0.16 m) chemical properties (pH, organic C, total N, available P, CEC and exchangeable cations) of Alfisols, Entisols/Inceptisols, Ultisols and Oxisols as related to the period of observation. Data from Chapters 5 to 10, excluding soil data from forest plantations. Note that the scales of the *y*-axes differ.

Most data points were found below the x-axis, indicating that rates of change were negative. On average, the rates of change are largest in Alfisols followed by Ultisols and Oxisols but there were more short-term data from Alfisols. Rates of change are on average lowest in relatively young soils (Entisols/Inceptisols). As mentioned in the section 'Interpretation of the Results' (Chapter 4, page 123), soil fertility decline should be differently appraised for different soils. A slight acidification trend in soils overlying limestone is less of a problem than in strongly acid and highly weathered soils. It appears that rates of change are higher in very old and weathered soils (Oxisols, Ultisols) than in younger soils (Entisols/Inceptisols). The difference is probably related to mineralogy of the soils (see Table 11.2) and the fact that nutrients removed from younger soils are partly replenished by weathering. The higher rates of change in Oxisols and Ultisols should be more of concern because of the few weatherable minerals left in these soils.

Discussion

In the previous sections, changes in soil chemical properties under different land-use systems were summarized. This section discusses how these changes were brought about, what they mean and whether soil chemical properties are useful indicators of the changes in the land-use systems. In the next two sections, implications for the productivity and sustainability of the land-use systems are discussed.

The general pattern of soil fertility decline unravelled in this book is as follows: annual cropping > sugarcane > sisal = perennial crops. The sequence can be relatively easily explained. An important cause of the high rates of soil fertility decline under annual cropping is the lack of nutrient replenishment, as was indicated in recent studies (e.g. Pieri, 1989; Smaling, 1993; van der Pol and Traore, 1993; Henao and Baanante, 1999) and earlier studies (e.g. Djokoto and Stephens, 1961; Toure, 1964; Clarke and Street, 1967; Brams, 1971; Jones, 1972; Juo and Lal, 1977). As mentioned, these studies have mostly focused on nutrient balances in subsistence agriculture. There is good agreement between the nutrient balance studies for smallholder agriculture and the changes in soil chemical properties as presented in this book. Assessing soil fertility decline by using soil chemical data substantiates the nutrient balance studies aiming to address the same problem: quantifying the effects of permanent cropping on soils of the tropics.

For the different soil orders, the decline in soil fertility ranked approximately as follows: Alfisols > Oxisols = Ultisols > Entisols/Inceptisols. Based on stage of weathering of these soil orders, Oxisols and Ultisols are classified as advanced, Entisols/Inceptisols as early to intermediate, and Alfisols take an intermediate position (see Table 11.2). Rates of change in highly weathered soils (Oxisols, Ultisols) were higher than in the younger soils (Entisols/Inceptisols) and there was fair agreement between stage of weathering, soil orders and rates of change in soil chemical properties.

Case study – sisal plantations in Tanzania

The research on the sisal plantations has shown striking differences between soil orders. Soil fertility under sisal declined strongly in Oxisols and Ultisols, whereas few changes were found in Mollisols and shallow Inceptisols. These soils are able to resist the process of decline as nutrients removed by the crop are replenished by weathering of the underlying limestone rock and the deep uptake by plants. This replenishment has maintained high Ca and Mg levels and the pH has remained near neutral. Levels of exchangeable K had decreased because the weathering of primary minerals or the underlying limestone rock does not replenish K. Rates of decline in Oxisols were about the same for Type I and II data, and there was good agreement between the two data types. A balance based on nutrient inputs and nutrient outputs compared with a soil nutrient balance of sisal fields, confirmed the pattern unravelled from the Type I and II data.

The explanation for the decline in soil fertility is the absence of nutrient inputs. In some sisal areas plantation managers have relied on bush fallows for the restoration of the soil fertility (Fig. 11.4). The effects of long fallow periods are only marginal in highly weathered soils (Table 10.13) and soil fertility also declined with the fallow system.

Case study – sugarcane plantation in Papua New Guinea

On the sugarcane plantation in Papua New Guinea, soil fertility changes were measured in relatively young and fertile Fluvents and Vertisols. For both soil orders a similar degree of soil acidification and fertility decline was found when Type I and Type II data were compared. Loss of organic matter and the acidification in topsoil and subsoil are the most significant changes that have occurred. Similar to the soils under sisal, changes were caused by the lack of nutrient inputs with the notable exception of N. The significant pH decline coincided with a change in inorganic fertilizer policy resulting from a change in harvesting technique. Since 1989, Australian cutter–chopper–loader harvesters were used instead of preharvest burning. Urea fertilizers were replaced by sulphate of ammonia, whose acidity is twice that of urea-N, and that may explain the significant increase in soil acidity

Fig. 11.4. Slashing (brush-cutting) 15-year-old fallow vegetation on a sisal plantation in Tanga Region, Tanzania. The mulch is left to dry and decompose and is burned to clear the field. Sisal growers assume that the soil fertility is restored during the fallow so that no inorganic fertilizers, lime or manure need to be applied when a new crop is planted.

observed in the 1990s. Contributing causes are possibly the end of burning, by which no more pH-increasing ashes are returned to the soil, and the yearly addition of sugarcane trash, which increases the organic matter content. Although the trash-harvesting method may favour the organic matter content, in the young alluvial soils it has indirectly resulted in significant acidification.

Optimum sugarcane yields are obtained when the pH is about 6.0 (Coale and Schueneman, 1993) but in some of the plantations the pH had decreased below 5.5, the point at which Al becomes soluble. Also cation availability is decreased at a lower pH because the increase in protons displaces cations from the exchange sites, which are subsequently leached. The levels of exchangeable K significantly declined due to a combination of the large K removal by the sugarcane and possibly leaching losses. Since 1979, soil organic C levels have decreased by about 40%, but the current practice of trash harvesting is likely to increase soil organic matter. Such an increase would increase the pH buffering capacity and reduce the acidifying effects of sulphate of ammonia.

Forest plantations

Forest plantations can have favourable effects on the soil: trees improve soil fertility by reducing losses and optimizing nutrient gains (Sanchez, 1995). This occurs mostly when degraded soils are planted with trees and a number of studies have reported such effects. The increase in soil fertility is possibly due to the lower initial fertility, which is enriched during the planting of trees ('the fallow-enrichment phase') and the absence of nutrient removal during the crop cycle. The relative increase was larger in soils that are chemically poor (Psamments, Oxisols). Forest plantations differ from perennial crop plantations, where nutrients are constantly removed with the economic produce. In addition, many forest trees have mycorrhizal associations that enhance their nutrient uptake capacity (Bakarr and Janos, 1996; Haselwandter and Bowen, 1996). Although mycorrhizal associations also occur in some annual crops (Yost and Fox, 1979; Howeler *et al.*, 1987), the extensive root system and the prolonged period during which the trees can grow make mycorrhizal associations more beneficial for tree crops than annual crops.

The effects of forest plantations on soil chemical properties are species specific. For example, soil acidification under *Pinus* was common and it is generally found that conifers are more likely to enhance soil acidification than are hardwoods (Cannell, 1999). Species thus affect soil processes resulting in different rates of change in soil properties over time. The dynamics of soil nutrient content in forest plantations are smaller than in agricultural plantations as there is less frequent disturbance through removal, fertilizer applications, pruning etc. These dynamics affect the assessment of soil fertility decline in agricultural and forest plantations. The risk for soil fertility decline is, however, similar to agricultural plantations as nutrients leave the soil with the harvest of the wood. Much depends, however, on the harvesting technique and residue management of the leaves, twigs, small branches etc. Removal of all residues results in a much larger drain of nutrients and a larger potential for soil fertility decline.

In a number of studies, soil organic C decreased under forest plantations. On *Pinus* plantations it took at least 20 years before the declining trend in soil organic C was stopped, whereas other studies showed that there was no increase in soil organic C after 22 years. Plantations established specifically for C accumulation will need to use silvicultural systems that are modified to maximize C accumulation (Turner and Lambert, 2000). Recent research in the USA has shown that C sequestration may be limited in *Pinus* plantations because of water and nutrients becoming limiting factors (Oren *et al.*, 2001). So the notion that forest plantations are a good sink for C is not universally applicable, but it all depends whether forest plantations are compared with natural forest or Imperata grasslands.

Study methods

The issue addressed here is what are the effects of plantation cropping on soil chemical properties. Soil changes reflect the effects of the crop under the given set of pedo-climatic conditions, but more importantly it reflects the management of the soil and provides insight into the sustainability of the system. It also reflects uptake and immobilization of nutrients in the biomass of the crop. There is a whole range of factors that affect quantitative conclusions, which can be roughly subdivided in controlled and uncontrolled variation (Table 11.3). These sources of variation explain in part the range of values found for the different land-use systems, and these factors were discussed in Chapter 4.

Given the complex nature of agricultural and forestry plantations and the sources of variation, it is evident that a large number of observations are required before solid conclusions can be drawn. The studies on sisal and sugarcane have shown a solid relation between crop characteristics (nutrient demand and removal), soil management aspects and soil chemical fertility decline. In both studies, soil fertility decline was assessed using measured changes in soil chemical properties and although this has limitations, it proved to be a robust method. Also the analysis per soil order proved a useful entrance point for grouping and analysing the data.

Table 11.3. Controlled and uncontrolled variation in studies on soil changes of agricultural and forest plantations.

	Source of variation	Factor
Controlled variation	Method of soil data collection	Sampling procedures, depth and time
	Method of soil analysis	Analytical technique
	Crops	Oil palm vs. cocoa, mycorrhizas
	Soils	Depth, drainage, inherent fertility, etc.
	Climate	Rainfall, seasonality, temperature
	Management	Inputs, residue management, tillage
Uncontrolled variation	Field heterogeneity	Scale, boundary
	Temporal variability	Frequency of observation, soil property

Soil chemical data proved useful for the assessment of soil fertility decline under different land-use systems but in some studies aberrant results were found. Where the level of significance of correlations between a change in a soil chemical property and period of cultivation is low, it may be concluded that the soil test or the sampling procedure is at fault. This is similar to the value of soil tests for the estimation of fertilizer requirements, which is usually judged from statistical tests of significance on correlations between soil test values and some measure of yield response to fertilizer (Colwell, 1968). However, it remains difficult to link changes in soil properties to the nutrient availability for crop production, because soil chemical properties determined in the laboratory may not be correlated to crop growth.

Differences in rates of change between soil orders are related to the fact that the same analytical methods are used and may give erroneous results. As mentioned, it makes little sense to determine the CEC at pH 7 in strongly acid soils whose mineral fraction is dominated by kaolinite and sesquioxides. Thus, in different soils different soil analytical methods should be used, but that would make comparisons difficult.

Soil chemical data and the rates of change can be used in nutrient-balance studies. The loss of nutrients expressed in kg/ha can be translated into inorganic fertilizer requirements and into economic terms (replacement costs, etc.). This is not straightforward, as nutrient-use efficiency needs to be taken into account and not all losses can be offset by inorganic fertilizers, but in order to express soil nutrient decline in kg/ha, soil chemical data are a crucial first step.

These are point data

This book has summarized studies from over 50 countries in the tropics, and in a large number of studies the soil chemical fertility had declined. Most of the data are from plots and extrapolation would be required to arrive at spatially referenced conclusions: this system on those soils in that region/country/continent/climatic zone depletes the soil chemical fertility and if trends continue, crop yields may be drastically reduced within 5 years.

Scaling is a problem in many areas of agricultural research as data collected at one scale (plot) need to be extrapolated to landscape, watershed, regional, national or global scale (upscaling), and vice versa; data obtained at the regional scale need to be interpolated to the plot or field scale (downscaling). Questions arise of how processes involved at different scales, and observations made across a range of heterogeneous scales should be compared. Most studies so far have been concerned with upscaling and very few studies have looked at downscaling (Bierkens *et al.*, 2000).

Assessing soil fertility decline is scale dependent, for example the soil fertility may decline at the plot scale, but no net changes may occur at a smaller scale (paddock, watershed) (Fig. 4.3). No scaling rules have been developed, and similar observations have been made for upscaling soil erosion measurements at a plot scale (Lal, 1994; den Biggelaar *et al.*, 2001). Nevertheless, many soil erosion studies have produced alarming facts about the extent, rate and impact of erosion and many have produced wrong statements. These wrong statements are usually made in studies involving the extrapolation of measurements taken at one scale for estimates at an entirely different scale (Stocking, 1995).

Upscaling the information on soil fertility decline is, among other things, hampered by unresolved scaling rules and the complexity of the systems under study. In the previous chapters it was shown that soil fertility decline takes place in the tropics, but quantitative statements on the extent and impact require additional studies for which this book has provided data and some of the boundary conditions of how data should be collected, processed and interpreted.

Effect on Crop Production

About 95% of the increase in global human population takes place in the tropics, of which sub-Saharan Africa is the fastest-growing continent (see Chapter 2, 'Malthus and his Followers', page 10). There is also firm evidence that the soil fertility is declining in the tropics, and in particular in sub-Saharan Africa. The question arises of whether the soil fertility decline is a factor that affects crop productivity and the feeding of the growing population.

Declining soil fertility (namely nutrient levels) may result in nutrient deficiencies because of nutrient depletion and imbalances, or an increase in toxic substances for plant growth (Al or Mn toxicity due to severe soil acidification). Secondary effects are the degradation in soil physical properties (e.g. structural degradation from resulting mechanical stress, which is more likely because of soil organic C decline) and the increased susceptibility to pests and diseases and competitive loss to weeds and parasites (e.g. striga). So a decline in soil fertility affects the productivity of a land-use system and dwindling crop yield is the most noticeable result.

Effects on the yield of annual crops

Since the 1960s, yields of most annual crops have doubled or tripled as a result of the Green Revolution. Improvements include the use of fertilizers and lime, irrigation and drainage, and the effects of these amendments outweighed possible soil physical and chemical degradation (Greenland *et al.*, 1997). In the past decade, there has been a decreasing trend in the growth rates of cereal production (Gruhn *et al.*, 2000). It is unlikely that many of the new crop varieties have reached their genetic plateau and it has been advocated that biophysical factors including soil fertility decline contribute to the stagnating growth rates in crop yield (Sanchez *et al.*, 1997). It was shown in the section 'Changes in Different Land-use Systems' (page 317–318) that the decline in soil fertility under annual cropping was larger than in the other land-use systems, which may thus have a significant effect on crop productivity.

Countless studies have investigated the effects of soil amendments (lime, inorganic fertilizers, etc.) on crop yield, but data to support the effects of soil fertility decline on yields of annual crops are limited. There is ample evidence, however, for low crop productivity in many tropical regions. Given the spectacular yield increases that are sometimes obtained with soil amendments, one could argue that declining soil fertility should result in a yield decrease. The timescale at which the yield decline occurs is much longer than the time needed for a yield increase following soil amendments. Detailed studies such as yield decline in lowland rice (Cassman *et al.*, 1997; Dawe *et al.*, 2000) are useful as they unravel the various factors involved in yield trends.

We have investigated trends in sweet potato yield in the humid lowlands of Papua New Guinea and compared this with changes in soil chemical properties (Table 11.4). The sweet potato was not fertilized. Although the observational period was only 2 years, marketable tuber yields at Hobu (Typic Eutropepts) decreased from 18 Mg/ha in the first season after the fallow to 13 Mg/ha in the fourth season (–28%). Sweet potato yields at Unitech, where soils were less fertile (Typic Tropofluvents), decreased from 9 to 6 Mg/ha (–33%). Only soil

Table 11.4. Changes in soil chemical properties under unfertilized sweet potato cultivation (sampling depth 0–0.15 m) at two locations in the humid lowlands of Papua New Guinea. Values are the arithmetic mean of four plots. Type I data, modified from Hartemink *et al.* (2000).

Site	Sampling time	pH H_2O (1:5 w/v)	Organic C (g/kg)	Total N (g/kg)	Olsen P (mg/kg)	Exchangeable cations ($mmol_c$/kg) Ca	Mg	K
Hobu	Before planting	6.2	69.9	6.0	10	268	61	12.2
(Eutropepts)	After four seasons[a]	5.8	71.3	5.9	6	227	59	8.4
	Difference	$P < 0.01$	ns	ns	ns	ns	ns	ns
Unitech	Before planting	5.9	22.1	1.9	30	247	39	11.2
(Fluvents)	After four seasons[a]	5.9	21.7	1.7	24	224	45	8.6
	Difference	ns	ns	ns	ns	ns	ns	$P < 0.05$

[a] One season is about 170 to 190 days; four seasons is about 2 years.
ns, Not significant.

pH at Hobu and exchangeable K at Unitech changed significantly over the same period and it was concluded that other factors (seasonal rainfall, nematodes, weevils) were important in causing the decline and interseasonal variation in sweet potato yields.

Effect on food production

If soil fertility decline is a serious problem, growth rates of crop yields decline, and if population keeps on increasing (Chapter 2), food availability is affected. Malthus would have called these agronomic factors the 'checks' on the growing population. Now, how much evidence is there that declining soil fertility affects food production? Soil analysis and pot experimentation might yield a positive relationship between the soil chemical fertility and decreased crop productivity, but studies at farm and plot level are required and appropriate scaling rules have to be found to quantify and predict the effects of soil fertility decline on the world food production.

Since the exact rate and extent of soil fertility decline cannot be assessed for different regions, it remains a challenge to quantify its possible effect on the loss of crop productivity. There is again a parallel with soil erosion studies. Quantitative data on the effects of erosion on crop yield have been used to estimate the loss of crop productivity at the continental level (Lal, 1995). Although the data were few, it was estimated that yield reductions for sub-Saharan Africa in the 1980s and 1990s due to erosion ranged from 2% to 40% (mean 6.2%). If accelerated erosion continues unabated, yield reductions by the year 2020 may be 14.5% (Lal, 1995). A similar upscaling approach was used for investigating the impact of soil erosion on crop yields in North America. Spatially referenced data of soil orders were linked to erosion rates to estimate the productivity loss in each soil order. A low rate of erosion-induced productivity loss was found, reflecting the impact of improved technology, especially of the increased use of fertilizers and other amendments and improved crop varieties (den Biggelaar *et al.*, 2001). A comparable study approach could be used to investigate the effects of soil fertility decline on crop productivity, whereby detailed studies relating soil fertility decline to crop yields are extrapolated to the national and supra-national level.

In spite of the difficulties in quantifying the effects of soil fertility decline on food production, some of the striking figures for sub-Saharan Africa are as follows: population grows at around 2–3% (Chapter 2); average yield reductions due to soil erosion are 6% (Lal, 1995); many of the soil nutrients decrease at rates faster than 3% per year (Chapter 5); and food production in sub-Saharan Africa only grows at a rate of less than 2% (Mwangi, 1997). Economics play a major role, but it is likely that these factors are linked.

Effects on plantation crop yields

The term Green Revolution has been commonly reserved for the dramatic increase in food crop production, but it could also be used for the increase in perennial crop production. Coffee, cocoa and tea yields tripled between the 1930s and 1980s, and yields of most perennial crops almost doubled between the 1950s and 1990s (Chapter 3). The most spectacular yield increase occurred in oil palm in Asia. The main technical advances that have contributed to the higher yields are: improved planting material and nursery techniques; improved pest and disease control; soil conservation measures and leguminous ground covers; chemical weed control; improved diagnosis of crop nutrient requirements; and field experiments leading to better use of inorganic fertilizers (Webster and Watson, 1988). Although plantation crop yields could be further increased as both the attainable and the potential production levels have not been reached, yield increases caused by technical advances have outweighed yield losses caused by the decline in soil nutrient levels and the increase in soil acidity.

There are no extensive data-sets that could be used to investigate changes in soil chemical properties in relation to plantation crop yields. Possible losses due to soil fertility decline may be small relative to the total value and production of these crops and the impressive yield increases of the past decades. Losses may be masked by interannual yield variation that arises from weather, pests and diseases and other factors. The limited effects of soil fertility decline can be expected, on average, to provide plantation managers with little incentive to adopt sound soil fertility management practices in the short term, which was also observed for erosion-induced crop losses in high-external-input agricultural systems in the USA (den Biggelaar *et al.*, 2001). If the cost of maintaining soil fertility is higher than the value of extra production, the necessary measures are not adopted. This study has, however, shown that the maintenance of soil fertility should also be a matter of concern under plantation cropping in order to sustain and improve crop yield levels.

Effects on sisal yields

There is circumstantial evidence that the decline in production of the Tanzania sisal industry is caused by the decline in yields. Average annual sisal yield from five plantations declined by 0.5 Mg/ha (–30%) between 1968 and 1990. There were fewer leaves to harvest and leaves were on average shorter. From one plantation in the Tanga Region, leaf length data were available and the fraction of long leaves decreased from above 80% in the early 1970s to less than 40% in the late 1980s (Chapter 10).

There were no systematic data to investigate the effects of the declining soil fertility on sisal yields. The only data-set available is from soil samples taken in fields with different fibre-yield levels (Table 11.5). The highest yields are obtained in sisal fields where the

Table 11.5. Sisal yields and soil chemical properties of three sisal fields on Oxisols. Type II data from one plantation in the Tanga Region, Tanzania. Modified from Hartemink and Wienk (1995).

Sisal yield (Mg fibre/ha/year):	2.3		1.8		1.5	
Sampling depth (m):	0–0.20	0.30–0.50	0–0.20	0.30–0.50	0–0.20	0.30–0.50
pH H_2O (1:2.5)	6.5	5.3	5.4	5.2	5.0	4.9
Organic C (g/kg)	16	8	19	6	15	5
Total N (g/kg)	1.1	0.5	1.6	0.7	1.2	0.4
Available P (Bray I) (mg/kg)	5	1	4	< 0.5	3	1
Exchangeable Ca ($mmol_c$/kg)	46	22	19	12	6	6
Exchangeable Mg ($mmol_c$/kg)	17	9	6	3	3	2
Exchangeable K ($mmol_c$/kg)	7	4	2	1	1	< 0.5
CEC ($mmol_c$/kg)	93	73	111	70	64	50
Base saturation (%)	79	51	25	23	16	17
Exchangeable Al ($mmol_c$/kg)	0	3	7	6	11	13
Al saturation (% CEC)	0	4	6	8	18	27

soil is near neutral and relatively high in exchangeable Ca and K. Lower yields are obtained when the pH is lower accompanied by lower levels of exchangeable cations and increased exchangeable Al. So it is very likely that productivity is affected in the strongly acidified soils with low levels of exchangeable cations (Chapter 10). Overall, it seems that the significant decline in soil fertility has contributed to the fall in sisal production in Tanzania. Economic factors such as the world market price for sisal fibre may have been one of the motives for the lack of nutrient inputs (Fig. 11.5).

Effects on sugarcane yields

In the case study on soil changes under sugarcane in Papua New Guinea (Chapter 9), it was found that plantation yields greatly varied in the 1980s but more or less stabilized in the 1990s at low levels because of the planting of varieties with a low yield potential. Although soil fertility had significantly declined and also leaf nutrients had declined over time, it was not possible to relate these changes to the yield variation (Fig. 11.6). Yield data were only available for the whole plantation whereas soil chemical data were available per block (field). Given the fact that yields at the plantation were largely controlled by the sudden and catastrophic outbreak of pests

Fig. 11.5. Poor growth of sisal in the coastal plain of Tanzania, which is partly caused by the weeds but also because these soils (Oxisols, Psamments) are low in fertility and had been permanently cropped with sisal for more than 60 years without nutrient inputs.

Fig. 11.6. Unravelling the effects of sugarcane cultivation on soil properties in the field. Relating soil data to crop yield patterns is another cup of tea.

and diseases, it is difficult to assume a prominent role for the changes in soil fertility in the productivity at the plantation. Papua New Guinea is the centre of origin of sugarcane and the sugarcane at the plantation has a wide range of pests and diseases. Therefore, the incidence of pests and diseases is possibly not related to the soil nutrient status and its changes. Once the pests and disease problems are solved, the decline in soil chemical properties is likely to become a factor affecting the productivity of the soils.

There are data from other sugarcane-producing countries that suggest that growth in production is stagnating. For example, the Australian sugar industry is on a production plateau to which changes in the farming system and the phenomenon of yield decline are believed to have contributed (Garside *et al.*, 1997). Changes in the farming system included the removal of assignment restrictions, introduction of chopper harvesters, improved varieties, green cane trash blanket, herbicide usage, major diseases, and increased used of inorganic fertilizers. From the Australian study it was not possible to apportion the relative importance of each, and it was concluded that there were several biophysical causes for the yield plateau (Garside *et al.*, 1997). The literature review on soil changes under sugarcane in Australia showed, however, that in some areas soil fertility had declined under sugarcane.

Soil acidification is a common trend that occurred in soils under sugarcane in Papua New Guinea, and in many other studies reviewed in Chapter 8. Although soil acidification decreases the availability of some nutrients and increases the possibility of cation leaching, it is uncertain what the effects of soil acidification are on sugarcane production (Sumner, 1997). Some sugarcane-producing areas are on very acid soils giving, nevertheless, high yields (Blackburn, 1984). However, where soil nutrient levels are limiting yields, an increase in soil acidity will almost certainly affect production and so the effects are site specific.

Sustainable Land Management

Sustainability problems occur through under use of resources (soil degradation) or over use of resources (environmental effects of nutrients and pesticides) (Kropff *et al.*, 1997). In this book, the emphasis has been mostly on the under use of resources, i.e. the reduction in soil nutrients as a consequence of cropping without adequate nutrient inputs. Crop yields are key indicators for the assessment of land management but other indicators and the determination of threshold values at which crop productivity is affected are also important. Indicators and threshold values are discussed below in relation to the two case studies (Chapters 9 and 10).

Indicators

Sustainability, although a dynamic concept, implies some sort of equilibrium or steady state (O'Callaghan and Wyseure, 1994). The analysis in this book has shown that many soil chemical properties significantly change with time, and it can be argued that land-use systems in which significant soil fertility decline takes place are not sustainable in the long term. Such a conclusion demands a rigorous assessment of soil fertility decline and a set of soil chemical properties that can be used as indicators. This book has used a set of basic soil chemical properties (pH, organic C, total N, CEC and exchangeable cations) to investigate change under cropping. Each of these shows a degree of natural variation that is affected by soil management and the cropping system. The advantage of these properties is that most soil laboratories use the same methods for their determination and data from the past can be used. A disadvantage is that short-term changes cannot be measured accurately so that little understanding is gained about abrupt processes that bring changes. For long-term assessment, basic soil chemical properties are valuable in combination with well-quantified nutrient-balance studies.

In the two case studies, there were only a limited number of soil chemical properties available from the plantation records that could be used as indicators. Other data would have been helpful, such as total P or total cation contents. There is, however, a cost to the collection of such data, and for the general assessment presented here, it is improbable that the extra costs would have been justified by the extra information obtained. For the plantation management, obtaining spatial information on the changes in soil properties is probably more useful so that action can be taken to avoid further deterioration and to remediate the affected soils.

A key indicator of sustainable land management is crop yield, and trends in perennial crop yields were discussed in the previous section. Productivity is a key component of sustainable land-use systems and any system in which yields are declining is not sustainable. It can be concluded that sisal production in Tanzania has been unsustainable and the significant changes in soil chemical properties show that the management of the plantation has not sustained the resource base on which the production depends. On the sugarcane plantation in Papua New Guinea, no clear pattern of yield decline was found. However, the significant decrease in soil chemical properties indicates that soil management in the past two decades has not been sustainable and will not sustain yields indefinitely. The observed changes reflect the way in which the soils were used, including continuous cultivation with acidifying N fertilizers and the failure to use P and K fertilizers.

Threshold values

Soil chemical properties in highly weathered soils under sisal in Tanzania had reached values at which growth and production is affected. Many of the upland soils had acidified significantly and as sisal prefers a neutral to slightly alkaline soil reaction, the strongly acid Oxisols and Ultisols pose severe limitations for optimum production. For the sugarcane in Papua New Guinea, the story is somewhat different. These soils were much younger and had been under cultivation for shorter periods. Soil chemical properties had changed significantly but did they reach levels that affect the sugarcane? The pH levels in 1996 were about 5.8. Although the optimum pH for sugarcane is about 6.5 (Yates, 1978), the crop is successfully grown on soils with pH 4, as in Guyana, to soils with pH over 7, as in many parts of Barbados. It is therefore unlikely that the current pH levels affect sugarcane production. Levels of available P (Olsen) were on average over 25 mg/kg, which are high levels for sugarcane (Blackburn, 1984). Also the exchangeable cations remained at favourable levels for sugarcane cultivation. This suggests that the soil chemical properties had not reached threshold values.

An additional way to investigate whether threshold values were reached was by the analysis of tissue sample data from the sugarcane, which reflects nutrient availability. It was shown that all major nutrients were significantly lower in the sugarcane leaves in the 1990s compared with the 1980s (Chapter 9). The number of samples below the critical nutrient concentration increased significantly in the 1990s and more than two-thirds of the leaf samples were deficient in N, about one-fifth deficient in P, and nearly one-half were deficient in K. Although P and K levels in the soil were still favourable, the increase in leaf nutrient deficiencies provides evidence that nutrient availability was reduced. In addition to the soil chemical data, leaf analytical data seem to be a useful indicator in the assessment of sustainable land management.

Which land-use systems are more sustainable?

A matter raised in the first chapters of this book is whether plantations are a more sustainable form of land use than annual cropping systems. The analysis of soil changes suggests that rates of change in soils of agricultural plantations are lower than in soils under annual cropping. The difference is related to soil type, soil management (level of nutrient inputs) and crop characteristics (trees vs. annual crops). The detailed case studies on sisal and sugarcane showed clearly that soil fertility decline is a problem in areas with low inherent fertility soils (sisal plantations) as well as in areas with inherently fertile soils (sugarcane plantation). From a partial assessment point of view (El-Swaify, 2000), it can be concluded that both sisal in Tanzania and sugarcane cultivation in Papua New Guinea are not sustainable under current management systems. Also if a partial assessment were made, many of the annual cropping systems in which soil fertility has declined would be found to be unsustainable. As the rates of change in soil chemical fertility of annual cropping systems were much higher, these systems are less sustainable than perennial crop systems and possible effects on crop productivity are expected to be larger.

References

Bakarr, M.I. and Janos, D.P. (1996) Mycorrhizal associations of tropical legume trees in Sierra Leone, West Africa. *Forest Ecology and Management* 89, 89–92.

Bierkens, M.F.P., Finke, P.A. and de Willigen, P. (2000) *Upscaling and Downscaling Methods for Environmental Research.* Kluwer Academic Publishers, Dordrecht, The Netherlands.

Blackburn, F. (1984) *Sugar-cane.* Longman, Harlow.

Brams, E.A. (1971) Continuous cultivation of West African soils: organic matter diminution and effects of applied lime and phosphorus. *Plant and Soil* 35, 401–414.

Cannell, M.G.R. (1999) Environmental impacts of forest monocultures: water use, acidification, wildlife conservation, and carbon storage. *New Forests* 17, 239–262.

Cassman, K.G., Peng, S.B. and Dobermann, A. (1997) Nutritional physiology of the rice plants and productivity decline of irrigated rice systems in the tropics. *Soil Science and Plant Nutrition* 43, 1101–1106.

Clarke, W.C. and Street, J.M. (1967) Soil fertility and cultivation practises in New Guinea. *The Journal of Tropical Geography* 24, 7–11.

Coale, F.J. and Schueneman, T.J. (1993) Sugarcane response to limestone and gypsum application on acidic sandy soils. *Journal of American Society of Sugarcane Technologists* 13, 73–86.

Colwell, J.D. (1968) Assessments of factors limiting the significance of soil tests for fertilizer requirements. In: *9th International Congress of Soil Science.* ISSS and Angus and Robertson, Adelaide, pp. 709–718.

Dawe, D., Dobermann, A., Moya, P., Abdulrachman, S., Singh, B., Lal, P., Li, S.Y., Lin, B., Panaullah, G., Sariam, O., Singh, Y., Swarup, A., Tan, P.S. and Zhen, Q.X. (2000) How widespread are yield declines in long-term rice experiments in Asia? *Field Crops Research* 66, 175–193.

den Biggelaar, C., Lal, R., Wiebe, K. and Breneman, V. (2001) Impact of soil erosion on crop yields in North America. *Advances in Agronomy* 72, 1–52.

Djokoto, R.K. and Stephens, D. (1961) Thirty long-term fertilizer experiments under continuous cropping in Ghana. II. Soil studies in relation to the effects of fertilizers and manures on crop yields. *Empire Journal of Experimental Agriculture* 29, 245–258.

El-Swaify, S.A. (2000) Is sustainable agriculture a fallacy? *Journal of Soil and Water Conservation* 55, 2–3.

Garside, A.L., Smith, M.A., Chapman, L.S., Hurney, A.P. and Magarey, R.C. (1997) The yield plateau in the Australian sugar industry. In: Keating, B.A. and Wilson, J.R. (eds) *Intensive Sugarcane Production – Meeting the Challenges Beyond 2000.* CAB International, Wallingford, UK, pp. 103–124.

Greenland, D.J., Gregory, P.J. and Nye, P.H. (1997) Introduction and conclusions. In: Greenland, D.J., Gregory, P.J. and Nye, P.H. (eds) *Land Resources: on the Edge of the Malthusian Precipice?* The Royal Society, London, pp. 861–867.

Gruhn, P., Goletti, F. and Yudelman, M. (2000) *Integrated Nutrient Management, Soil Fertility, and Sustainable Agriculture: Current Issues and Future Challenges.* IFPRI, Washington, DC.

Hartemink, A.E. and Wienk, J.F. (1995) Sisal production and soil fertility decline in Tanzania. *Outlook on Agriculture* 24, 91–96.

Hartemink, A.E., Poloma, S., Maino, M., Powell, K.S., Egenae, J. and O'Sullivan, J.N. (2000) Yield decline of sweet potato in the humid lowlands of Papua New Guinea. *Agriculture Ecosystems and Environment* 79, 259–269.

Haselwandter, K. and Bowen, G.D. (1996) Mycorrhizal relations in trees for agroforestry and land rehabilitation. *Forest Ecology and Management* 81, 1–17.

Henao, J. and Baanante, C. (1999) *Estimating Rates of Nutrient Depletion in Soils of Agricultural Lands of Africa.* IFDC, Muscle Shoals, Alabama.

Howeler, R.H., Sieverding, E. and Saif, S. (1987) Practical aspects of mycorrhizal technology in some tropical crops and pastures. *Plant and Soil* 100, 249–283.

Jackson, M.L. and Sherman, G.D. (1953) Chemical weathering of mineral in soils. *Advances in Agronomy* V, 219–318.

Jenny, H. (1941) *Factors of Soil Formation. A System of Quantitative Pedology.* McGraw-Hill, New York.
Jenny, H. (1980) *The Soil Resource. Origin and Behavior.* Springer-Verlag, New York.
Jones, E. (1972) Principles for using fertilizers to improve red ferralitic soils in Uganda. *Experimental Agriculture* 8, 315–332.
Juo, A.S.R. and Lal, R. (1977) The effect of fallow and continuous cultivation on the chemical and physical properties of an Alfisol in Western Nigeria. *Plant and Soil* 47, 567–584.
Kropff, M.J., Muchow, R.C., Rabbinge, R. and Aggarwal, P.K. (1997) Limits to intensive agricultural systems and opportunities for sustainable agricultural development. In: Keating, B.A. and Wilson, J.R. (eds) *Intensive Sugarcane Production – Meeting the Challenges Beyond 2000.* CAB International, Wallingford, UK, pp. 17–35.
Lal, R. (1986) Conversion of tropical rainforest: agronomic potential and ecological consequences. *Advances in Agronomy* 39, 173–264.
Lal, R. (1994) Global overview of soil erosion. In: *Soil and Water Science: Key to Understanding our Global Environment.* SSSA Special Publication No. 41, Soil Science Society of America, Madison, Wisconsin, pp. 39–51.
Lal, R. (1995) Erosion–crop productivity relationships for soils of Africa. *Soil Science Society of America Journal* 59, 661–667.
May, R. (2001) Risk and uncertainty. *Nature* 411, 891.
Mwangi, W.M. (1997) Low use of fertilizers and low productivity in Sub-Saharan Africa. *Nutrient Cycling in Agroecosystems* 47, 135–147.
Nye, P.H. and Greenland, D.J. (1960) *The Soil under Shifting Cultivation.* Commonwealth Bureau of Soils, Harpenden.
O'Callaghan, J.R. and Wyseure, G.C.L. (1994) Proposals for quantitative criteria in the management of sustainable agricultural systems. In: Syers, J.K. and Rimmer, D.L. (eds) *Soil Science and Sustainable Land Management in the Tropics.* CAB International, Wallingford, UK, pp. 27–39.
Oren, R., Ellsworth, D.E., Johnsen, K.H., Phillips, N., Ewers, B.E., Maier, C., Schafer, K.V.R., McCarthy, H., Hendrey, G., McNulty, S.G. and Katul, G.G. (2001) Soil fertility limits carbon sequestration by forest ecosystems in a CO_2-enriched atmosphere. *Nature* 411, 469–472.
Pieri, C. (1989) *Fertilité des Terres de Savanes.* Ministère de la Cooperation et CIRAD–IRAT, Paris.
Richter, D.D. and Markewitz, D. (2001) *Understanding Soil Change. Sustainability over Millennia, Centuries, and Decades.* Cambridge University Press.
Sanchez, P.A. (1995) Science in agroforestry. *Agroforestry Systems* 30, 5–55.
Sanchez, P.A., Villachica, J.H. and Bandy, D.E. (1983) Soil fertility dynamics after clearing a tropical rainforest in Peru. *Soil Science Society of America Journal* 47, 1171–1178.
Sanchez, P.A., Shepherd, K.D., Soule, M.J., Place, F.M., Buresh, R.J., Izac, A.M., Mokwunye, A.U., Kwesiga, F.R., Ndiritu, C.G. and Woomer, P.L. (1997) Soil fertility replenishment in Africa: an investment in natural resource capital. In: Buresh, R.J., Sanchez, P.A. and Calhoun, F. (eds) *Replenishing Soil Fertility in Africa.* SSSA Special Publication No. 51. Soil Science Society of America, Madison, Wisconsin, pp. 1–46.

Smaling, E.M.A. (1993) An agro-ecological framework for integrated nutrient management with special reference to Kenya. Doctoral thesis, Agricultural University, Wageningen.

Stocking, M. (1995) Soil erosion in developing countries – where geomorphology fears to tread! *Catena* 25, 253–267.

Sumner, M.E. (1997) Opportunities for amelioration of soil physical and chemical constraints under intensive cropping. In: Keating, B.A. and Wilson, J.R. (eds) *Intensive Sugarcane Production – Meeting the Challenges Beyond 2000.* CAB International, Wallingford, UK, pp. 305–326.

Toure, E.H.O. (1964) Maintenance of the fertility of savanna soils. Effects of annual or biennial mineral fertiliser applications under conditions of continuous cropping in the Sudan zone of West Africa. *African Soils* 9, 221–246.

Turner, J. and Lambert, M. (2000) Change in organic carbon in forest plantation soils in eastern Australia. *Forest Ecology and Management* 133, 231–247.

van der Pol, F. and Traore, B. (1993) Soil nutrient depletion by agricultural production in southern Mali. *Fertilizer Research* 36, 79–90.

Webster, C.C. and Watson, G.A. (1988) Tree crops of the humid tropics. *TAA Newsletter* 8, 11–19.

Yates, R.A. (1978) The environment for sugarcane. In: *Land Evaluation for Standards for Rainfed Agriculture.* World Soil Resources Report no. 49. FAO, Rome, pp. 58–72.

Yost, R.S. and Fox, R.L. (1979) Contribution of mycorrhizae to P nutrition of crops growing on an Oxisol. *Agronomy Journal* 71, 903–908.

Summary and Conclusions

12

The previous chapters have focused on soil fertility decline in annual and perennial cropping systems in the tropics. In most studies on this subject, nutrient balances were used as a tool to evaluate soil and crop management practices and the sustainability of the land-use systems. Moreover, the focus of most studies has been on low-external-input agriculture of subsistence farmers.

This book differs in two important ways from earlier studies on soil fertility decline: (i) the assessment of soil fertility decline is based on soil chemical data; and (ii) it focuses on plantation crops. A quantitative approach has been taken combining published data on soil fertility decline with detailed case studies from plantations in Tanzania and Papua New Guinea. In the case studies, existing soil chemical data were used to select new sampling sites and to determine whether soil chemical properties had changed. The data were used to calculate rates of change in soil chemical properties and to compare annual and various perennial cropping systems.

In this final chapter, the main conclusions are summarized. Conclusions are made on the rates of soil changes and the differences in rates between land-use systems, followed by conclusions on methods to quantify soil fertility decline, and the effects of soil fertility decline. The chapter ends with some comments on the need for and direction of future research.

Rates of Soil Fertility Decline

Changes in soil chemical properties under permanent cropping are reported from various parts of the tropics and under different land-use systems. There were differences between the land-use systems but the data presented in this book confirmed the results from other studies on soil fertility decline and nutrient mining in the tropics. The decline in soil fertility is caused by the inadequate use of inorganic fertilizers and other nutrient inputs, and the increased loss of nutrients by erosion, leaching and gaseous losses as compared to natural ecosystems. Soil fertility decline should be a major concern because of inadequate inorganic fertilizer, lime and manure applications, and the need to maintain and increase crop yields to feed the ever-growing population.

In some studies, the pattern in soil fertility decline was ambiguous. The absence of soil fertility decline under permanent cropping without nutrient replenishment is extremely implausible and the result might be due to errors in the sampling system employed or in the soil analytical procedures. It implies that methodologies for its assessment have not been used with rigour, but this can be partly understood as the systems under study – like all terrestrial ecosystems – are intractably complicated. Although the complexity is no excuse for a deficiency in scientific rigour, it may be an important factor explaining some of the variation and aberrant results reported in this book.

Differences Between Land-use Systems

There are differences between land-use systems in the rate of soil fertility decline, which is related to crop characteristics and land management aspects. Annual cropping systems showed the largest rate of decline in soil fertility followed by the rates in soils under sugarcane. Rates of change in perennial cropping systems, including sisal, were lower than the rates under annual crops or sugarcane. Under forest plantations, rates of change were variable and in a number of studies the soil chemical fertility increased with time.

Soil organic C declined on average in all land-use systems. The decline was largest in annual cropping systems and lowest in soils under perennial crops. Also the pH showed the largest decline under annual crops followed by the pH decline in soils under sugarcane cultivation. Although it is generally perceived that soil fertility decline is a problem in annual cropping systems of subsistence farmers, this book has provided firm evidence that soil fertility decline also takes place on agricultural plantations.

Sustainability implies a steady state with a certain degree of natural variation. It is clear that many of the annual cropping systems in which soil fertility declined are not sustainable from a soil chemical fertility point of view. As rates of change in soil fertility of annual cropping systems were on average much higher, it could be concluded these are less sustainable systems than cropping systems with perennials.

Quantification of Soil Fertility Decline

In this book, soil fertility decline was assessed using a set of basic soil chemical properties (pH, organic C, total N, available P, CEC and exchangeable cations) from different periods at the same site, or from different land-use systems. The advantage is that the data are hard, provided sampling and laboratory analyses were done correctly and that samples were properly stored. Another advantage is that soil analytical procedures have been standardized in the past decades, and as most laboratories use the same methods, soil analytical data from the past can be used. The disadvantages are that soil chemical properties may hold no clear relationship with crop yields, that much time is required for the assessment, and that spatial and temporal variation requires large sampling schemes and abundant data before concrete conclusions can be drawn.

The set of basic soil chemical data proved sufficient for the overall purpose of this study. However, soil fertility decline studies should include bulk density data. Bulk density is relatively easy to measure and can have significant impact on the results of a study. With the bulk density, nutrient concentrations can be expressed in kg/ha, which can be used in nutrient-balance studies or economic studies on the costs of soil fertility replenishment. Bulk density is mostly relevant for total elemental analysis and not so much for the nutrients that are extractable (exchangeable, available) as the available amount may hold no relationship with the total nutrient content in the soil.

The detailed case studies on sisal and sugarcane plantations were primarily undertaken to investigate soil fertility changes under permanent cropping. The studies combined various data sources and types, which resulted in much more solid conclusions because of the deployed study methods and the amount of data. Soil maps were used to distinguish the major soil orders and for the selection of sampling sites. Old reports and routine sampling schemes from the past were used, but careful selection was required. In addition to comparing data from the past with recent observations, the investigation was strengthened by sampling different land-use systems (virgin vs. cultivated soils), semiquantitative nutrient balances, leaf nutrient data, and soil physical measurements. Quantification of soil fertility changes as well as its impact on crop productivity is important for determination of the long-term sustainability of the plantation industry. This study has shown what data can be used, what the limitations are of the data, how data should be processed, and how the results should be interpretated.

The Effects of Soil Fertility Decline

The study confirms that soil fertility decline occurs in many land-use systems in the tropics. There are incidental reports on the relationships between soil fertility decline and crop yields but there is no study available in which these relationships have been systematically investigated. It is likely that the decline in soil fertility under annual cropping systems is contributing to the stagnation in the growth of food production experienced in some parts of the world, as in sub-Saharan Africa. The data presented in this book were mostly point data and as soil fertility decline is scale-dependent, neither fertility-decline maps nor systematic generalizations can be applied.

This study has shown that soil fertility decline can be serious under plantation cropping, which will sooner or later affect production and thus reduce the export and income of a country. Plantation agriculture is a major contributor to the income of many countries in the tropics and provides hundreds of thousands of people with labour and income. Sustaining and improving the production capacity of agricultural plantations is therefore important, and a decline in soil fertility should be avoided.

The case study on sisal revealed that soil fertility decline is a contributing factor to the fall of sisal production in Tanzania. Many of the soils under sisal were already of low inherent fertility, which was further lowered by permanent cropping with sisal. Sugarcane production in Papua New Guinea is largely determined by the prevalence of pests and diseases, and soil chemical properties have not decreased to very low levels. The effects of soil fertility decline are masked by many other factors, but productivity could be adversely affected if soil fertility decline continues. Soil amendments are required to maintain and improve sugarcane productivity. There is a cost to these measures that may not be compensated for by an increase in crop production, but the costs to restore degraded soils may be higher than those required to maintain the soil in favourable condition for crop production.

Future Research

Much research in the tropics has focused on the alleviation of soil chemical fertility constraints. Although inorganic fertilizers, manure and lime were widely promoted in the past, since the 1980s integrated nutrient management systems have been advocated, in which organic and inorganic nutrient inputs are combined. Enormous progress has been in made in this area and technical solutions to soil fertility constraints are known. Fundamental research seems to advance less rapidly and the understanding and quantification of a number of soil processes is lacking. Also data availability remains behind the development of soil and crop models. Models have been very useful in conceptualization of soil and crop problems, but their greatest benefit is possibly that they have shown the gaps in basic knowledge and the dearth of data by which they may be validated.

In order to differentiate beliefs from facts and figures, rigorous and detailed studies on soil fertility decline and its effect on crop yield and the world food production are required. The lack of widespread and documented evidence of crop yield decline makes soil fertility decline an enigma that requires rigorous research efforts to unravel the impact of the many factors involved. Both long-term field experiments and models in which the available data are aggregated should be used. This is essential for proper assessment and spatial quantification of nutrient decline. This book has provided data and boundary conditions that could be used in future studies on soil fertility management and crop productivity in the tropics.

Author Index[1]

[1]The index refers to authors and where their work is mentioned in the text; only first authors are indexed and the full list of authors per reference is given in the literature section at the end of each chapter. Institutional publications (e.g. from FAO or IFPRI) are not listed in this index.

Subject Index[1]

[1] In this index no Latin names for plants are used and the common names are listed, e.g. sisal instead of *Agava sisalana*, cocoa instead of *Theobroma cacao*, teak for *Tectona grandis*. For some plants genera or species name is given (e.g. *Albizzia*, *Leucaena*). Only soil orders (soil taxonomy) are indexed and no suborders or lower levels; e.g. Alfisols are listed but not Tropudalfs.

Printed and bound by CPI Group (UK) Ltd, Croydon, CR0 4YY

16/09/2026

14963431-0005